Practical Cytodiagnosis

By

YOSHIO ISHIZUKA, M. D.

Associate Professor of Obstetrics and Gynecology,
Nihon University School of Medicine, Tokyo;
Director of Obstetrics and Gynecology, Chief of Cytology Laboratory,
Kosei Chuo General Hospital, Tokyo

Under the supervision of

KUNIO OOTA, M. D.

Professor of Pathology,
Faculty of Medicine, University of Tokyo

and

KAZUMASA MASUBUCHI, M. D., F.I.A.C. (hon.)

Chairman of Department of Gynecology, Director of Cytology Laboratory
and School of Cytotechnology, Cancer Institute Hospital,
The Japanese Foundation for Cancer Research, Tokyo

IGAKU SHOIN LTD. TOKYO

J. B. LIPPINCOTT COMPANY

PHILADELPHIA AND TORONTO

ISBN 0-397-58120-3

Sole distributor in the United States and Canada
J. B. LIPPINCOTT COMPANY, East Washington Square, Philadelphia Pa. 19105, U. S. A.

Printed and Bound in Japan

FOREWORD

Diagnostic cytology, introduced into clinical medicine in 1941 by the late Dr. G. N. Papanicolaou, has made an enormous contribution. It has not only made possible the diagnosis of hospital cases but also has offered a unique tool in detecting pre-symptomatic cancers. Those engaged in diagnostic cytology are members of a global team helping to save patients and combat a dreaded enemy.

The task of diagnostic cytology is, however, not easy. In order to master the art, one must be thoroughly acquainted with both the physiological and the pathological conditions of the organs involved. Mere knowledge of the general features of a cancer cell is not sufficient.

To those who are starting the study of diagnostic cytology this book will be helpful in differentiating the various pathological conditions which can be indicated by the cells appearing on the smear slides. To those who have had more experience, the book will serve as a guide in systematizing knowledge and in recognizing the need for an even more profound understanding of organ pathology.

Dr. Y. Ishizuka is an experienced and respected gynecologist and has devoted himself to teaching the art of diagnostic cytology. He was trained in the Cytology Laboratory of the Strang Clinic of the Memorial Hospital for Cancer and Allied Diseases in New York, a laboratory which was founded by the father of clinical cytology, Dr. G. N. Papanicolaou.

May 1972, Tokyo

KUNIO OOTA, M.D.
KAZUMASA MASUBUCHI, M.D.

PREFACE

Since the cell is a basic unit of tissue, diseases of the human body are expressed most directly at the cellular level. Therefore, both normal and abnormal states of the organs should be reflected accurately in the morphology of cells derived from the organs concerned. Thus, training of our professional eyes for the recognition of such minute cytologic changes has become of the utmost importance in improving our diagnostic ability.

It was almost a quarter of a century ago that Dr. George N. Papanicolaou first suggested the practical use of cytology. His basic concepts gradually developed and, as further knowledge accumulated, enabled one to perform diagnoses on various organs affected with a disease from a wider point of view. Cytodiagnosis has acquired increasing accuracy as a diagnostic tool and it is for this reason that the cytodiagnosis cultivated by Dr. Papanicolaou has now prevailed throughout the world.

Cytodiagnosis, however, is by no means an easy technology to learn. It is difficult to develop skill in the cytodiagnostic practice without the accumulation of experience over a long period of time. The author recognized the need for adequate interpretations of the difficulties involved in cellular description and the need for numerous explanatory pictures and illustrations to aid those beginning to study cytodiagnosis.

Cytological details, however, should be studied by color pictures as realistic as possible. Therefore, in the course of writing this book I decided to compile as many color photographs as possible. I of course included only those cases whose diagnoses were histologically established, and I believe that the photographs of the cells provide a good selection and are appropriate for the given diagnoses of the cases presented. Needless to say, there may be manifold variations in addition to those included in the selections. It is, however, apparent that it would be impossible to compile all the variations of normal and abnormal cells. I have, therefore, provided representative photographs of cells for each type of disease.

The photographs of the cells were prepared mostly by magnifications of 20×10 for low and 100×10 for high power. The photographs are primarily based on materials from the Cytology Laboratory at Kosei Chuo General Hospital, Tokyo, the Cancer Institute Hospital in Tokyo, and University of Tokyo. All the specimens were stained by Papanicolaou's method. Most of the photomicrographs and illustrations were taken or drawn by the author.

It is impossible to identify abnormal cells without understanding normal ones. Even in those cytodiagnostic cases handled by special medical services, the majority of specimens are in the normal or benign range. Therefore, one must study normal cytology before cancer cytology. For this reason in each separate section of the book a considerable number of pages are devoted to the explanation of epithelial cells, normal or benign. For convenience such items as the criteria for malignant cells and radiation effects on cells are included under the heading of "cytologic examination of the female genital organs".

This book should be convenient for physicians and/or cytotechnologists who have just begun the routine work of cytodiagnosis or screening, since it does not stress the theoretical aspects which often confound the beginner. To this end, all documentary records were omitted.

Since this book represents a new type in the cytological field, undoubtedly there will be different interpretations regarding the cytological presentation. The author welcomes any criticism and/or correction from the reader.

I sincerely hope that the book will make a contribution to the field of cytodiagnosis and thereby help patients all over the world.

Finally, I want to express my sincerest gratitude to Professor Kunio Oota, University of Tokyo, and to Dr. Kazumasa Masubuchi of the Cancer Institute Hospital, who kindly guided me to publish this book.

I am also deeply indebted to Dr. Leopold G. Koss, Memorial Hospital for Cancer and Allied Diseases, New York, and to Dr. Herbert Derman, City of Kingston Laboratory, Kingston, N. Y., for my initial introduction to diagnostic cytology.

I wish to thank Professor Chiaki Sawasaki, Chairman of Department of Obstetrics and Gynecology, Nihon University School of Medicine, Tokyo, for his unfailing support of my work.

I also owe my gratitude to Mrs. Grace R. Durfee and Mr. Andrew Ricci at the Memorial Hospital for Cancer and Allied Diseases, New York, for their generous instruction.

Dr. Masayoshi Takahashi and Dr. Yoshitaro Sakai kindly contributed slides from their collections.

The author wishes to acknowledge the gracious co-operation of his wife, Mrs. Shigeko Ishizuka.

Dr. Yukihiro Shigenaga, Dr. Kiichi Matsuzawa and Mrs. Satoko Iwabuchi provided invaluable assistance in the collection and preparation of materials.

I am particularly indebted to my publisher, IGAKU SHOIN LTD. who allowed me to compile a large number of color photomicrographs in this book.

May 1972, Tokyo

YOSHIO ISHIZUKA, M. D.

CONTENTS

FUNDAMENTALS OF CYTOLOGY

Every tissue is composed of cells. In a sense, they are the basic units of the living organism. The study of the cell is called "cytology". There are numerous types of cells, and they differ from one another in size, shape and function. Their basic structure, however, is the same. They consist of a round or oval structure in the center called the nucleus, which is surrounded by cytoplasm.

Nucleus

The life cycle of the nucleus is divided into two phases : the interphase and the mitotic phase.

Interphase (Fig. 1)

Metabolic activity increases during this phase; therefore, it is not accurate to call it the "resting phase". In a normal cell that has been properly fixed and stained, several nuclear components are observed during this phase.

1. Nuclear membrane, 2. Nucleolus, 3. Chromatin granules,
4. Chromocenter, 5. Sex chromatin, 6. Nuclear sap.

1) Nuclear membrane

The nuclear membrane with the optical microscope appears as a single layer but with the electron microscope, an inner and outer layer are revealed. Furthermore, numerous micropores about 1 micron in diameter can be seen and exchange of material is believed to take place through these openings between the nucleus and the cytoplasm. Under optical microscope the nuclear membrane often looks thickened, due to accumulation of chromatin inside the membrane.

2) Nucleolus

A nucleolus is an eosinophilic body in the nucleus. It is clearly seen in some types of cells and only vaguely in others. Sometimes more than two are present. Because it differs from the chromocenter, it is not demonstrated with the Feulgen reaction. It consists of RNA, protein, lipid and a small amount of DNA. Because it contains protein, it is eosinophilic and is stained with basic dyes. The outline of the nucleolus is distinct and regular in the interphase, but it becomes irregular and larger in the pathologic state.

3) Chromatin granules

Chromatin forms a granular pattern after ordinary methods of fixation and staining, and the fineness or coarseness of the granules depends on the type of cell. Chromatin accumulation underneath the nuclear membrane is sometimes called the "chromatinic membrane" and, under optical microscope, it often seems that the nuclear membrane itself is thickened.

4) Chromocenter

This is a dense clumping of chromatin substance, and is sometimes called the "false nucleolus" or the "karyosome". The size and shape differ according to the type of cell. It differs from the nucleolus by containing a large amount of DNA and stains dark purplish-red with the Feulgen reaction.

5) Sex chromatin

This is a plano-convex body (convex on one side and flat on the other), about 1 micron in diameter. It is seen in many cells of the human female and is located on the inner aspect of the nuclear envelope. It is strongly basophilic, and the Feulgen reaction reveals that it consists mainly of DNA. Sex chromatin is rarely observed in the nuclei of cells from males.

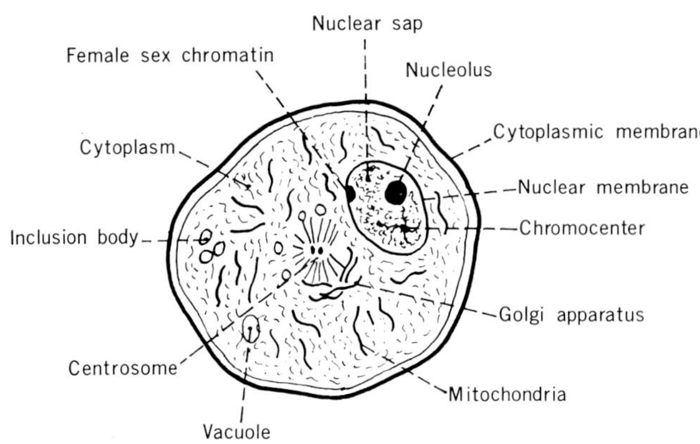

Fig. 1 The structure of a cell.

6) Nuclear sap

This is a medium in the gel state which fills the nucleus, and nuclear structures are thought to be suspended in it. It stains poorly.

Mitotic phase (Fig. 2)

This is the phase of cell division. Cells proliferate by division, and malignant tumor tissue which proliferates suddenly and rapidly consists of cells that divide more frequently.

In cell division, both nuclear and cytoplasmic divisions take place. First, the chromatin substance in the nucleus is evenly distributed, then division of the nucleus and the cytoplasm takes place. Thus, the cell is divided into two cells. Mitosis can be divided into four phases ; prophase, metaphase, anaphase and telophase.

1) Prophase

In the interphase, chromatin is in the form of minute granules and is distributed evenly throughout the nucleus. In the mitotic phase, these granules condense to form rod-shaped chromosomes. The number of chromosomes differs among animals, but it is constant in all cells of animals of the same genus. In humans the number of chromosomes is 46. At the onset of mitosis, 23 pairs of chromosomes are aligned.

The nuclear membrane and the nucleolus dissolve with the formation of chromosomes. If only one centrosome is present, it divides into two. When there are two centrosomes each moves to opposite poles of the cell, the radiating cytoplasmic threads form a spindle shaped structure between them.

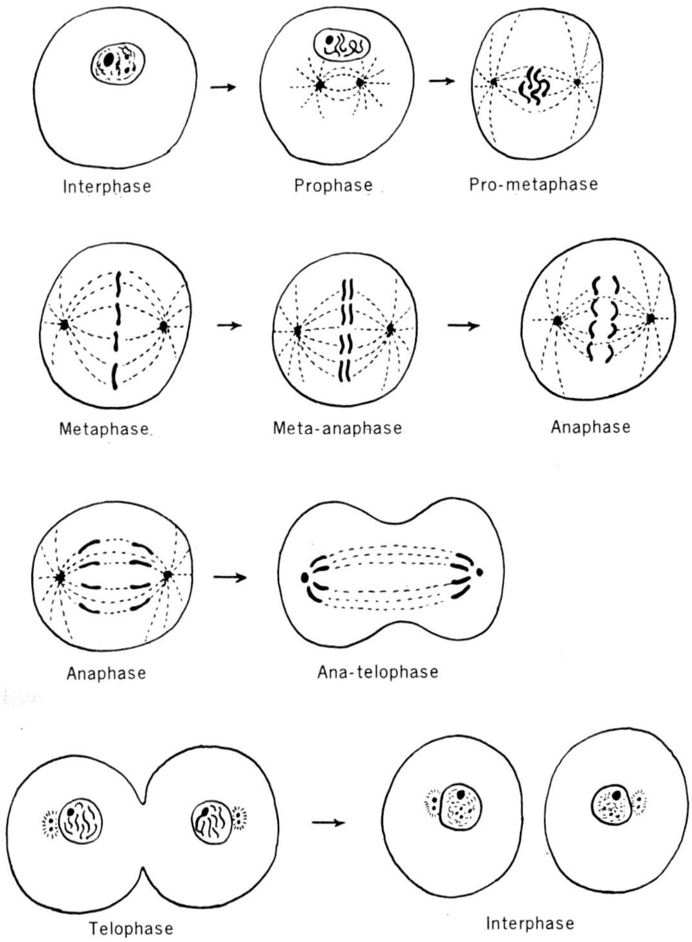

Fig. 2 Course of nuclear division.

2) Metaphase

In this phase, each centrosome moves to an opposite extreme of the cell, and fusiform threads radiate from both poles. Chromosomes line up at the equatorial plane and are attached to the spindle fibers.

3) Anaphase

Chromosomes split lengthwise, and the halves are drawn by the adherent threads to the centrosomes at opposite poles.

4) Telophase

The cytoplasm is constricted in the middle and finally divides into two cells of the same size. The nuclear membrane again reappears around the chromosomes, and the nucleoli return to the nuclei.

The number of chromosomes is the same as before in both cells, and they gradually change into the original chromatin granules.

Thus the interphase of the new cell begins.

Cytoplasm (Fig. 1)

1) Cytoplasmic membrane

This is a lipoprotein boundary membrane surrounding the cell. It gives shape to the cell according to the equilibrium between external and internal conditions. Owing to the cytoplasmic membrane, the cell is able to change its shape fairly freely under various conditions.

2) Mitochondria

These are minute, rod or filament-shaped particles and widely distributed in the cytoplasm. They participate in cellular enzyme reactions, especially oxidation, and metabolism.

3) Golgi apparatus

It is located in the cytoplasm, in a network or vacuolar structure comparatively near the nucleus. Its function is not well known, but it is closely connected with the synthesis of special substances in cellular differentiation and secretion. It is most clearly revealed with silver impregnation.

4) Centrosome

This is a short cylindrical structure located near the nucleus, and is sometimes called the "centriole". In the interphase, it appears under optical microscope as one or two minute globular granules. It is closely related to cilia and other structures on the cellular surface. In the mitotic phase, it splits into two, and each half becomes the center around which chromosomes gather.

5) Cytoplasmic inclusions

These are not normal, basic structures, but intracytoplasmic substances which are produced intracellularly or taken in from the outside. Therefore, their existence is often not peculiar to certain kinds of cells. The shape may be globular or indeterminate, and their size is not uniform.

Inclusions are variable, and may be a fragment of a destroyed cell, droplets of carbohydrate, fat, protein, a substance containing a hydrolytic enzyme (lysosome), pigment of various types and an inorganic substance like dust. Even a viral particle is occasionally seen as an inclusion.

6) Secretory granules and chromophile substances

These are functional cytoplasmic products observed in the cytoplasm of a glandular cell with definite secretory function or in other certain specialized cells. Since RNA and protein granules are frequently observed in an actively proliferating cell, such as a carcinomatous one, they appear basophilic.

7) Structures which appear according to cellular functions

These are special, minute, linear or tubular structures which appear inside, sometimes outside, the cytoplasm according to the functional differentiation of the cell. Tonofibrils, neurofibrils, myofibrils and cilia belong to these structures.

Tonofibrils are observed mainly in epithelial cells (*e.g.* squamous epithelial cells), and at present are thought to be related to keratinization. Cilia, which are significant when found in cytological examinations, are observed in endocervical or bronchial respiratory cells. It may transport substances in a definite direction or may eliminate harmful material.

Physiological functions of cells

All cells, regardless of type, have the following functions and characteristics in common.

1) Metabolism

Various chemical reactions take place within a cell, and these reactions are collectively termed metabolism.

Such metabolic reactions occur constantly in the cytoplasm. Some of them are anabolic while others are catabolic.

2) Absorption

A cell must absorb necessary substances from outside for its metabolism.

3) Excretion

Metabolism produces various by-products, some of which are of no value or may be even harmful to the cells when allowed to accumulate. Thus they are removed from the cell.

4) Secretion

Secretion is classified into external secretion (*e. g.* mucus) and internal secretion (*e. g.* various hormones). Intracellularly produced substances are secreted from the cell to fulfill their functions.

5) Movement

A living cell is able to make partial or whole movement. The cytoplasm flows and the shape of the cell changes during the movement.

6) Conduction

A stimulation given to a cell is conducted from the stimulated part to other parts of the cell. Every cell, more or less, has the ability to conduct such impulses, but a nerve cell has special structure as well as function for conduction.

7) Irritability

A cell is very sensitive to changes in its environment, and external factors greatly influences cellular vitality. Such external factors vary from physical factors like heat, light, and pressure, to chemical ones, such as pH and others.

CYTOLOGIC EXAMINATION OF THE FEMALE GENITAL ORGANS

I. Anatomy of the Female Genital Organs

Vulva

The vulva consists of the mons pubis, perineum, labia majora, labia minora, hymen, vestibule and clitoris. The urethral orifice and introitus vaginae are located medially. The external surface of the labia majora is covered with skin, but the inner surface is covered with mucous membrane composed of stratified squamous epithelium. This mucous membrane covers the vagina and the portio vaginalis. This is one of the most important areas for cytological examination of the female genital organs (Fig. 3).

Vagina

The vagina is a tubular organ about 10 cm long connecting the vulva with the uterus. In a multipara it is comparatively distended and widened, but in a nullipara the vaginal walls are in contact with each other having numerous wrinkles and are highly distensible. The portio vaginalis protrudes into the internal end of the vagina and forms a blind end. The internal surface of the vagina and the surface of the portio vaginalis are covered with a stratified squamous epithelium (Fig. 6), and the cell debris exfoliating from these areas may accumulate mainly in the posterior vaginal fornix, the recess surrounded by the vagina and the posterior part of the portio vaginalis. These cells, as well as endometrial cells as described later, often show definite cyclic changes according to the ovarian cycle. The thickness of the stratified squamous epithelium differs with age. In childhood, there are fewer cellular layers and the epithelium is generally thin, but after the menarche, and in the childbearing age, the epithelium thickens and becomes stratified. In such periods, the glycogen content may be, more or less, increased. After the menopause, it becomes thinner again and atrophic (Fig. 4).

Uterus

The uterus is a pear-shaped muscular organ whose size in the normal state is about $8 \times 5 \times 2.5$ cm. Fallopian tubes are connected to the superior lateral margins of the uterine fundus. The uterus consists of the cervix, the lower third up to the internal orifice, and the body, the upper two-thirds. The cervical portion protruding into the vagina is the portio vaginalis (Fig. 5).

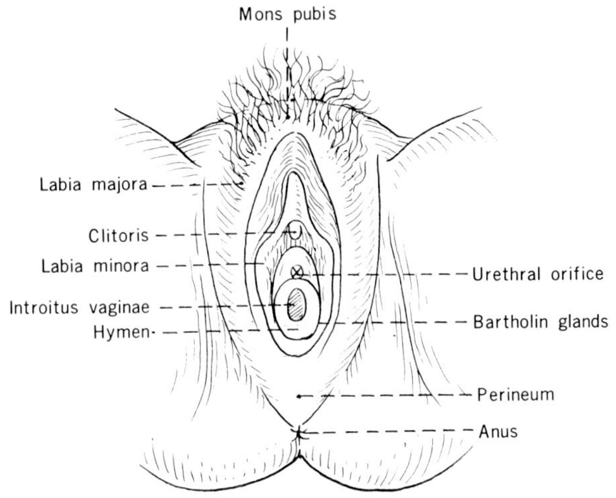

Fig. 3 Female genital organs.

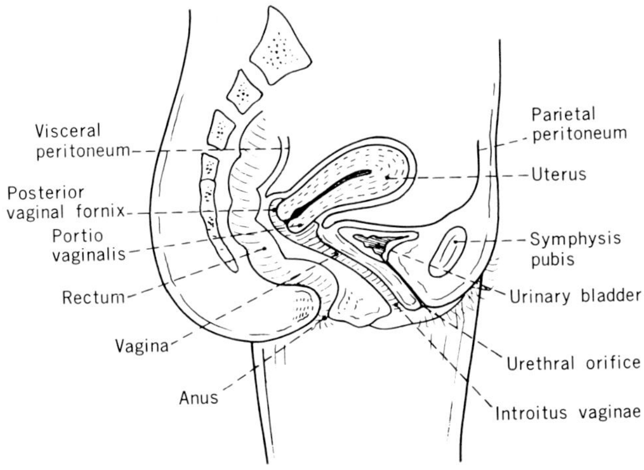

Fig. 4 Female genital organs (sagittal section).

Uterine cervix

The portio vaginalis of the uterine cervix is covered with a stratified squamous epithelium resembling the mucous membrane of the vaginal wall. The cells, however, contain less glycogen than those of the vaginal wall. The internal surface of the cervix is covered with a single layer of mucus-producing columnar epithelium consisting of mucous cells. This is called the endocervix and it extends into an underlying stroma of connective tissue, forming the endocervical mucous glands (Figs. 7 and 8).

The junction between the stratified squamous epithelium of the portio vaginalis and the mucus-producing columnar epithelium of the endocervix is histologically distinct, and is called the squamocolumnar junction (Fig. 7). When this junction is everted, the mucous membrane of the cervix can be clinically seen by speculum examination of the portio vaginalis, but it cannot be seen macroscopically when the junction is inside the cervix.

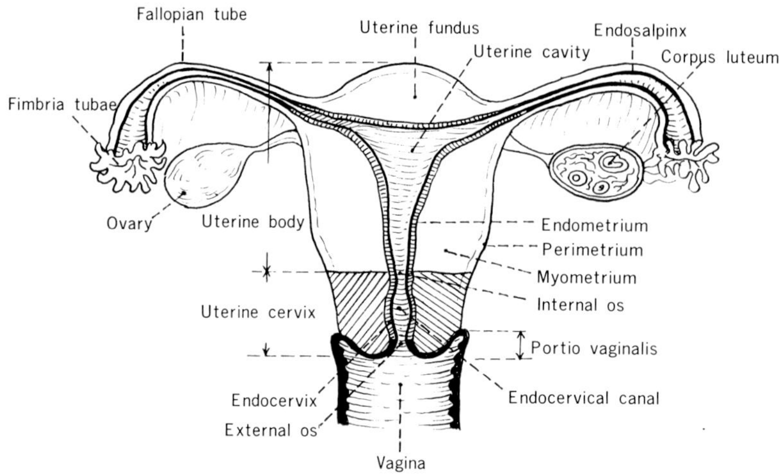

Fig. 5 Uterus and adnexa.

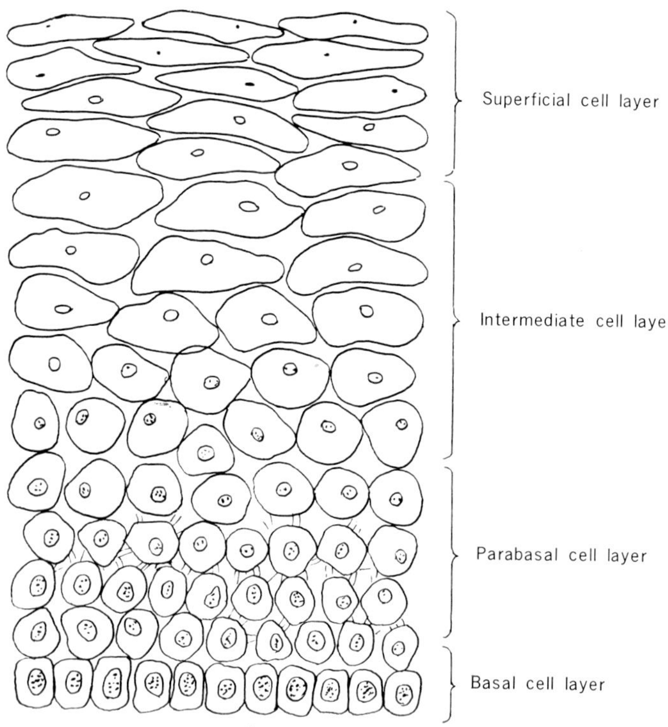

Fig. 6 Stratified squamous epithelium.

Histologically, the endocervical columnar epithelium shows periodic changes according to the ovarian cycle; but unlike the changes in endometrial cells, it is difficult to demonstrate endocervical columnar epithelial cell changes in exfoliated cells.

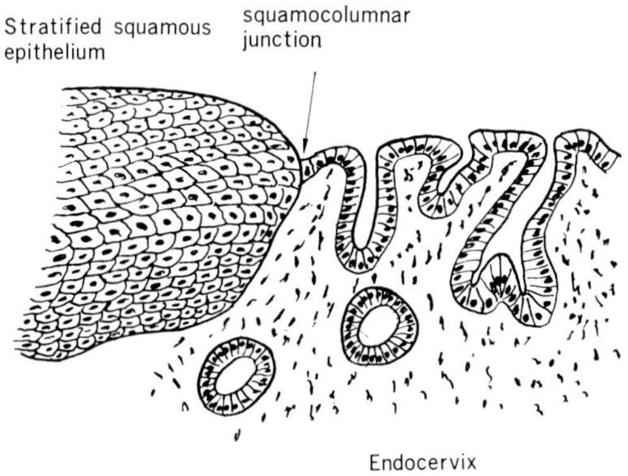

Fig. 7 Squamocolumnar junction.

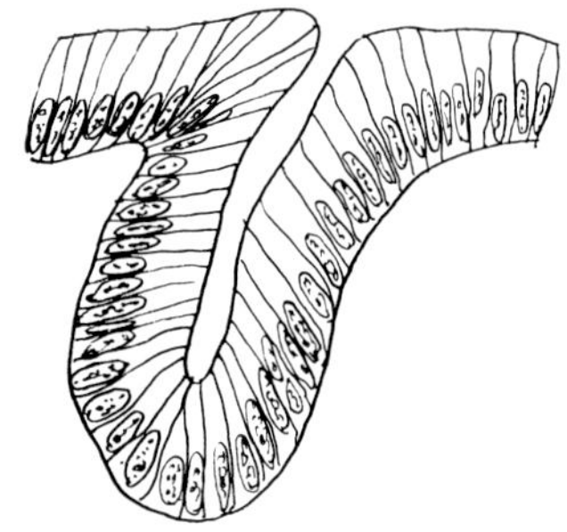

Fig. 8 Endocervical columnar epithelium.

Uterine body

Inside the body of the uterus is a narrow triangular space called the uterine cavity, and the inner surface is lined with endometrium. This lining membrane consists of a single layer of columnar epithelium and an underlying glandular epithelium which opens to the surface. Gaps in the glandular epithelium are filled with stroma, the connective tissue consisting of short fusiform cells. In the normal state, this endometrial epithelium, especially the glandular epithelium, shows remarkable

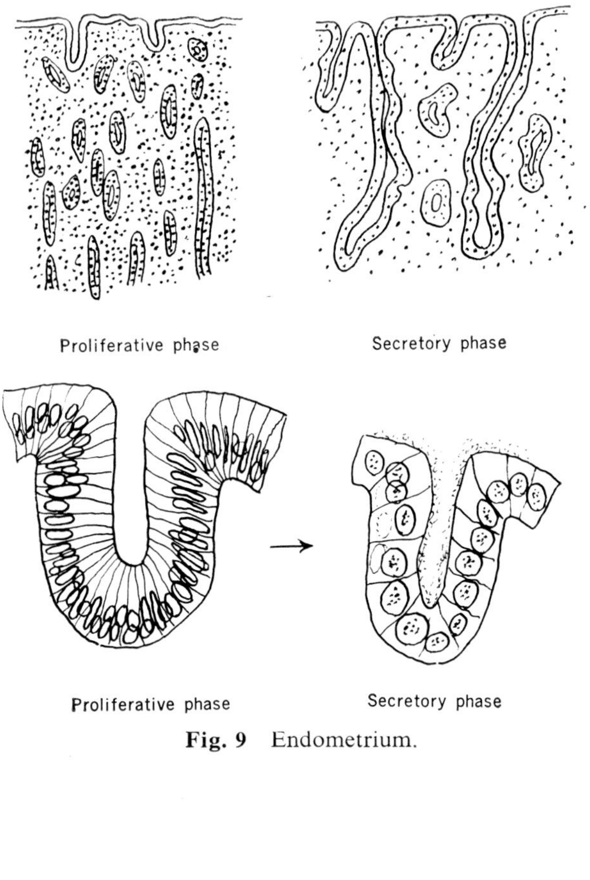

Fig. 9 Endometrium.

cyclic changes according to the ovarian cycle, and secretes glycogen during the childbearing age (Fig. 9).

Fallopian tubes

The tubes, two in number, are slightly tortuous, tubular organs. Each tube is 8 – 12 cm in length and 3 – 5 mm in diameter. It is connected to the lateral superior margin of the uterine fundus distally and, at the other end widens to form the fringe like fimbria which opens into the abdominal cavity. The internal cavity of Fallopian tube is covered with a single layer of columnar epithelium called the endosalpinx, consisting of ciliated cells and non-ciliated secretory ones, among which peg cells are scattered. The ciliated and secretory cells may show cyclic changes, but cells of the endosalpinx are only accidentally found in the ordinary vaginal smear (Fig. 5).

Ovaries

The ovaries are paired ovoid organs about 4×2×1 cm and are situated on each side of the uterus close to the fimbria. The external surface is covered with germinal epithelium. The ovary is connected to the vagina only through the endocervical canal, uterine cavity and Fallopian tube, and is, therefore, hardly an object of the vaginal cytologic examination in the normal state. In ovarian tumor, however, observation of the tumor cells in ascites may help the diagnosis (Fig. 5).

II. Normal Cells from Epithelium of the Female Genital Organs

Stratified squamous epithelial cells of the vagina and the portio vaginalis

These cells are classified into superficial cells, intermediate cells and deep cells (parabasal and basal). It seems highly unlikely that anyone would be able to diagnose a cell as a true basal cell because the basal cell is rarely exfoliated in a normal state and, beside that, no one can tell in a vaginal smear if a cell definitely originates from the stratum germinativum.

General characteristics (Fig. 10)

1) As a whole cell
   1. Size : 10 – 40 or 50 microns in diameter. The more superficial cells are larger than the deeper cells.
   2. Shape : Comparatively round or rather elliptical. The superficial cells may be polyhedral.

2) Nucleus
   1. Number : Usually one, but sometimes multinucleated in inflammation.
   2. Location : Approximately at the center of the cell.
   3. Size : Medium. Smaller in the more superficial cells and larger in the deeper cells. The nucleus is round, pyknotic, and small in the superficial cell, while the nuclear-cytoplasmic ratio is much larger in the deep cell.
   4. Shape : Round or elliptical. As stated above, rounder in the more superficial cells and more elliptical in the deeper cells. The nuclear border is comparatively distinct in the deep cell.
   5. Chromatin : Chromatin is more homogeneous in the more superficial cells, while it is rather unevenly distributed and coarsely granular in the deeper cells. Chromocenters are observed in the latter.
   6. Nucleolus : Usually not observed.
   7. Vacuolation is not observed unless in cellular degeneration.

3) Cytoplasm
   1. Size : Larger in the more superficial and smaller in the deeper cells.
   2. Shape : Polygonally round or elliptical in the superficial and rounder or more elliptical in the deeper cells.
   3. Border : The border is often wrinkled in superficial cells. This tendency is especially marked in the secretory phase of the menstrual cycle.
   4. Thickness : Thicker in the deeper and thinner in the more superficial cells.
   5. Generally homogeneous and transparent but sometimes vesiculated.
   6. Staining reaction : Generally, eosinophilic and stained reddish in the superficial, and basophilic and stained bluish in the deeper cells. This staining reaction, however, changes with various factors.
   7. Vacuolation : Usually not observed, but sometimes observed in degeneration or in the deeper cells.

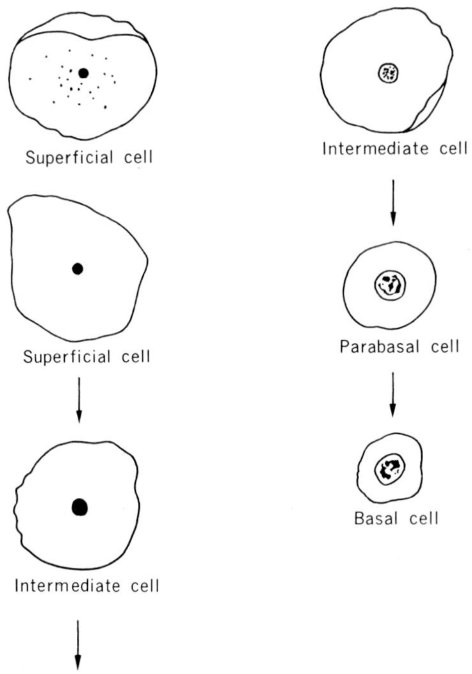

Fig. 10 Stratified squamous epithelial cells of the vagina and the portio vaginalis

8. Cytoplasmic inclusions : Fragmented nuclear substances in the form of extremely minute granules that stain deeply may be observed in the cytoplasm of the superficial cells.
9. Occasionally a narrow, perinuclear halo can be present around the nucleus by inflammation.

1. Superficial cells

a. Low magnification

Large cells with small and pyknotic nuclei.

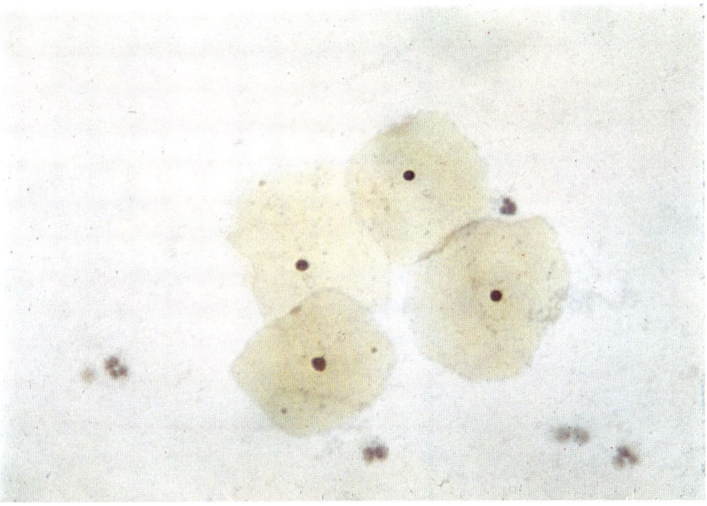

b. High magnification

1. Nucleus: Round or elliptical. Pyknotic and homogeneous.
2. Clear border.
3. Perinuclear halo is observed around the nucleus.
4. Cytoplasm: Abundant, polygonal, homogeneous, thin and somewhat translucent.
5. The border is folded.

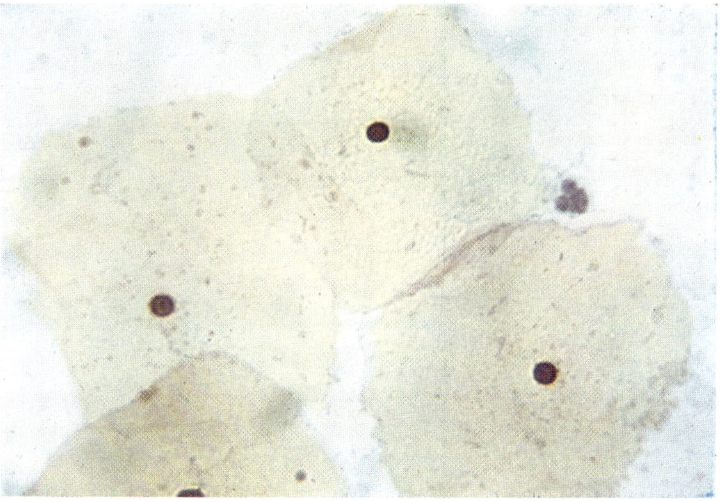

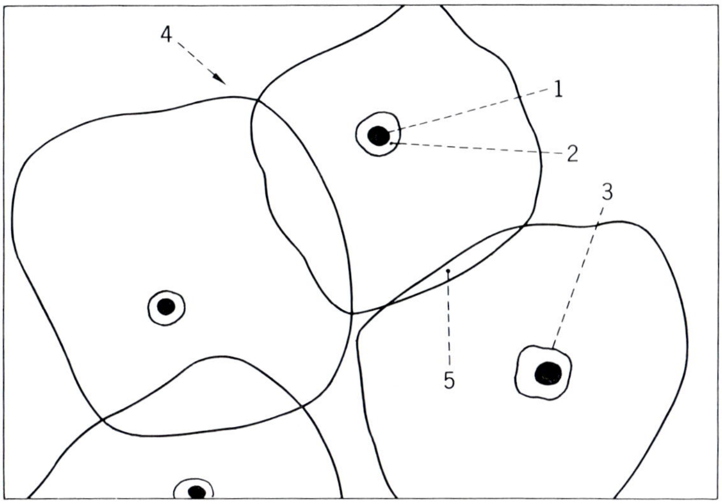

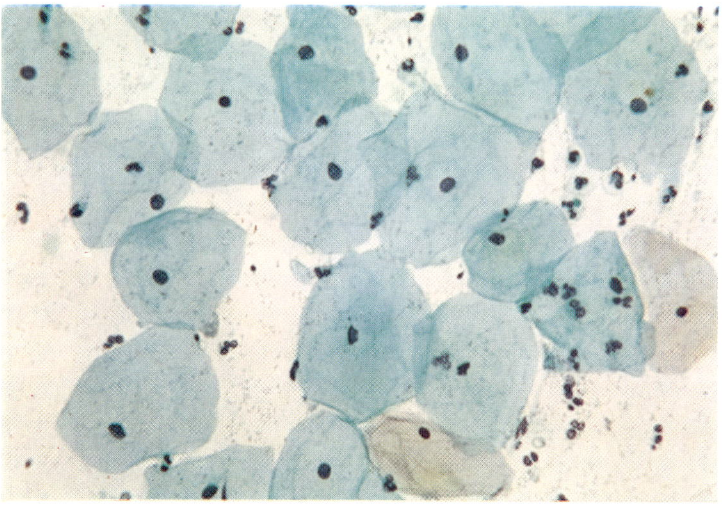

2. Superficial cells

a. Low magnification

Numerous large cells are observed, and the nuclei are slightly larger than those described previously.

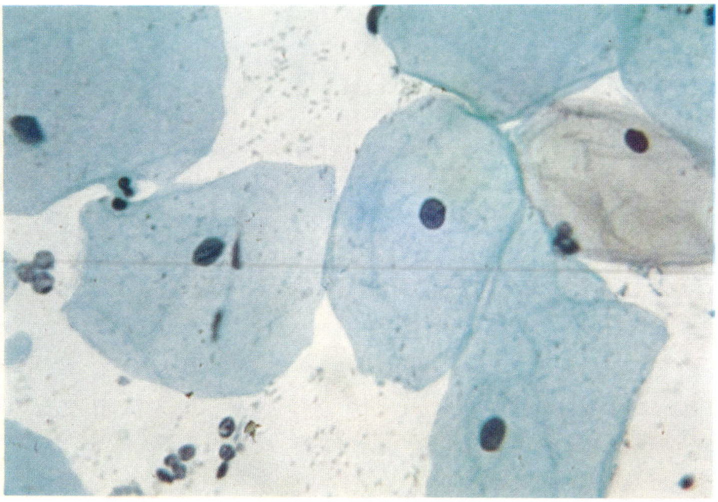

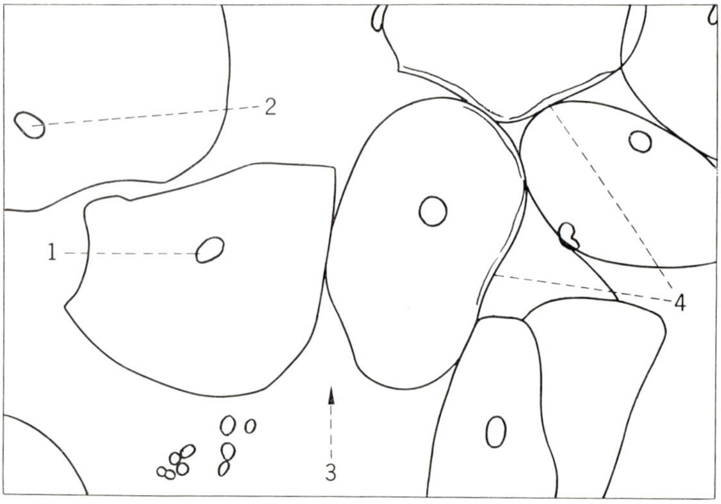

b. High magnification

1. Nucleus: Round or elliptical and pyknotic. Slightly larger than the nuclei of those described previously. Homogeneous.
2. A little less chromatin. The nuclear structure is sometimes observed.
3. Cytoplasm: Abundant, thin and transparent. Sometimes rather polygonal in shape.
4. The border may be wrinkled.

3. Intermediate cells

a. Low magnification

As a whole cell: Nearly as large as a superficial cell.

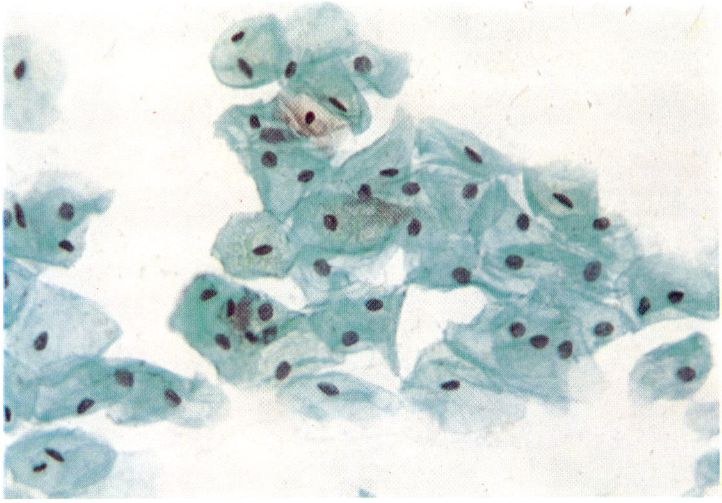

b. High magnification

1. Nucleus: Larger than that of superficial cells. Round, or more elliptical.
2. Less chromatin than in superficial cells. Lightly stained and finely granular.
3. Nuclear border is distinct. No chromocenter, nucleolus or vacuole is observed.
4. Cytoplasm: Variously shaped and slightly thicker than that of superficial cells. The border is distinct.
5. Wrinkling of cell border is observed. Vacuolation is not noticed.

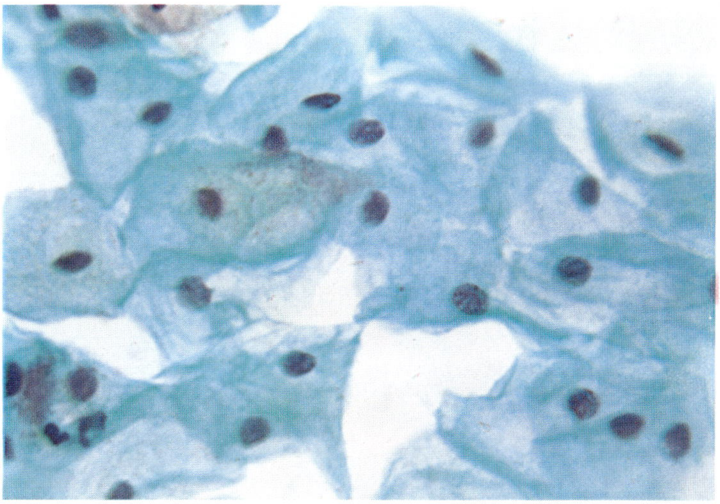

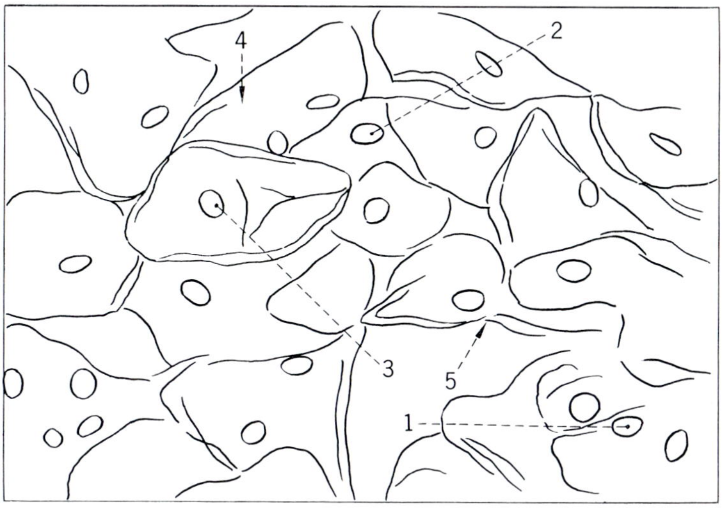

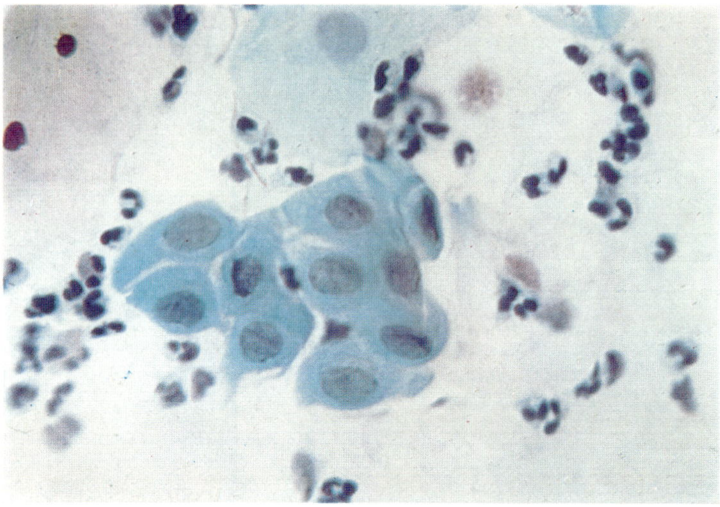

4. Parabasal cells

a. Low magnification

Among numerous, scattered leukocytes, a group of cells with comparatively large nuclei is observed.

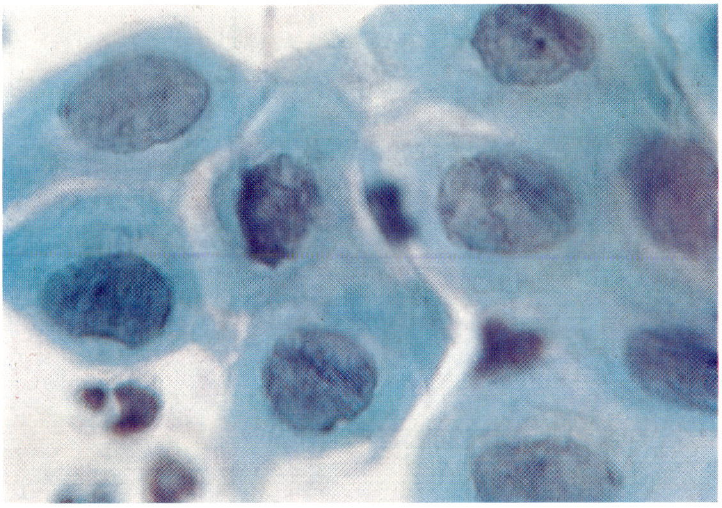

b. High magnification

1. Nucleus : Large round or elliptical. Nuclear-cytoplasmic ratio is larger than in superficial cells.
2. Nucleus is occasionally located eccentrically, but usually at the center.
3. Chromatin granules are more distinct.
4. Cytoplasm : Less abundant than in superficial cells and partially thickened. The shape is round, elliptical or frequently polygonal.
5. Sometimes there is perinuclear halo surrounding the nucleus.

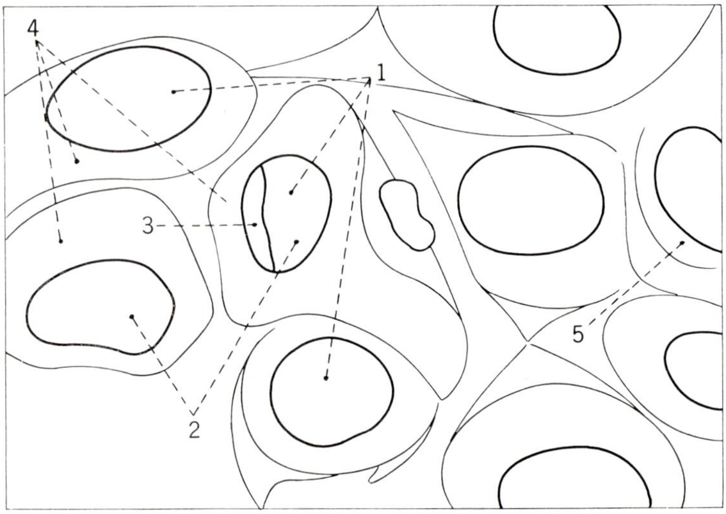

5. Parabasal cells

a. Low magnification

Every cell is much smaller than superficial or intermediate cells, but the nucleus is disproportionately large.

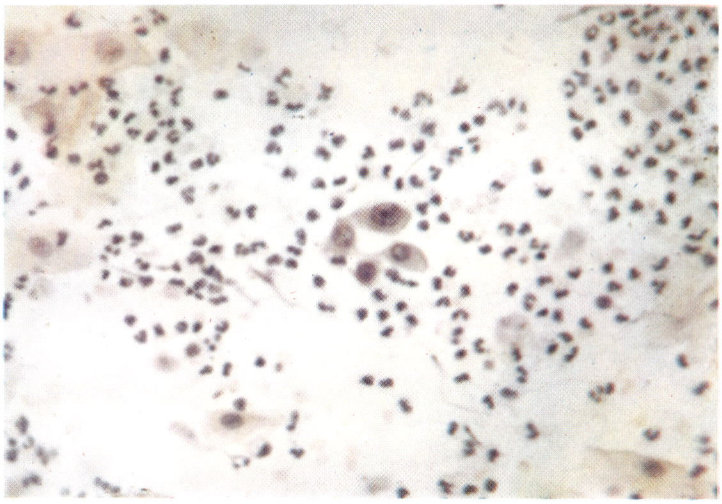

b. High magnification

1. Nucleus : Round or elliptical, much larger than that of the superficial cell, and fairly uniform in size.
2. Sometimes located near the border of the cell, but usually at the center. Chromatin granules are clearly observed and a chromocenter is seen.
3. The nuclear membrane is distinct. The nucleus of the basal cell is larger than that of the parabasal cell. No nucleolus is observed.
4. Cytoplasm : Less abundant than in superficial cells. The border is clear and appears even thicker.
5. Shape : Round or elliptical.
6. Sometimes polygonal as if the original shape in the tissue remains.
7. Partially thickened.
8. Inflammatory cells, like leukocytes, are often observed in the background.

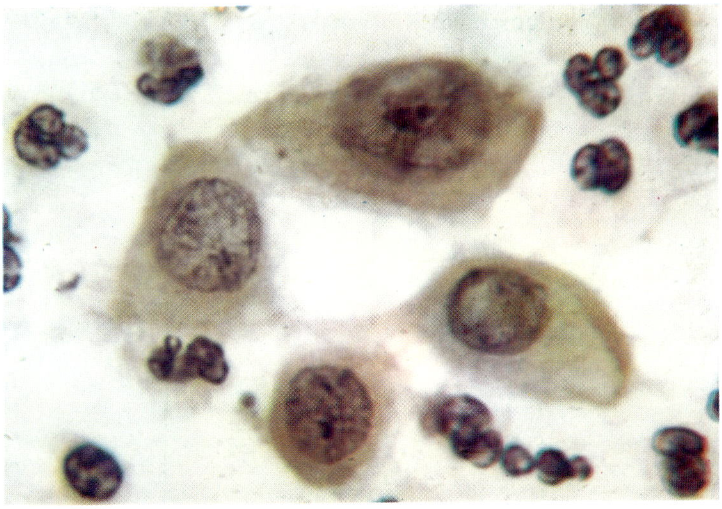

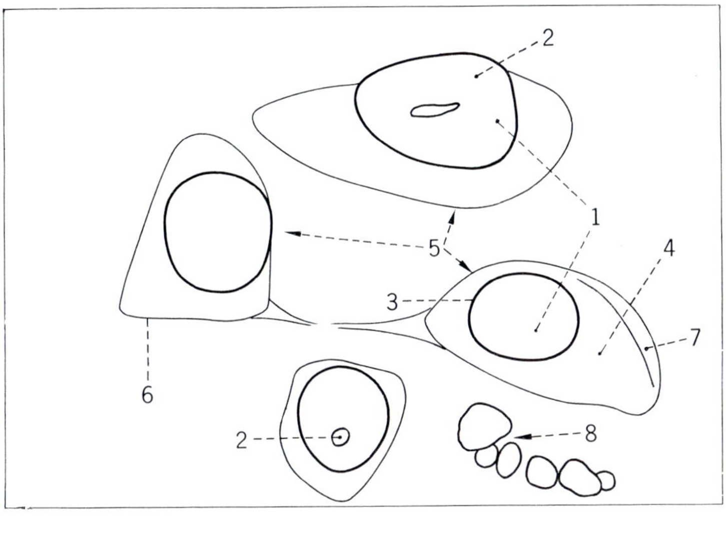

Cyclic changes of vaginal cells — General characteristics (Figs. 11–17)

1) Early follicular phase

Mainly large and small basophilic intermediate cells with vesicular nuclei. Scattered leukocytes.

2) Middle follicular phase

Large intermediate cells and superficial cells. The nuclei are comparatively small and fairly dense. Variously stained.

3) Late follicular phase

Numerous superficial cells. Small and pyknotic nuclei. Eosinophilic cytoplasm. A few leukocytes.

4) Post-ovulatory phase

The superficial cells observed in the late follicular phase have slightly decreased turgor. Scattered intermediate cells. The border of the cytoplasm is folded. A few leukocytes.

5) Middle to late luteal phase

Mainly intermediate cells. Cells adhere to each other and are disorganized. The border of the cytoplasm is folded. A fairly large number of leukocytes. Döderlein bacilli. Cytolysis and stripped nuclei.

6) Menstrual phase

Mainly intermediate cells. Vesicular nuclei. Numerous erythrocytes and leukocytes. Cell debris.

The above changes roughly represent that, in the normal menstrual cycle, superficial cells with small and pyknotic nuclei increase in the middle phase of the cycle and then gradually decrease, and intermediate cells with slightly larger nuclei increase in the latter half of the cycle. It is, of course, necessary to examine the karyopyknotic and eosinophilic indices also. It is often very difficult to determine the definite date of the cycle from an individual vaginal smear. This is because the length of the cycle differs with individuals. For instance, the typical cytological feature of the 14th day of the normal 28-day cycle can appear on the 7th or 8th day of the 21-day cycle. On the other hand, the cytologic presentation of the 14th day of a prolonged cycle may correspond to that of the 10th day of the normal cycle. Furthermore, the hormone level (Fig. 18) differs individually, and inflammatory changes often complicate an evaluation of the cytological pattern of the vagina. Therefore, it is quite risky to estimate the date of the cycle from the vaginal cell picture. The basal body temperature chart (Fig. 19) or other gynecological observations should always be taken into consideration together with vaginal cytology.

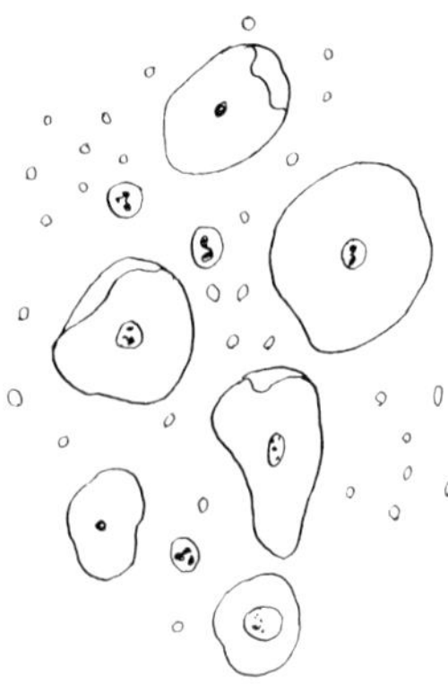

Fig. 11 Early follicular phase.

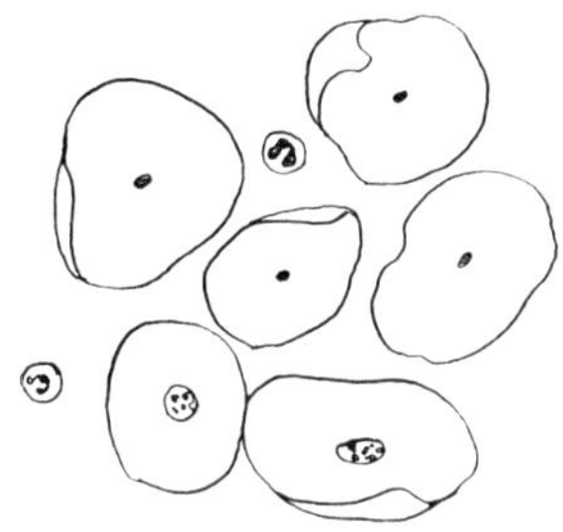

Fig. 12 Middle follicular phase.

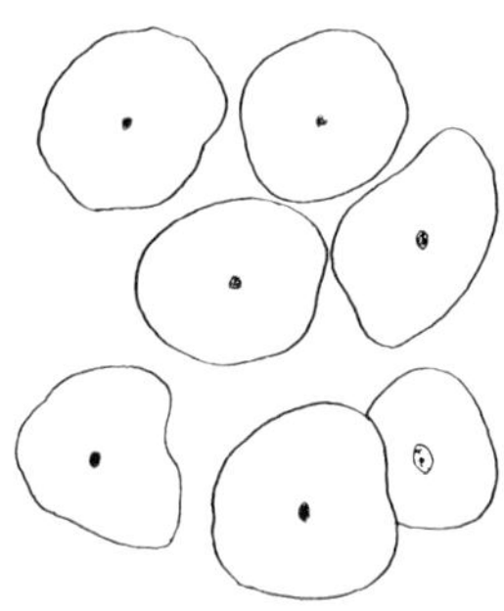

Fig. 13 Late follicular phase.

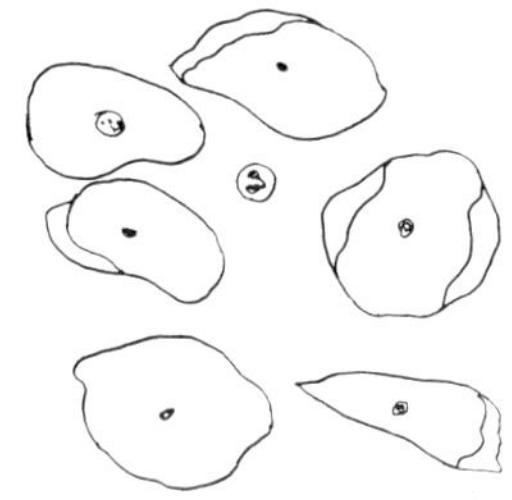

Fig. 14 Post-ovulatory phase.

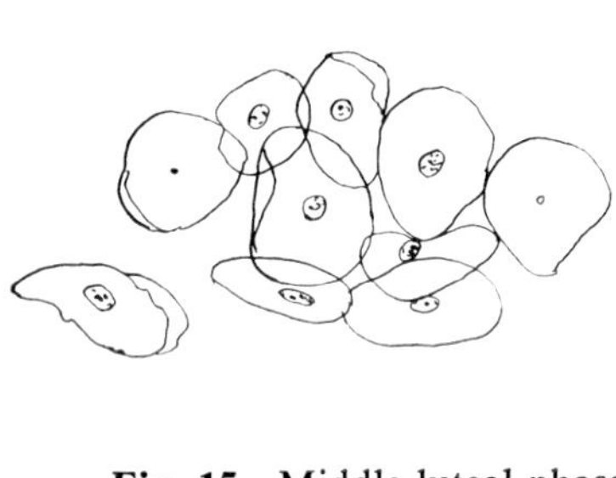

Fig. 15 Middle luteal phase.

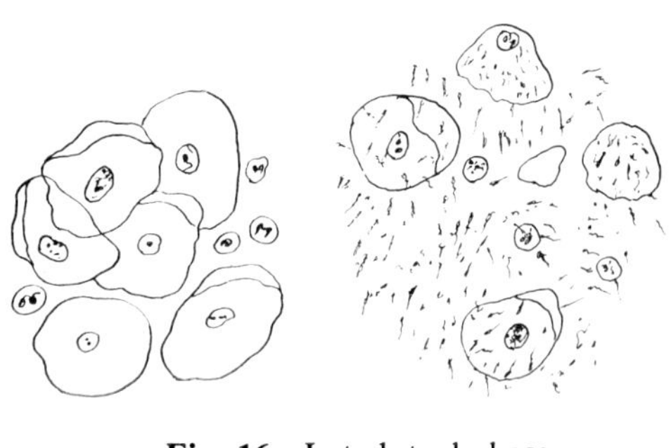

Fig. 16 Late luteal phase.

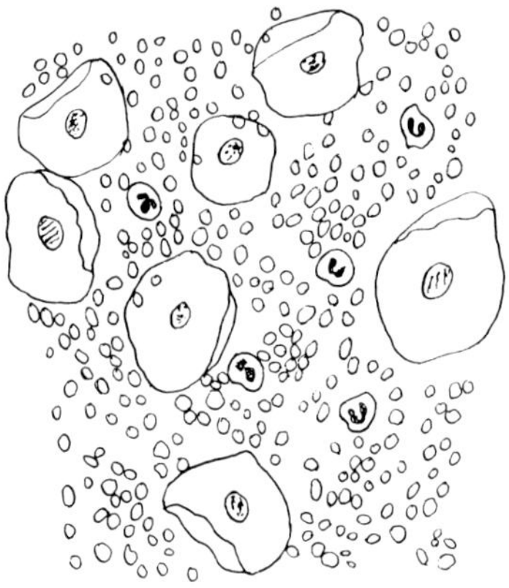

Fig. 17 Menstrual phase.

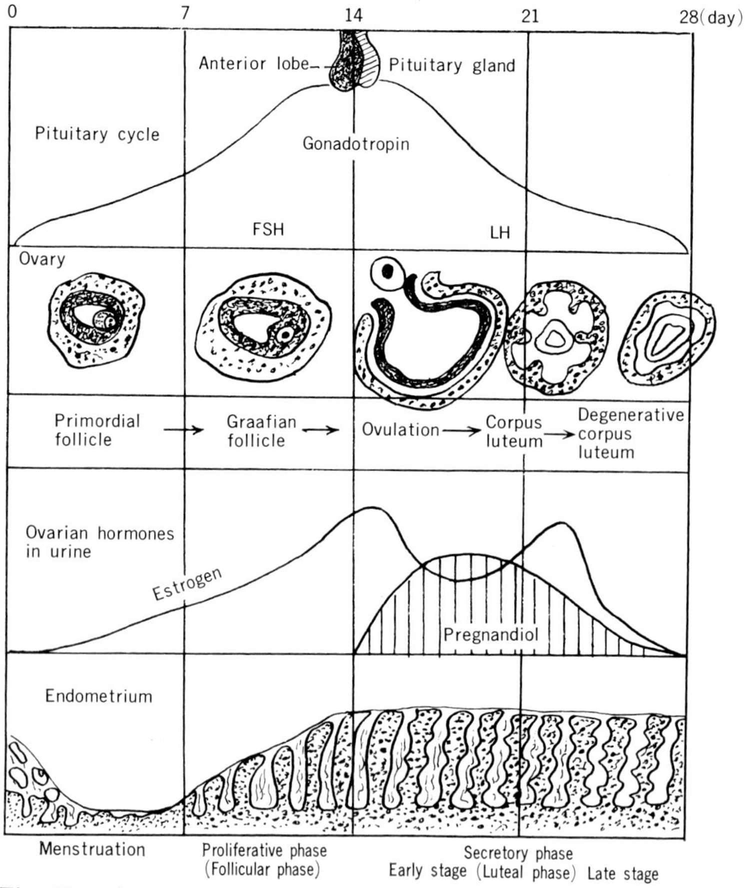

Fig. 18 Diagrammatic representation of endocrine relationships in the normal menstrual cycle.

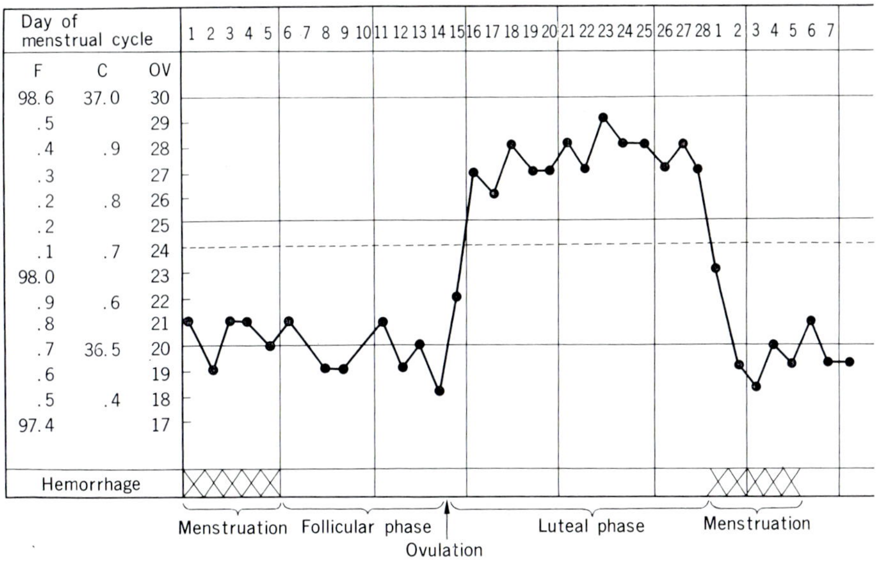

Fig. 19 Normal basal body temperature.

6. Early follicular phase
Intermediate cells are dominant. The nucleus is rather large and vesicular.

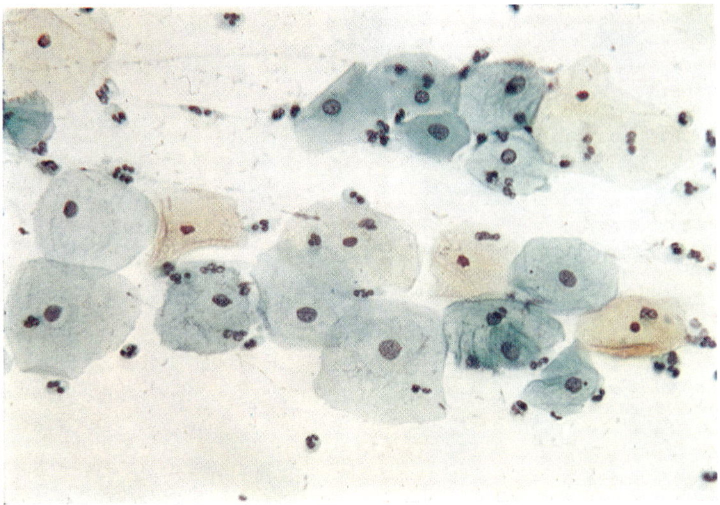

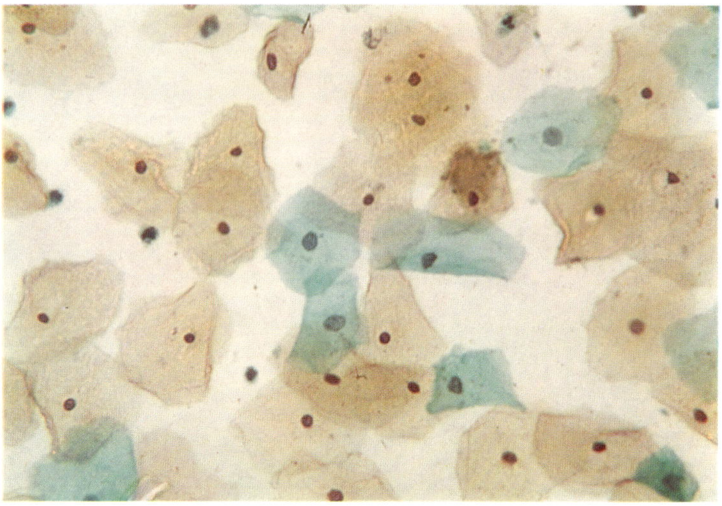

7. Middle follicular phase
Superficial cells with pyknotic nuclei appear.

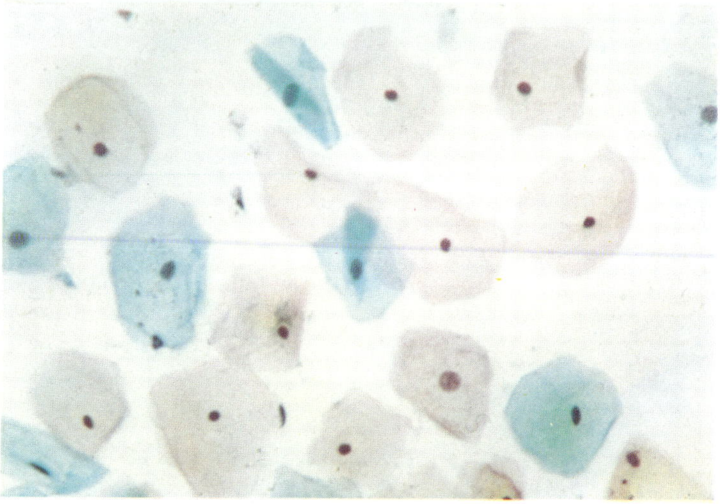

8. Late follicular phase
The majority are superficial cells with markedly pyknotic nuclei.

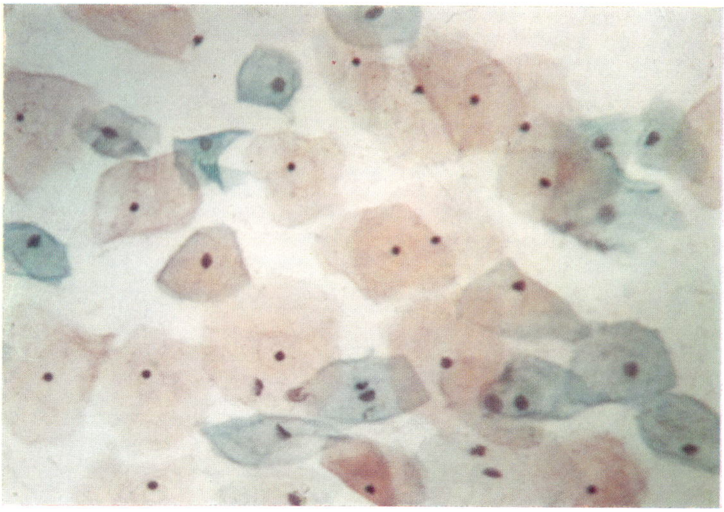

9. Post-ovulatory phase
The cytoplasm of superficial cells are slightly decreased in turgor.

10. Middle luteal phase
Intermediate cells with slightly larger nuclei predominate. The cells are grouped together.

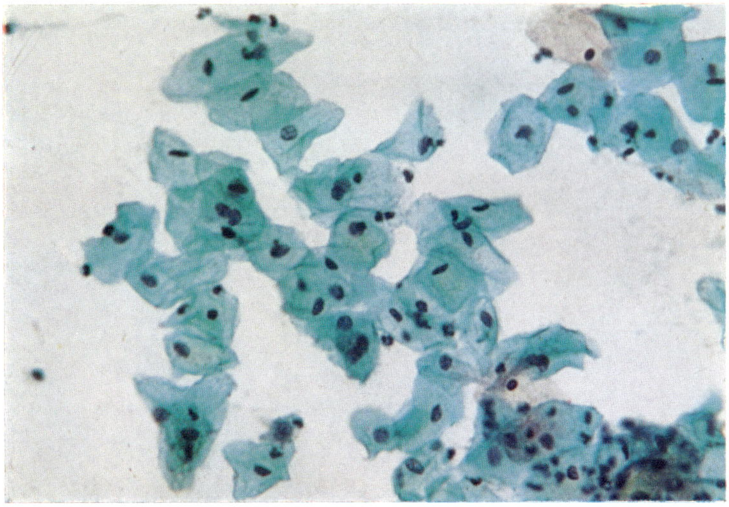

11. Late luteal phase
Numerous leukocytes and Döderlein bacilli can be seen. Therefore, cytolysis is prominent.

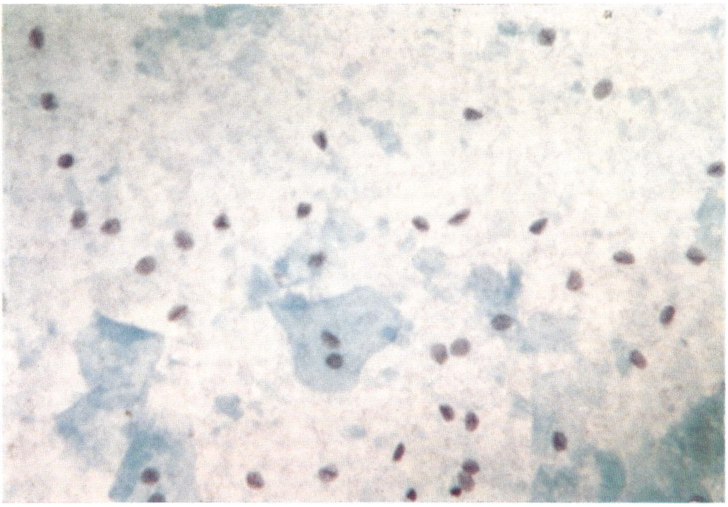

12. Menstruation
Intermediate cells are scattered among numerous erythrocytes and leukocytes.

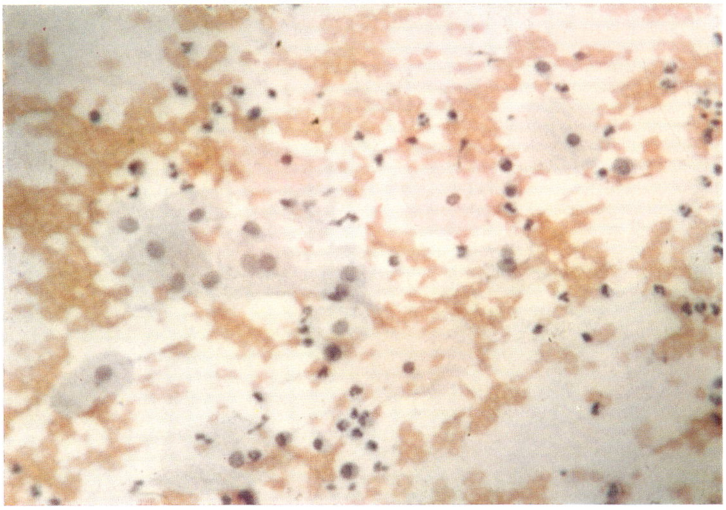

Endocervical cells

General characteristics (Fig. 20)

1) As a whole cell

The cells appear as an elongated ellipse or goblet-shaped, as large as a parabasal cell, and the large and rather elliptical nucleus is eccentrically located near one end. The cells are often seen in sheets. Sometimes stripped nuclei without cytoplasm are observed.

2) Nucleus

1. Round or elliptical.
2. Large for the size of the cell.
3. The border is distinct and the nuclear membrane looks thick.
4. Chromatin is finely granular and occasional clumps of chromatin are observed.
5. No nucleolus or vacuole is seen.
6. Sometimes more than two nuclei overlap.

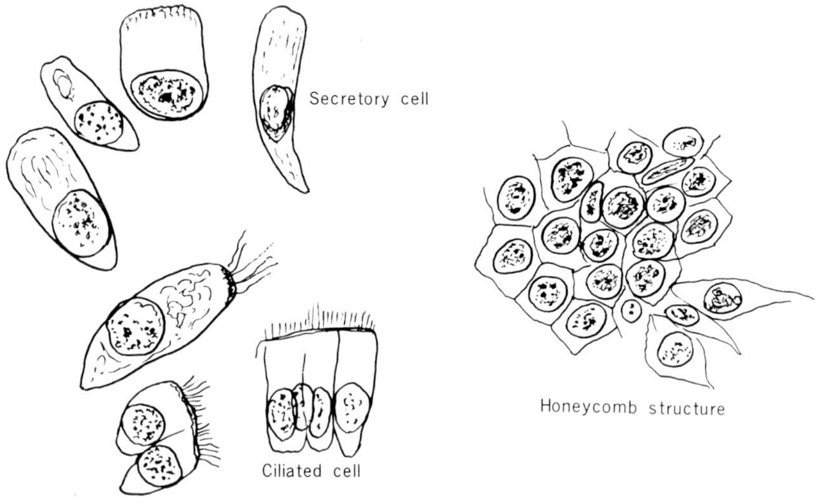

Fig. 20 Endocervical cells.

3) Cytoplasm

1. Shape : Characteristically long elliptical or goblet-shaped.
2. In the ciliated cell, the border of cilia is slightly thickened and distinct. The other borders are sometimes indistinct, and especially in degeneration, they have a lacy appearance.
3. Basophilically stained Slightly vacuolated.
4. A honeycomb structure can be seen when the cells gather on end in sheet.

13. Endocervical cells

a. Low magnification

Among scattered leukocytes and three of squamous cells, tall, columnar, and elliptical cells are drawn up in a cluster.

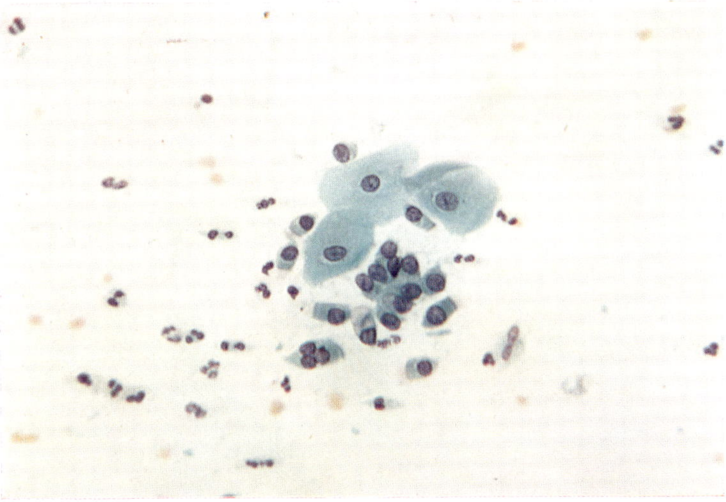

b. High magnification

1. Nucleus: Round or elliptical. Chromatin is finely granular. Clumps of chromatin which may be called chromocenters, are observed here and there. No nucleolus or vacuole is observed.
2. The border of the nucleus is quite distinct and has an increased amount of chromatin.
3. Cytoplasm: Columnar and elliptical with a lacy appearance.
4. Vacuolation in the cytoplasm.
5. Goblet-shaped cytoplasm. There occurs a prominent transverse line at the lumenal end. This is flat and may look thickened.
6. Numerous cilia are seen perpendicular to this terminal plate. They usually stain pinkish.

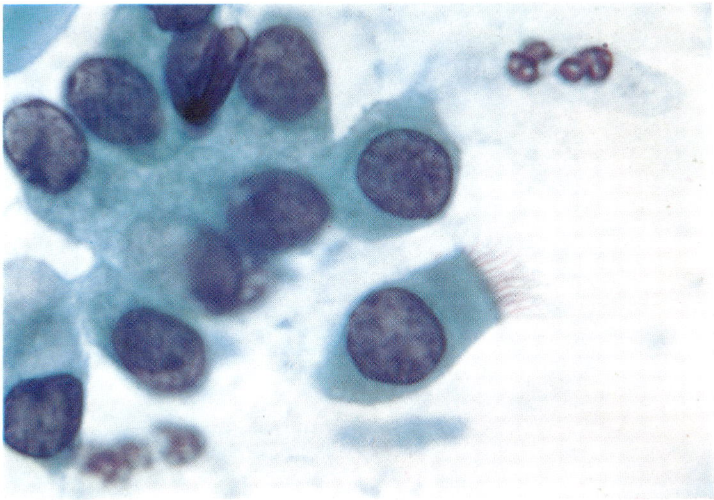

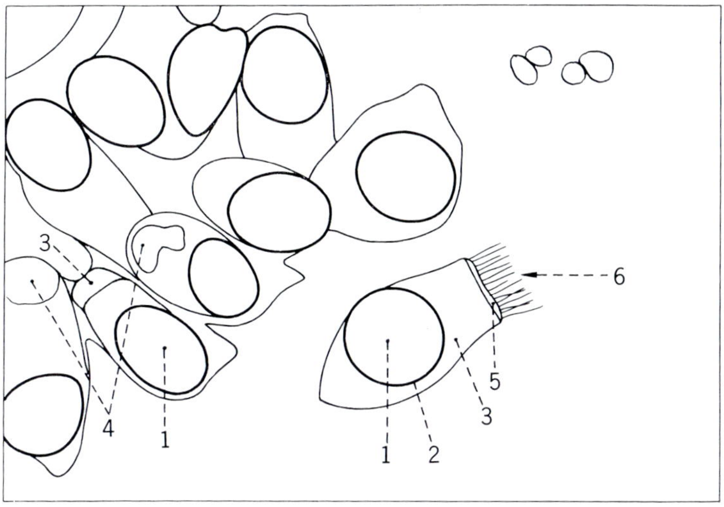

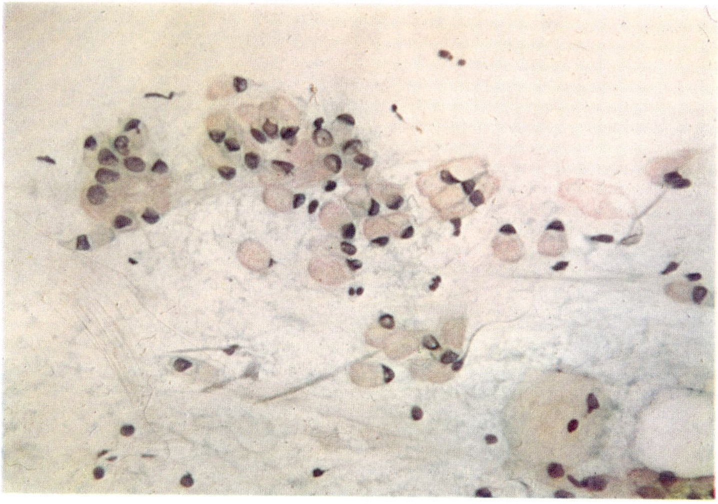

14. Endocervical cells

a. Low magnification

Clusters of columnar and goblet-shaped cells are observed.

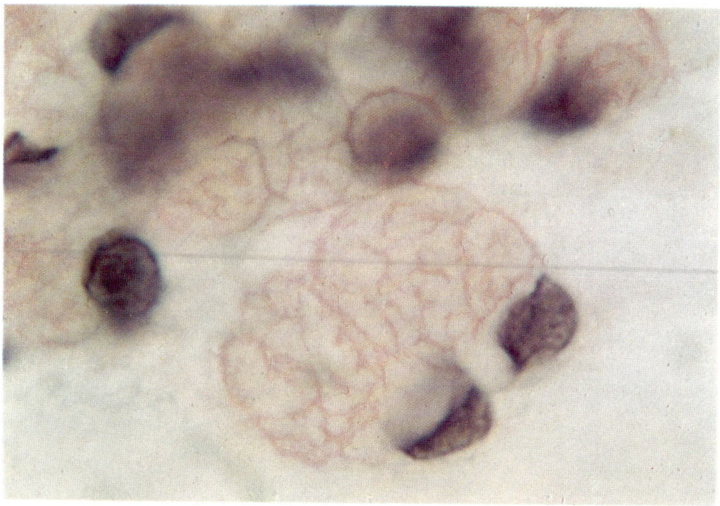

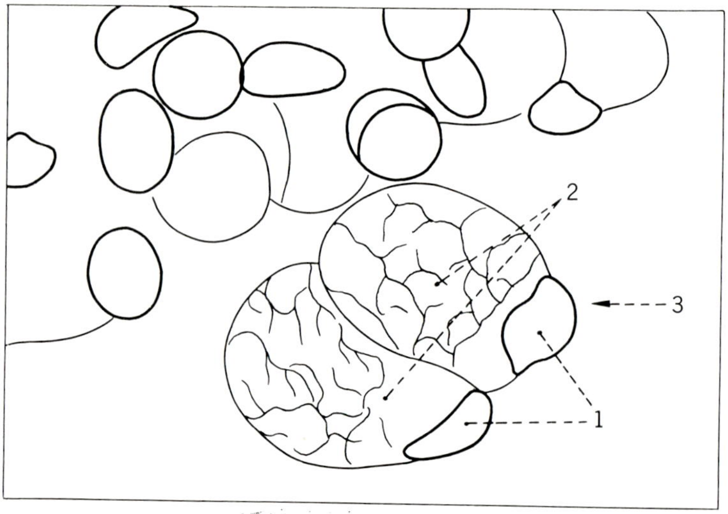

b. High magnification

1. Nucleus: Generally round with evenly distributed chromatin. Chromocenters are noticed. The nuclear membrane is thick and the border is distinct.
2. Cytoplasm: Goblet-shaped, faint and lacy.
3. The nucleus lies in the proximal part of the cell or basally.

15. Endocervical cells

a. Low magnification

Among scattered squamous epithelial cells, a group of cells in a honeycomb structure is observed.

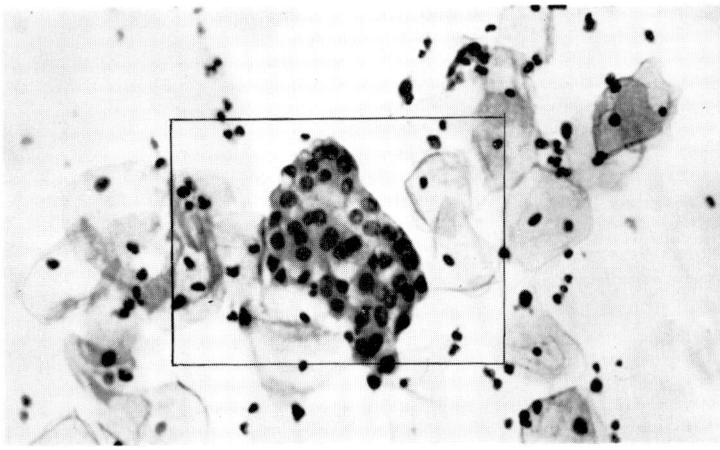

b. High magnification

1. Nucleus: Round or elliptical and uniform in size. Coarse chromatin granules are observed.
2. Cytoplasm: Vesicular. The border of each cell is fairly distinct; thus a honeycomb structure is formed.
3. Squamous epithelial cells.
4. Leukocytes.

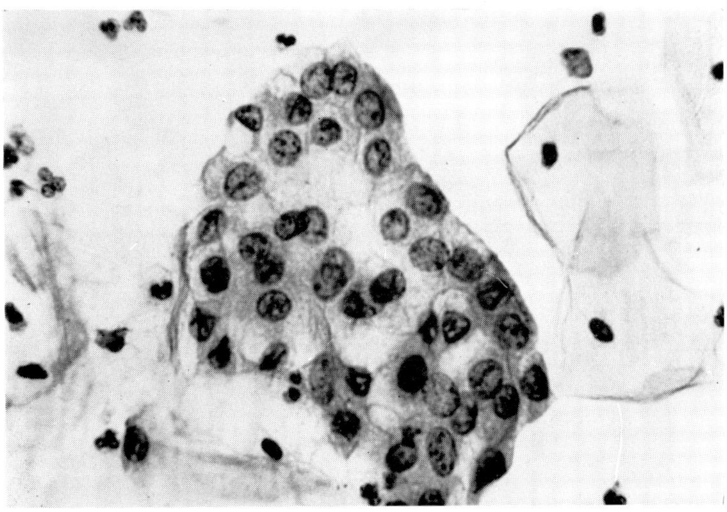

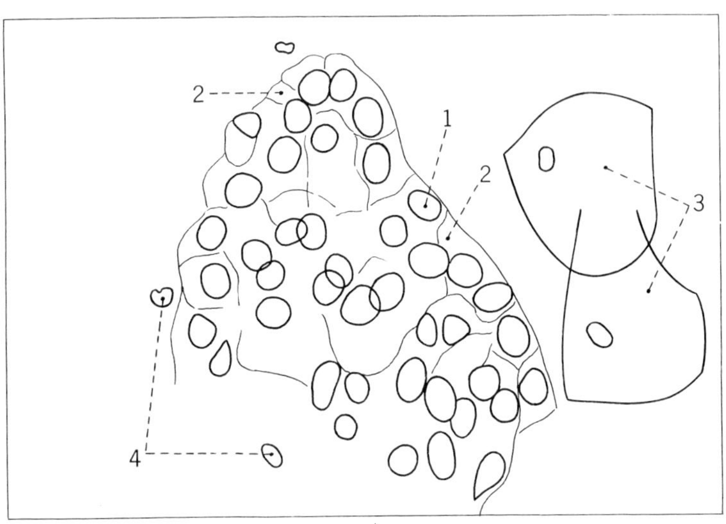

Endometrial cells

General characteristics (Fig. 21)

The endometrial cells observed in the vaginal smear are ciliated columnar epithelial cells originating in the inner lining of the uterine cavity, glandular cells and stromal cells. It is difficult to distinguish glandular cells from superficial epithelial cells in the vaginal smear. Cilia are often indistinct.

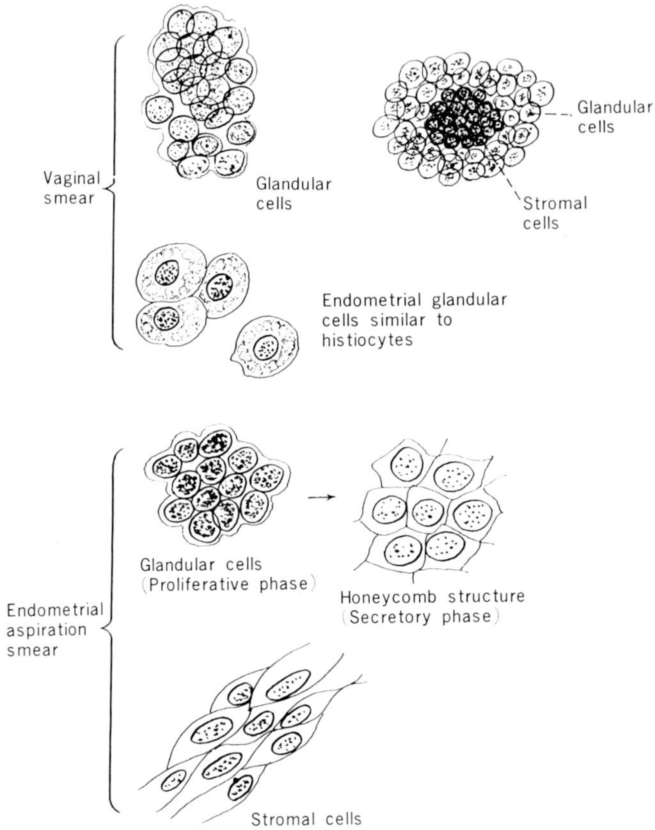

Fig. 21 Endometrial cells.

1) Glandular cells

Glandular cells may appear in the smear during menstruation or until about the 12th day of the menstrual cycle. These cells may be scattered or in groups overlapping each other. Sometimes they are in a doughnut shape surrounding stromal cells. Cells taken directly from the endometrium by curettage or aspiration are better preserved than those exfoliated into the vagina. Accordingly, their shapes are quite different in their appearance (*cf.* Fig. 21 and Plates 18–19).

Nucleus :

1. Round and uniform in size.
2. About 1-3 times larger than leukocytes.

3. The border is distinct.
4. Generally, finely granular (in the secretory phase) or coarsely granular (in the proliferative phase).
5. Small clumps of chromatin are often observed.
6. A nucleolus is sometimes seen.
7. Sex chromatin is occasionally noticed.
8. No vacuoles.
9. Mitosis is rarely noticed in the vaginal smear.

In the vaginal smear, it is difficult to distinguish cyclic changes in the nuclei of cells that naturally exfoliate as these cells may have degenerated. However, the changes are clearly seen in those of cells directly obtained by aspiration or curettage.

Cytoplasm :

1. Cell borders are indistinct when cells are in a group.
2. The cytoplasm looks lacy and delicate.
3. Sometimes it may be eosinophilically stained.
4. Fairly distinct large vacuoles are noted in the degenerated cytoplasm, and they contain leukocytes and other substances.

2) Cyclic changes

Observed in curettage or aspiration smears.

Table 1

	Proliferative phase (Plate 18)	Secretory phase (Plate 19)
As a whole	Fairly dense in a cluster	Loose in a cluster
Nucleus	1-3 times larger than leukocytes. Round, elliptical and occasionally polygonal. Chromatin is coarsely granular. Deeply stained. Chromocenters are observed. The border is rather indistinct.	Larger than in the proliferative phase. Rounder. Finely granular. Lightly stained. Distributed evenly. Chromocenters are clearly seen. The border is distinct and chromatin is often increased at the border.
Cytoplasm	Less abundant. The border is indistinct. Dense.	More abundant. The border is usually distinct. Honeycomb structure. Lacy in appearance.

3) Stromal cells

Identification of stromal cells in the vaginal smear is difficult, but it is not so difficult in curettage or aspiration smears after the examiner gains appearance.

As a whole cell : Usually rather elongated.

Nucleus : Elliptical, elongated and varies in size. Chromatin is coarsely granular and small clumps are scattered throughout. The smaller ones are all equally deeply stained. The border of the nucleus may be distinguishable, but is not very distinct.

Cytoplasm : Elongated and rather dense. Lightly stained basophilically. The border is indistinct.

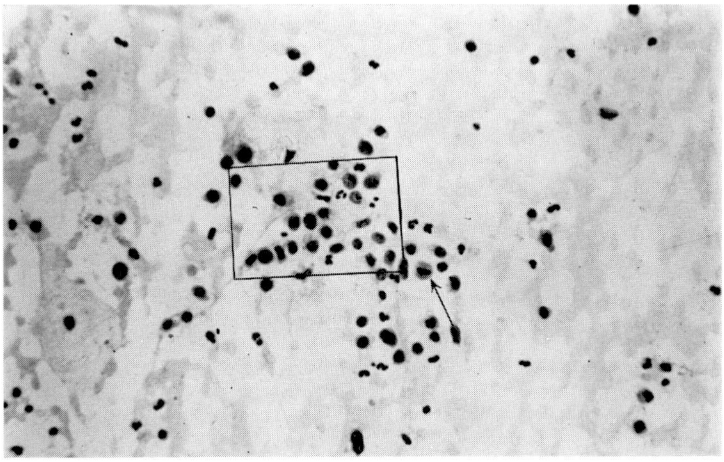

16. Endometrial glandular cells (Vaginal smear)

a. Low magnification

Fairly large round cells are scattered among a few squamous epithelial cells and numerous degenerated erythrocytes. Mitotic figure is noticed in the cell with an arrow. This cytologic illustration represents the proliferative phase.

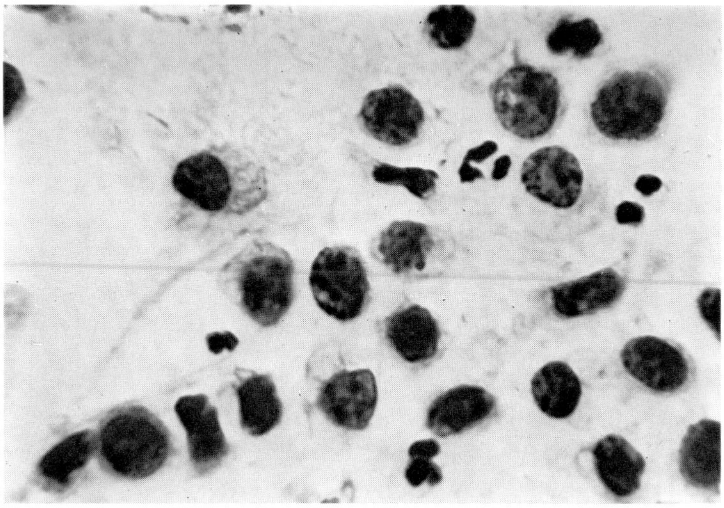

b. High magnification

1. Nucleus : Usually round or elliptical, but occasionally rect angular. Chromatin is coarsely granular. About 2–3 times larger than a leukocyte.
2. Cytoplasm . Lacy with indistinct borders.
3. The cytoplasm is almost lost due to degeneration.

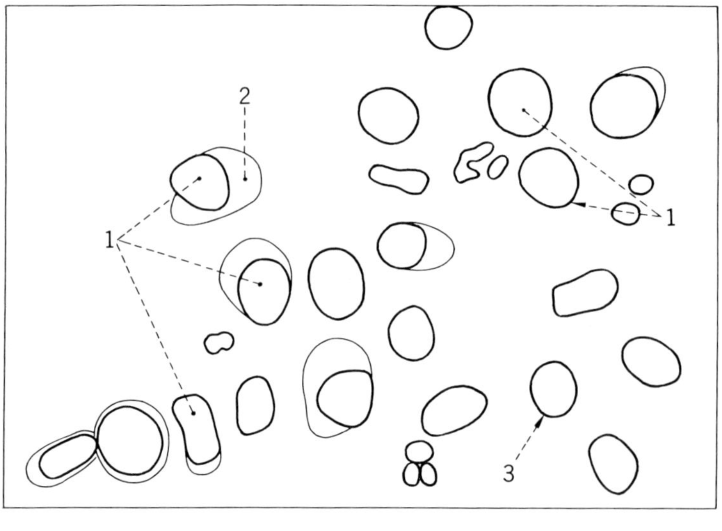

17. A cluster of endometrial glandular and stromal cells (Vaginal smear)

A doughnut-shaped cellular group is observed among numerous inflammatory cells.

1. The central dense cluster consists of stromal cells.
2. The surrounding cluster in a doughnut-shaped ring consists of glandular cells.
3. Inflammatory cells and cell debris are seen in the background.

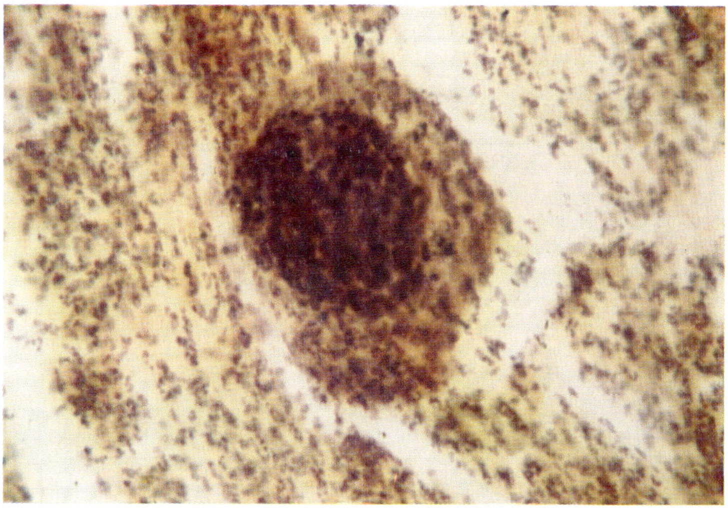

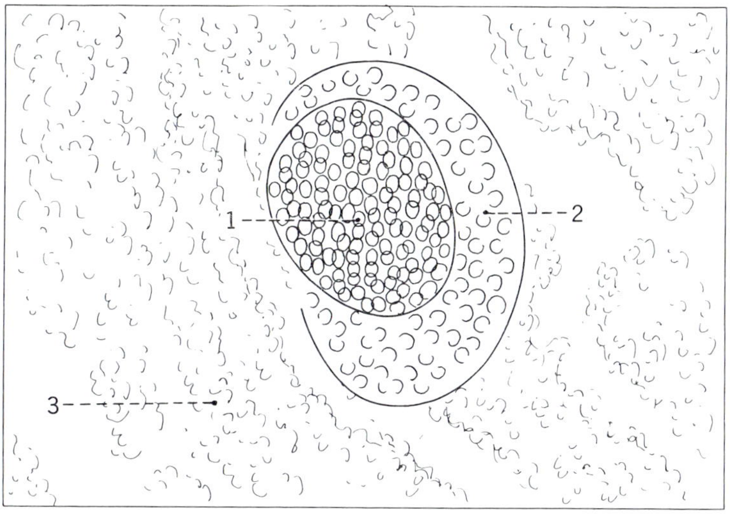

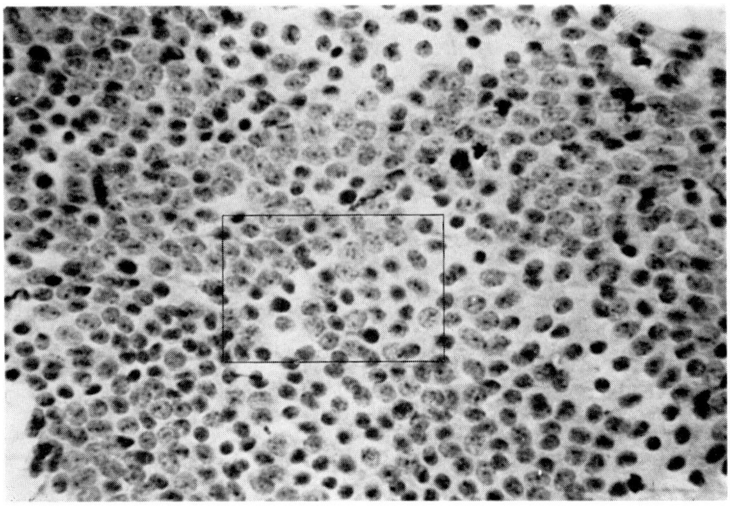

18. Endometrial glandular cells—Proliferative phase(Aspiration smear)

1. Nucleus: Coarsely granular and deeply stained.
2. Chromatin clumps and chromocenters are observed.
3. The border of the nucleus is less distinct than in the secretory phase.
4. Cytoplasm: Dense with indistinct borders. Generally densely distributed.

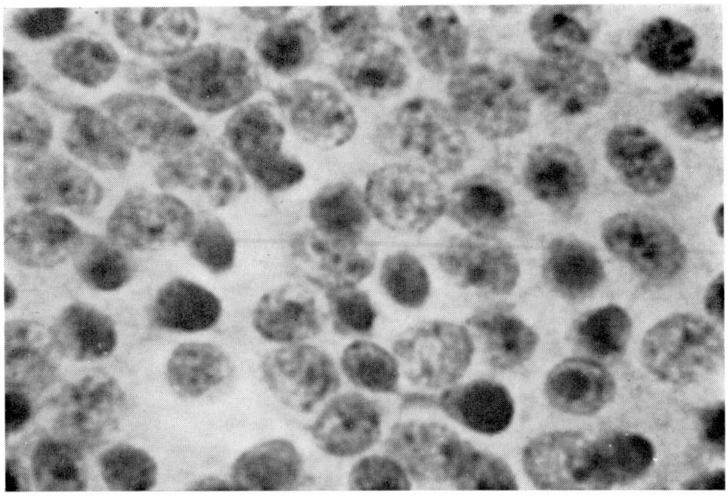

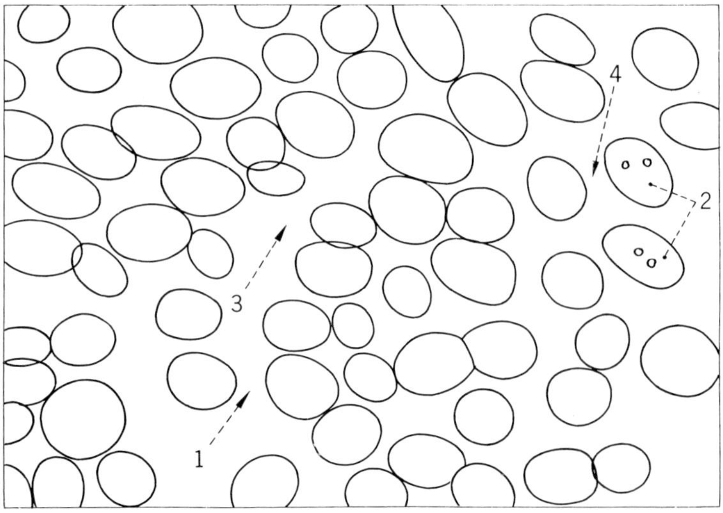

19. Endometrial glandular cells—Secretory phase(Aspiration smear)

1. Nucleus: Round and uniform in size. The border is distinct.
2. Chromatin is less abundant, lightly stained and is evenly distributed.
3. Chromocenters are observed.
4. Cytoplasm: More abundant and with quite distinct border. A honeycomb structure is seen.
5. Lacy and fine cytoplasm.

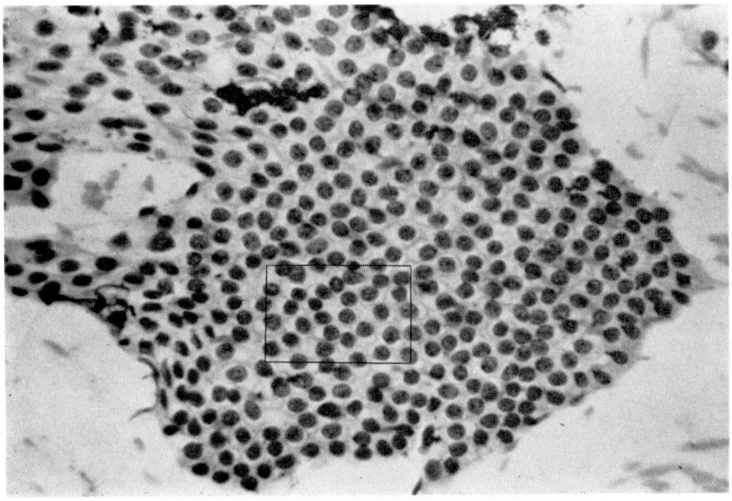

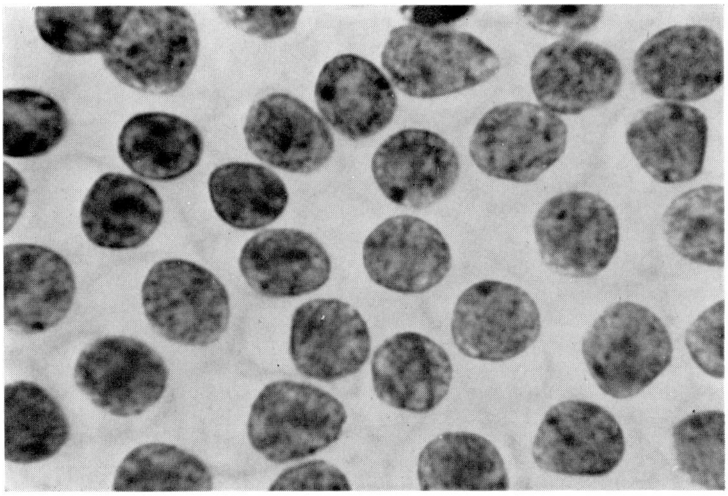

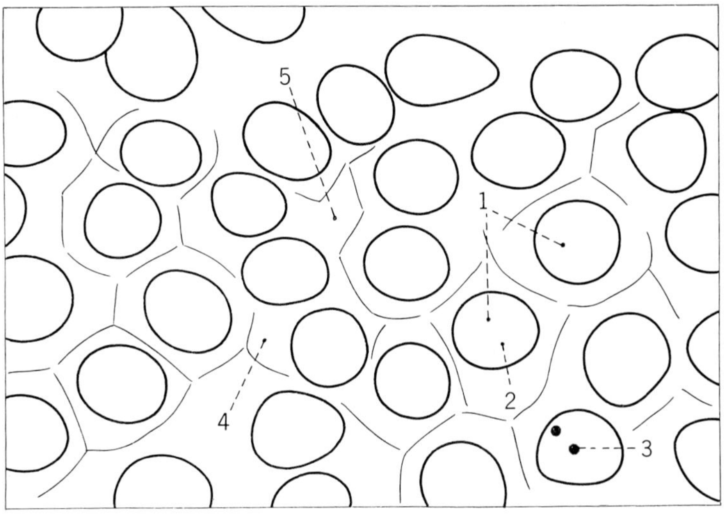

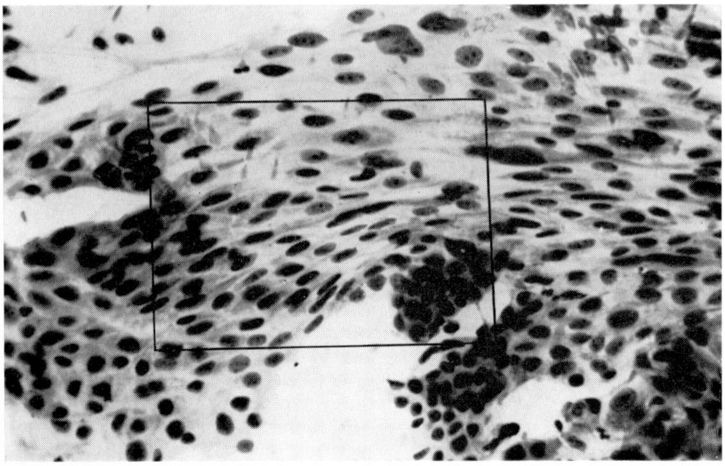

20. Endometrial stromal cells (Aspiration smear)

a. Low magnification

Elongated cells are aligned in a definite direction.

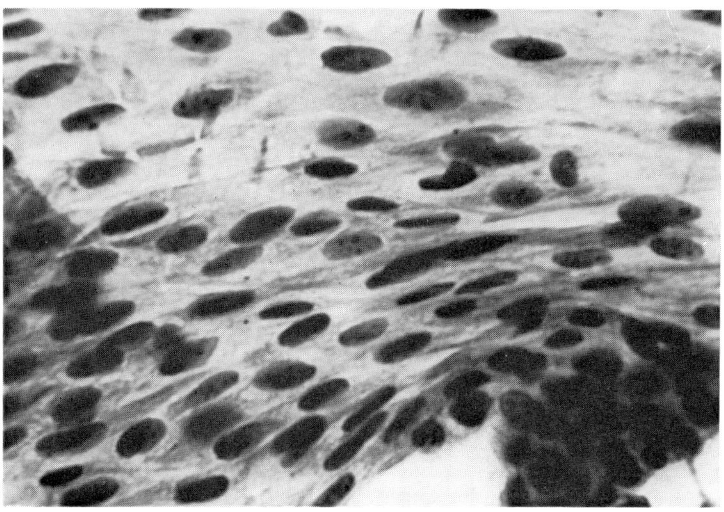

b. High magnification

1. Nucleus: Most are elliptical or quite elongated. Chromatin is evenly distributed and comparatively deeply stained.
2. Cytoplasm: The border is indistinct, but a long reticular structure is vaguely recognized. Rather thick and even.
3. A cluster of glandular cells: Note that their nuclei are rounder and they are more closely grouped together than stromal cells.

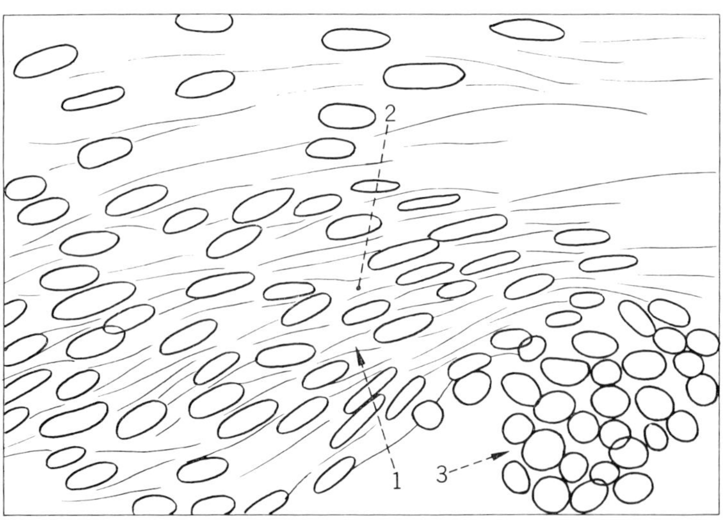

Differences between endocervical and endometrial cells

It may be difficult to make a distinction between these two types of cells in vaginal smears, especially because endometrial cells vary according to the phase of the menstrual cycle.

Table 2

	Endocervical cells	Endometrial cells
As a whole	Larger. A little looser when in a group. Characteristically columnar-shaped.	Slightly smaller. Denser and often overlapping each other.
Nucleus	Slightly larger. Oval or round; frequently multiple. The chromatin granules are finer and more lightly stained Eccentrically located in the columnar cells.	Smaller. Irregular; usually single. Coarsely granular, clumped and more deeply stained in the proliferative phase
Cytoplasm	More abundant. Lightly stained. Basophilic, or sometimes eosinophilic. The border is more distinct showing honeycomb structure or columnar shape. Occasionally ciliated.	Less abundant. Rather deeply stained. Slightly eosinophilic, but not always definite. Often indistinct due to degeneration.

III. Other Components in the Vaginal Smear

Blood cells — mainly polymorphonuclear leukocytes (Fig. 22)

Large numbers of polymorphonuclear leukocytes are often seen in vaginal smears. They are especially numerous in cervicitis, but it is difficult to establish a definite relationship between the number of leukocytes and the degree of inflammation. The cytological significance of leukocytes in smears is that they can be used as a standard to determine: 1) the sizes of other cells, 2) the density of the nuclei or cytoplasm of other cells.

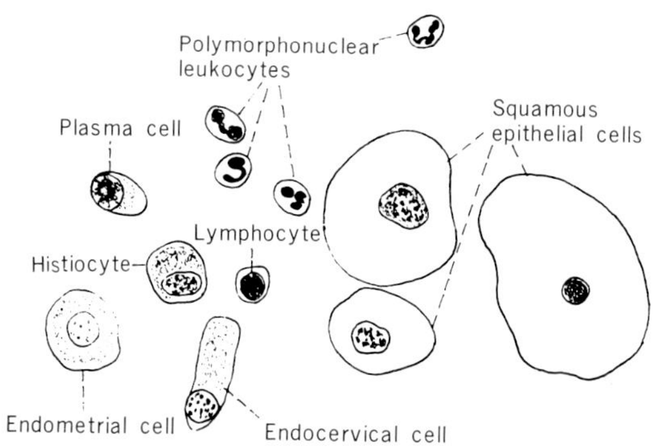

Fig. 22 An approximate comparison of the leukocyte with other cells in size and density.

Histiocytes (Fig. 23)

1) As a whole cell

Histiocytes are classified into three types: large, small and giant cell type. They are not uniform in size and are generally round or elliptical in shape. The nucleus is often arranged eccentrically in the cytoplasm.

2) Nucleus

1. The nucleus is typically kidney-shaped, but it is frequently round or elliptical.
2. Each cell has one nucleus, but in a giant histiocyte there may be several round, eccentric nuclei.
3. The nuclear border is fairly distinct.
4. Chromatin is finely granular and small clumps are seen here and there. There is no nucleolus.

3) Cytoplasm

1. The border of a well preserved histiocyte is usually distinct, but the border of a degenerate one is indistinct.

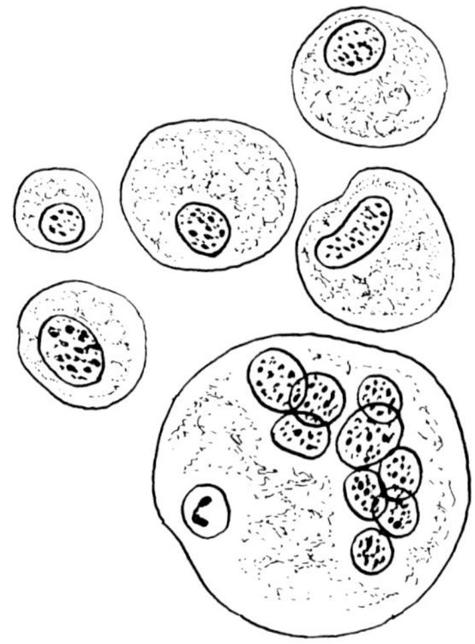

Fig. 23 Histiocytes in vaginal smears.

2. The cytoplasm may vary in shape or size.
3. The distinctive features of the cytoplasm are finely foamy, basophilic and faintly stained.
4. In giant histiocytes the cytoplasm is relatively homogeneous and rather dense. Phagocytized leukocytes or vacuoles may occasionally be observed.

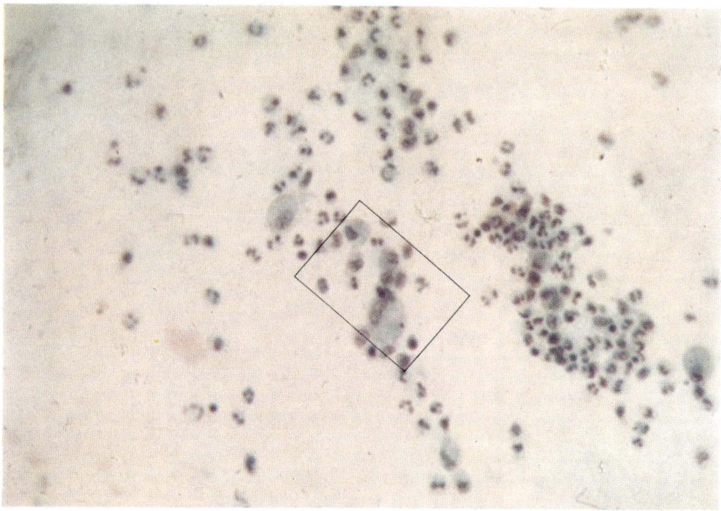

21. Large and small histiocytes

a. Low magnification

Round and vesicular cells with eccentrically located nuclei.

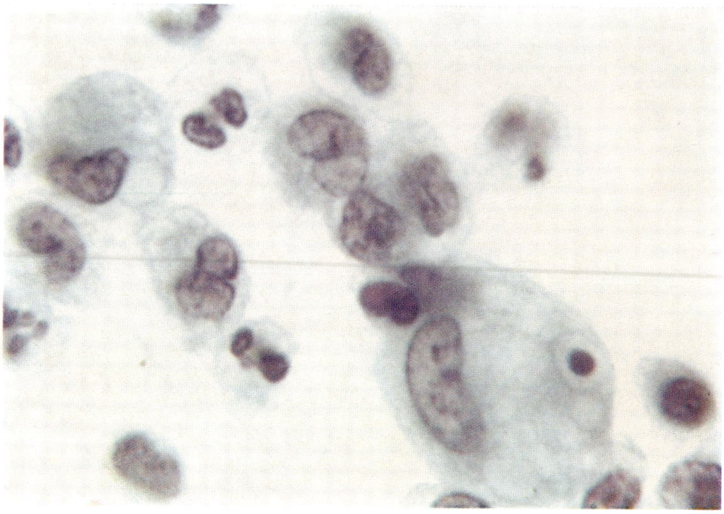

b. High magnification

1. Typical kidney-shaped nuclei with distinct borders, and small clumps of chromatin.
2. A cytoplasmic vacuole.
3. An inclusion inside a vacuole.
4. Characteristic foamy cytoplasm with a fairly distinct border. Lightly stained and basophilic.
5. Small histiocytes. One of them appears double nucleated.
6. Polymorphonuclear leukocytes.

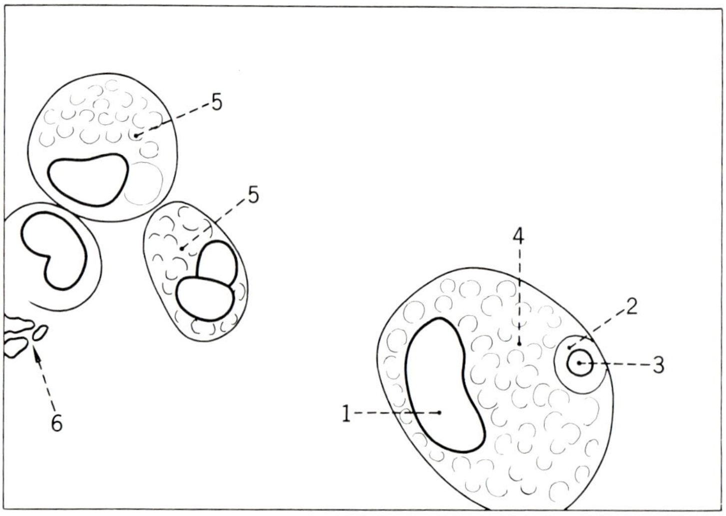

22. Giant-cell type histiocytes

a. Low magnification

There is a centrally-located multinuclear cell.

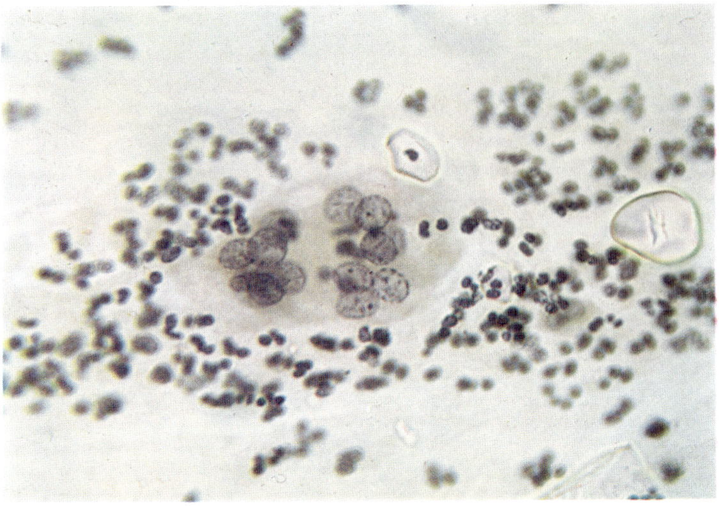

b. High magnification

1. Round and overlapping nuclei.
2. Distinct nuclear borders.
3. Small chromatin granules.
4. Comparatively homogeneous cytoplasm with vacuoles.
5. Inclusions within a vacuole.
6. Leukocytes.

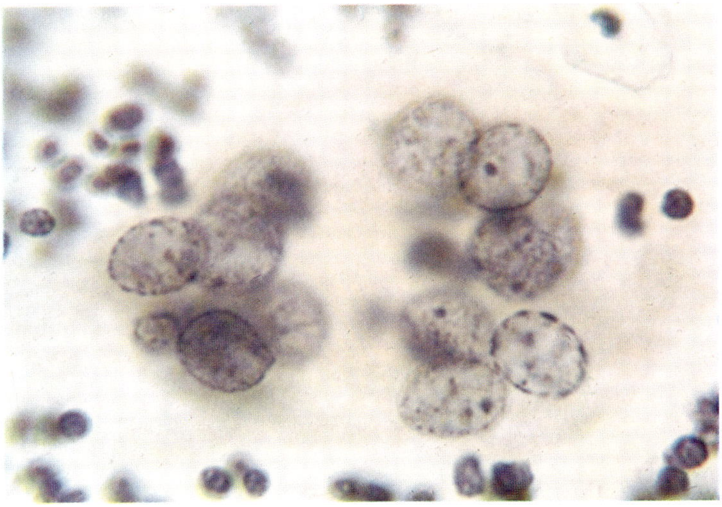

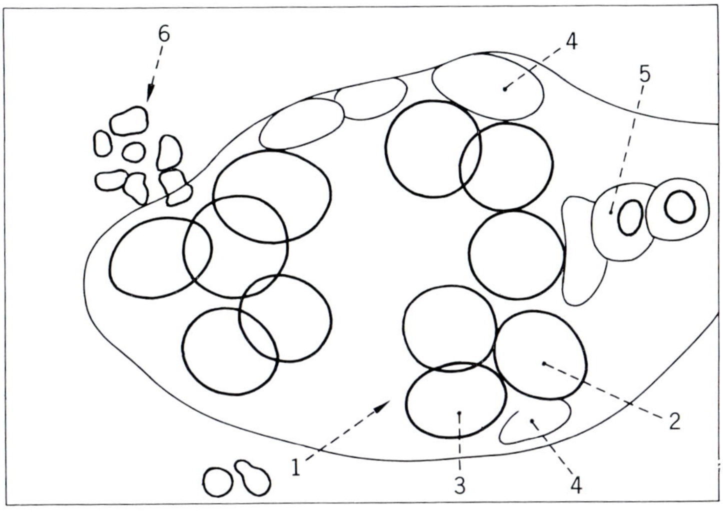

Trichomonas vaginalis

1) Schematical observation of trichomonas (Fig. 24)

This is a pear-shaped or elliptical protozoon with a nucleus and two granules located near the front end. Four long and locomotive flagella have their origin in one of the granules, and an undulating membrane extends to the rear end from the other granule. Sometimes, round degenerate-type cell without flagella and undulating membrane is seen.

It is difficult to distinguish the above details in the vaginal smear.

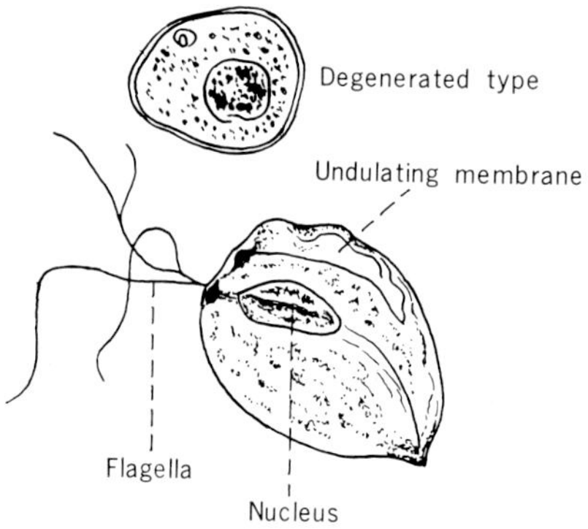

Fig. 24 Trichomonas vaginalis.

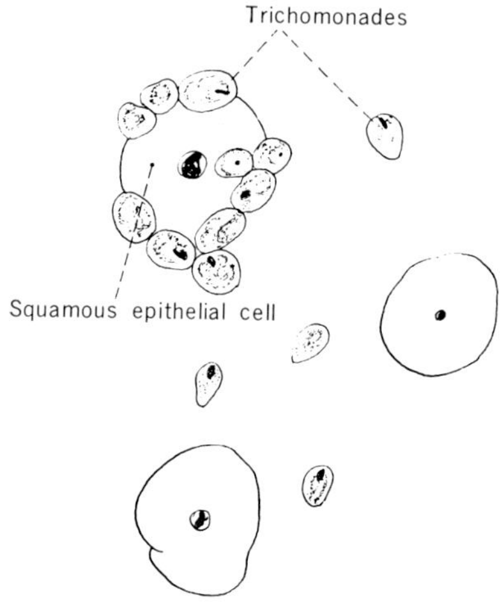

Fig. 25 Trichomonas in the vaginal smear.

2) How can it be seen in the vaginal smear? (Fig. 25)

1. About the same size as the nucleus of a parabasal cell.
2. Elliptical and slightly pointed at one end.
3. Indistinct border.
4. Very lightly stained. Usually light blue-gray.
5. On close observation, an elongated nucleus may be seen near the rounded end.
6. The cells are sometimes scattered, but often crowd around a superficial squamous cell.
7. Generally, the smear looks dirty and includes polymorphonuclear leukocytes, light blue-stained mucus and exudate. Degenerate epithelial cells are often present.

23. Trichomonas vaginalis

a. Low magnification

A number of trichomonads crowding around a squamous cell.

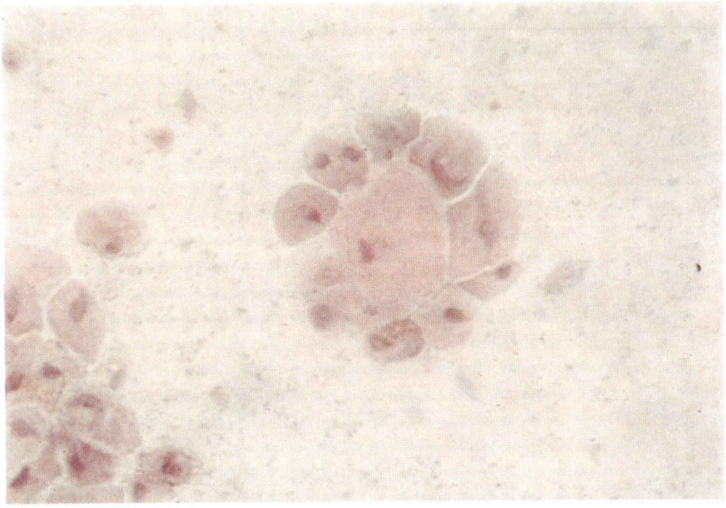

b. High magnification

1. Superficial squamous cell.
2. Trichomonads crowding around the squamous cell. Each individual organism can be easily distinguished one from another.
3. The nucleus of a superficial cell.
4. Nucleus of trichomonad.

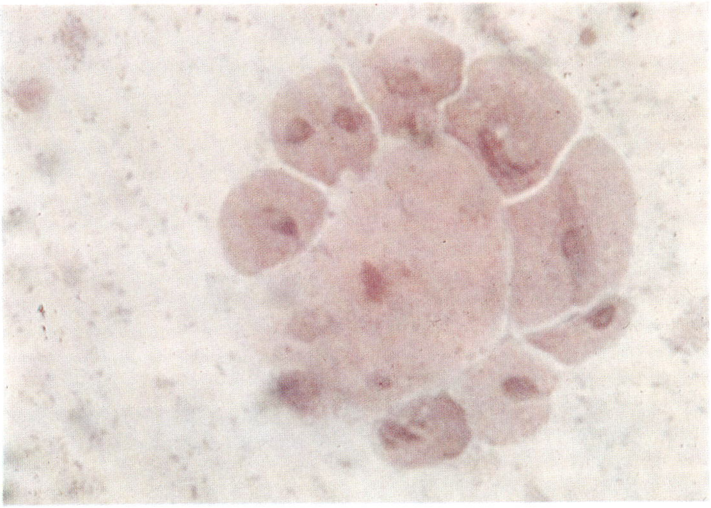

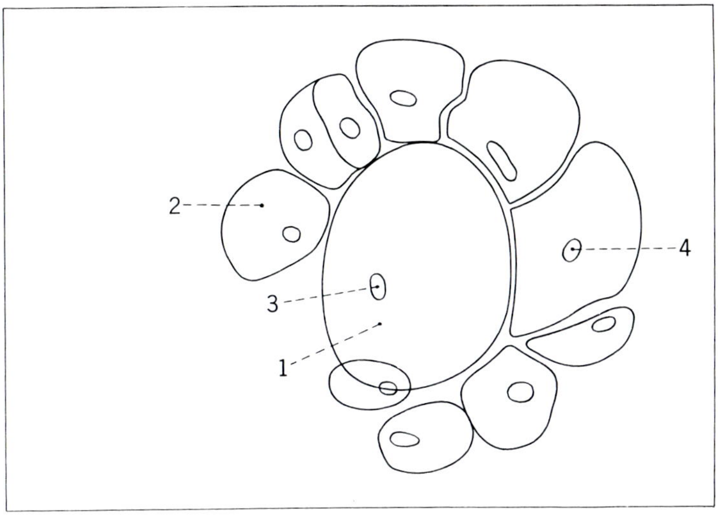

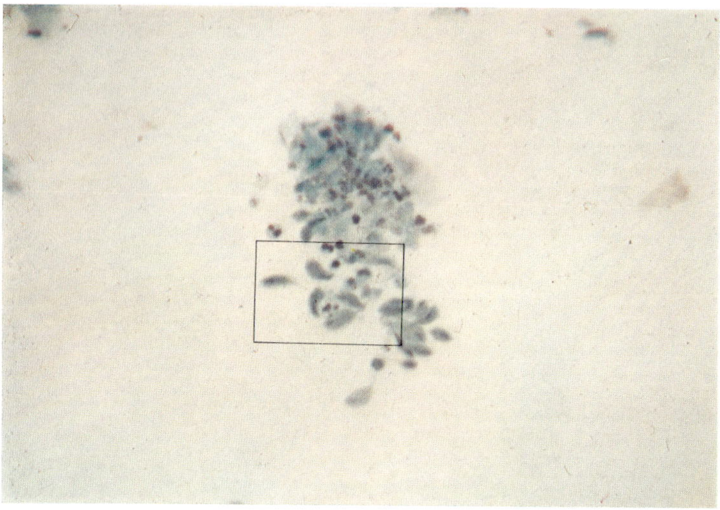

24. Trichomonas vaginalis

a. Low magnification

Light bluish stained trichomonads are seen in a cluster mixed with polymorphonuclear leukocytes.

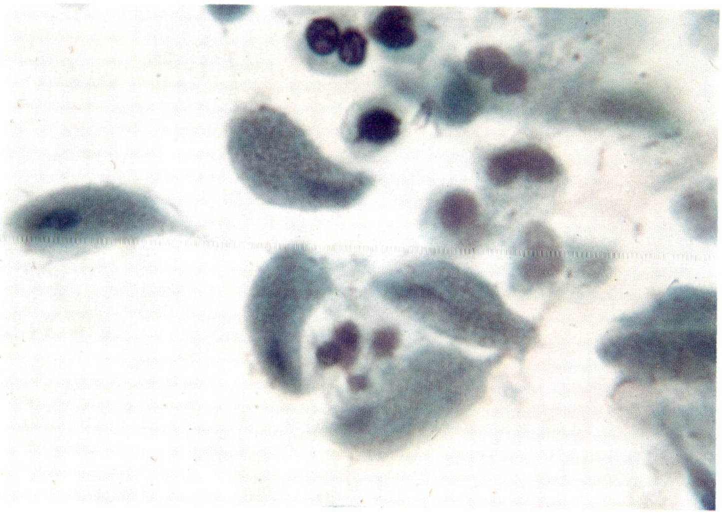

b. High magnification

1. Pear-shaped trichomonad showing an elliptical nucleus in the one end.
2. Faint and indistinct cytoplasm. Light basophilic stain.
3. Undulating membrane as well as flagella hardly can be recognized in the picture.

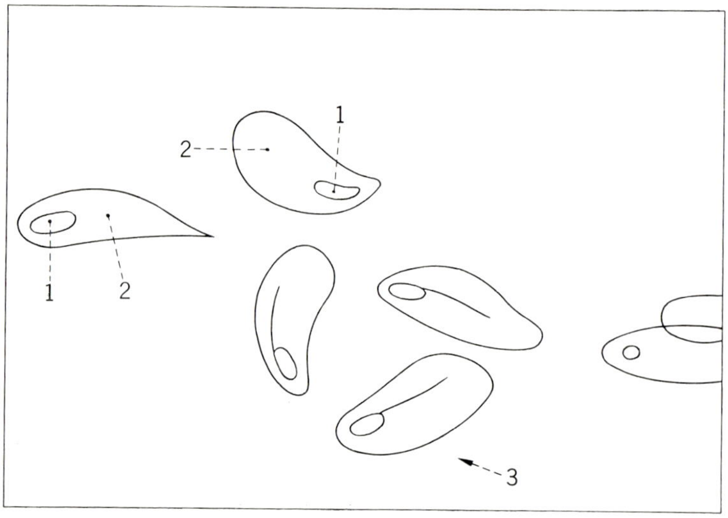

Candida albicans (Monilia)

1) Schematical observation of candida (Fig. 26)

Candida albicans is a genus of fungi with a pseudomycelium. They consist of gram-positive, oval, budding yeasts, measuring 2–3 × 4–6 microns and elongated, gram-positive hyphal cells.

2) How can it be seen in the vaginal smear?

With Papanicolaou's stains, the spores appear small and elliptical and look like sesame seeds. They are reddish-maroon and appear double refractile. The hyphae are rather lightly stained. It is fairly difficult to notice the fungi in a smear, but they are often revealed by Gram stain.

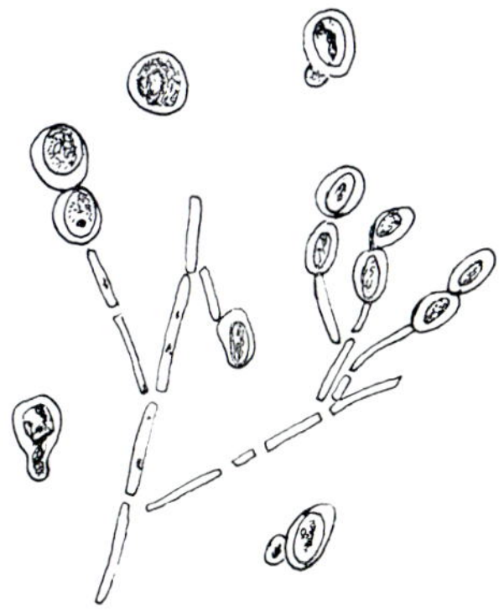

Fig. 26 Candida albicans.

25. Candida albicans

1. A long hypha.
2. Spores. All are reddish-maroon. Focusing up and down helps to reveal the borders.
3. Squamous cell.
4. Nucleus of the cell.

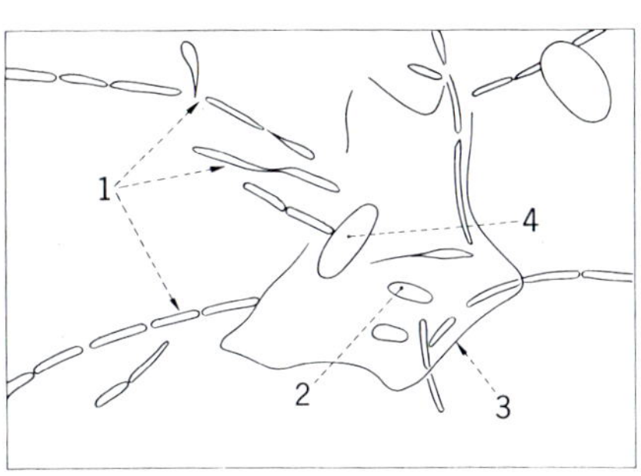

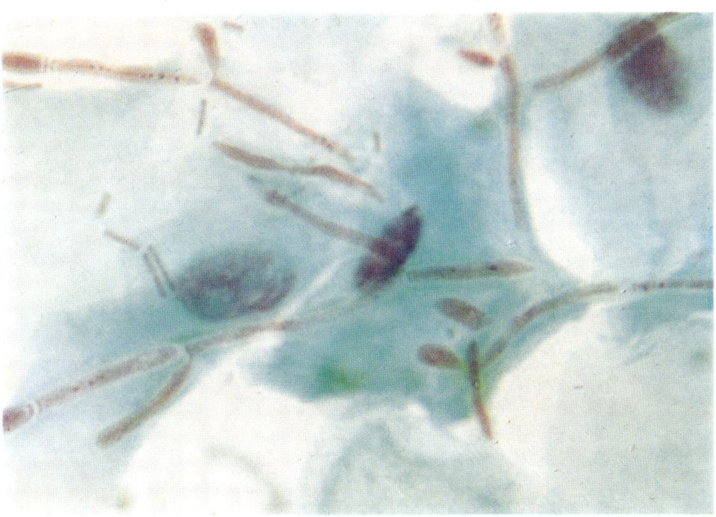

Leptothrix vaginalis

These are sometimes seen with trichomonas vaginalis in vaginal smears, but have little clinical significance. The examiner should not be confused by their hair-like threads.

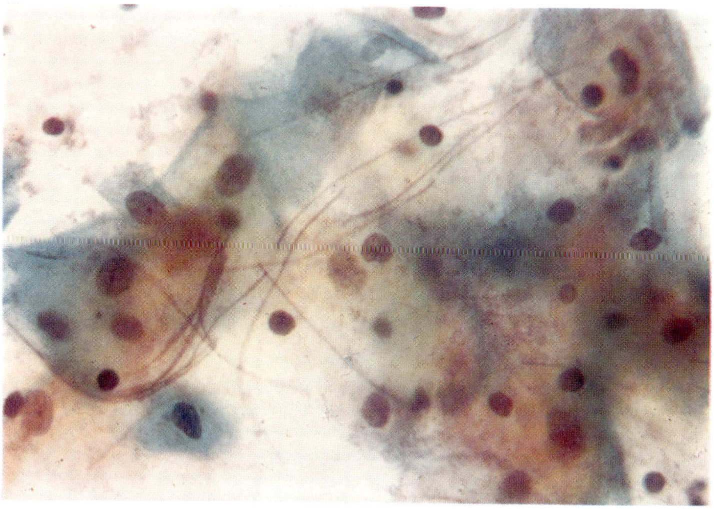

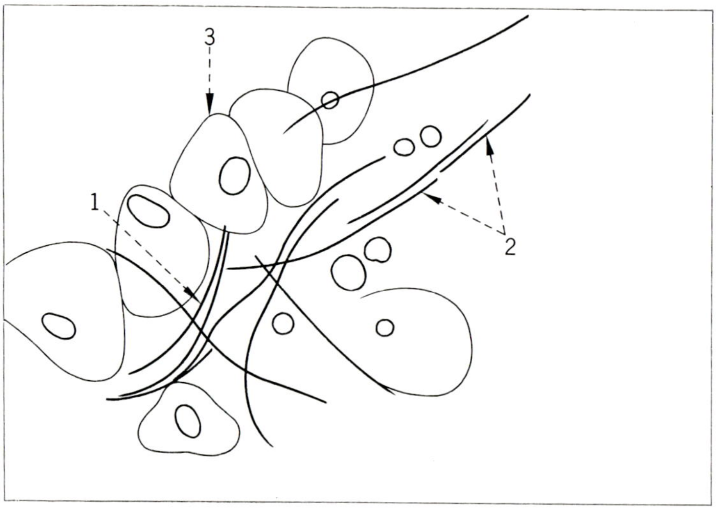

26. Leptothrix vaginalis

1. They appear individually or in a group. They are not divided into branches and have a hair-like appearance.
2. Closer observation reveals that they are segmented.
3. Squamous cells.

Spermatozoa

Although not diagnostically important, they are fairly often seen in vaginal smears.

The tails of well-preserved spermatozoa are clearly seen, but in degenerated ones only the heads are present. The top half of the head is lightly stained, but the bottom half is blue-black and the tail is stained pale pink by Papanicolaou's stain.

27. Spermatozoa

1. Tail : Stained pale pink.
2. Head.
3. Bottom half of the head is stained bluish-black by Papanicolaou's stain.
4. Top half shows light staining reaction.

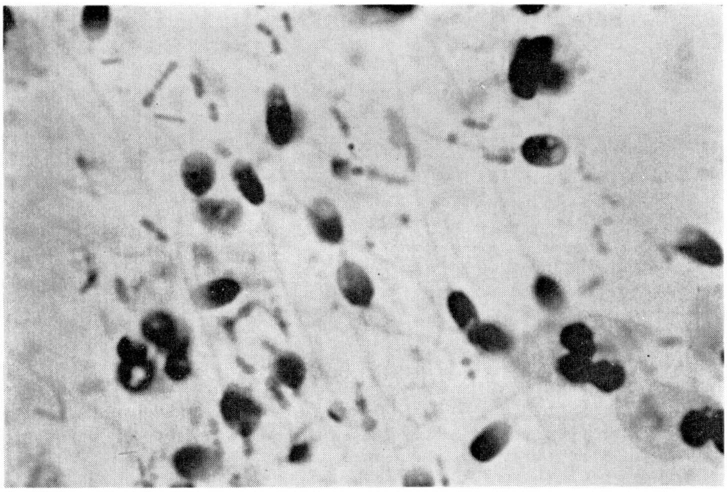

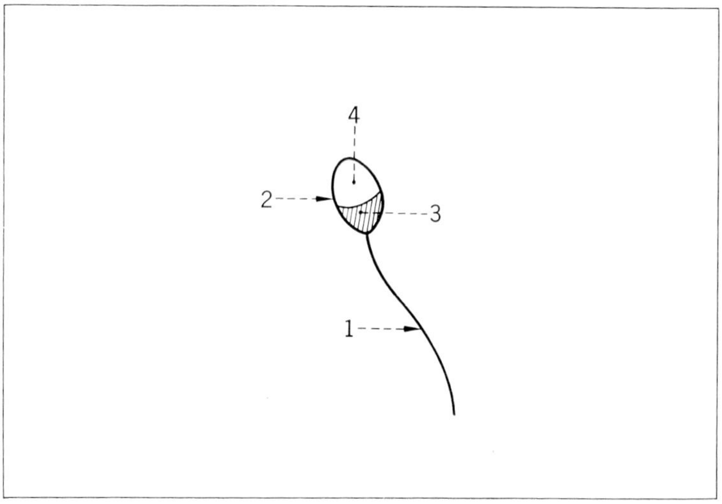

Döderlein bacilli

These are gram-positive bacilli often seen in ordinary vaginal smears and appear light blue with Papanicolaou's stain. They can live in exsistence of the glycogen of squamous cells, and it is likely that cytolysis can take place in the presence of the bacilli. They are quite common during the latter half of the menstrual cycle, which corresponds to the secretory phase.

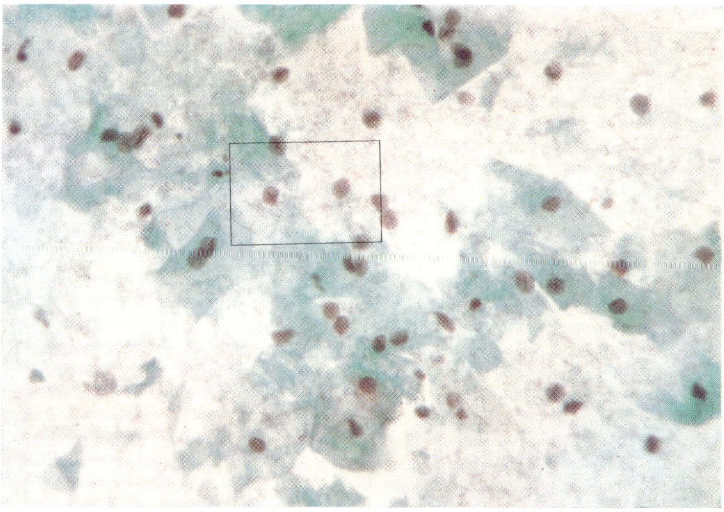

28. Döderlein bacilli
a. Low magnification
Numerous stripped nuclei and cells with their poorly preserved borders are seen. The whole picture looks untidy.

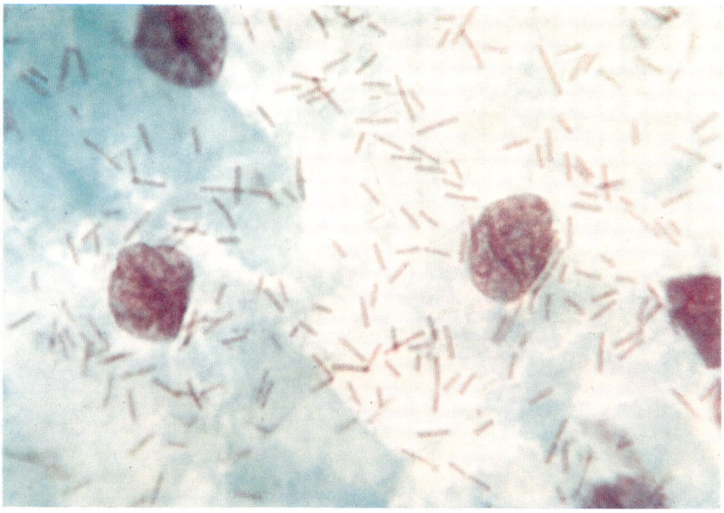

b. High magnification
1. Many characteristically slender, rod-shaped bacilli are seen.
2. A squamous intermediate cell lacking half its cytoplasm is also present.
3. Stripped nuclei are seen.

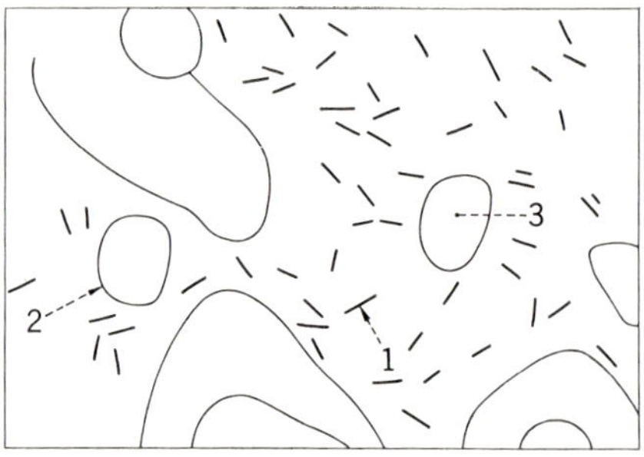

Tuberculosis

It is very difficult to diagnose tuberculosis of the genital organs from smears and cells from tuberculosis are seldom encountered. Langhans giant cells rarely appear in ordinary vaginal smears. However, tubercles or Langhans giant cells may be observed in cervical smears or endometrial aspiration smears taken during or shortly after menstruation from sterile patients in whom tuberculosis is suspected.

29. Tubercle (Endometrial aspiration smear)

1. An indistinct border. Homogeneous and cheesy cytoplasm.
2. Round or elliptical nuclei, several times as large as lymphocytes, are overlapping each other. The borders are not very clear but can be distinguished. Chromatin is comparatively evenly distributed and coarsely granular.
3. Numerous lymphocytes, plasma cells and epithelioid cells can be seen.

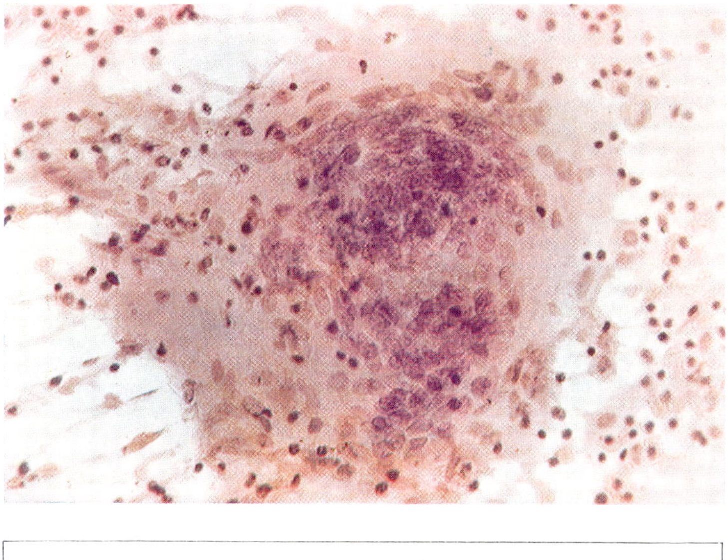

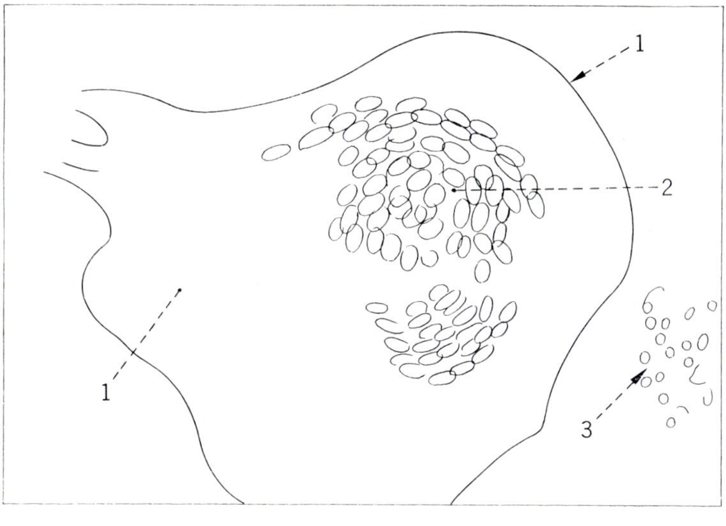

Viral infection — Herpes

Herpes infections of the vagina and the external genitalia in both sexes recently have become increasingly important in the cytodiagnostic field. Most striking feature of the disease is the appearance of intranuclear inclusion bodies. They may be found early in the second week of the viral infection. Prior to the appearance of the inclusion bodies in the nuclei of epithelial cells in smears the nuclei are slightly enlarged, become somewhat irregular in shape and look hazy. All changes may disappear within two weeks.

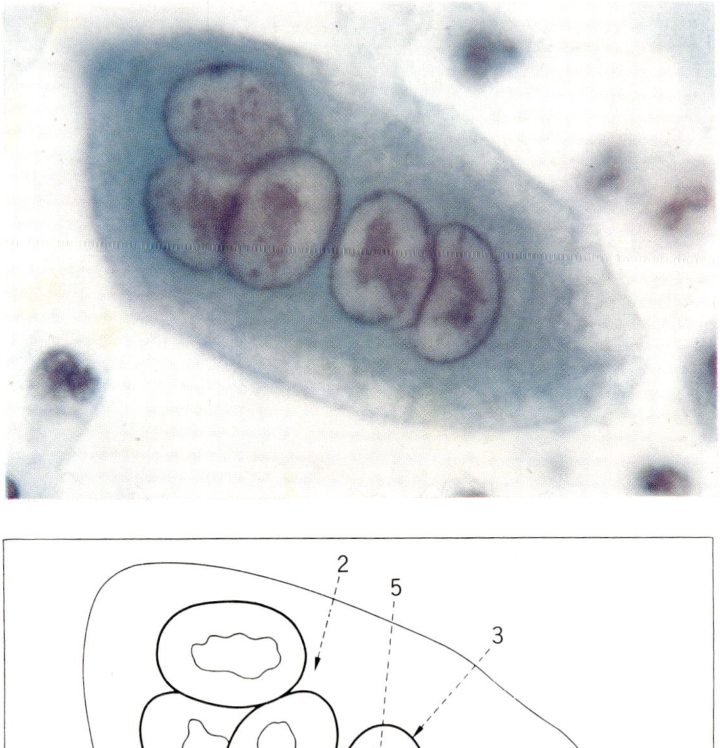

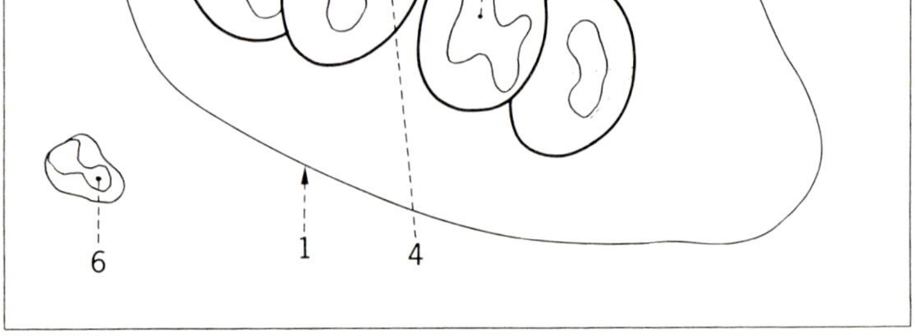

30. Intranuclear inclusion bodies due to herpes (Scraped smear from an ulcer in the labium major)

There are multinucleated giant cells in the inflammatory background.

1. Multinucleated giant cell, the border of which is quite definite.
2. The nuclei are large and mostly round or oval in shape, but occasionally kidney-shaped.
3. The nuclear border is very distinct and appears thickened.
4. The interior of the nuclei looks washed out and is faintly stained.
5. A prominent inclusion body stained pinkish is observed centrally. It is irregularly shaped and usually situated in the middle of the nuclei.
6. Polymorphonuclear leukocyte.

IV. Changes in the Vaginal Cell Pattern during Menopause

General characteristics (Fig. 27)

Smear types during the menopause

An atrophic cell pattern may easily be thought to occur after menopause, but this is not necessarily true. The vaginal epithelium sometimes may show cyclic changes even after the menopause, but only some years after the menopause complete atrophy may take place.

Three types of vaginal smears seen during the menopause

Vaginal smears during the menopause are roughly classified into three types: smears in early menopause when the squamous epithelium is still sufficiently proliferating, those in the crowded menopause when proliferation decreases, and these in the advanced or atrophic menopause when proliferation had ceased.

1) Early menopause

The vaginal smear is similar to that during the premenstrual period. It is usually seen in the stage when the menstrual flow is decreasing.

1. Intermediate and superficial cells predominate. They are rather small, paler stained and basophilic.
2. Nuclei are vesicular.
3. Döderlein bacilli may be abundant. Consequently, cytolysis is accelerated and some stripped nuclei are present.

2) Crowded menopause

1. Intermediate cells predominate. They are slightly smaller than in the early menopause and appear in a group.
2. The cytoplasm is lightly stained and basophilic. The nuclei are vesicular, uniform in size and easily recognized. The cytoplasmic borders are distinct.
3. Leukocytes and Döderlein bacilli are abundant. Some nuclei are stripped as the result of cytolysis.
4. Superficial and parabasal cells may be sometimes seen among intermediate cells. This kind of smear is called the mixed type.

3) Advanced or atrophic menopause

1. Smears show an atrophic picture consisting mainly of parabasal cells. The cells and nuclei vary in size.
2. Cytoplasm: Round or elliptical and uniform in size. Basophilic. The borders are distinct and even thicker. The entire cytoplasm is dense and homogeneous.
3. Nuclei: Round, vesicular and comparatively large.
4. Some cells are strongly eosinophilic and stained deep orange. The nuclei of these cells are small and pyknotic in their appearance.
5. The background is cluttered with leukocytes, old blood components or ruptured nuclear substances in fine thread-like structure formed by karyorrhexis. Stripped nuclei due to cytolysis are also observed.

6. Sometimes parabasal, intermediate and superficial cells are mixed.
7. Döderlein bacilli are usually not seen.
8. Small,eosinophilic,orange-colored angular cells are observed in some smears.

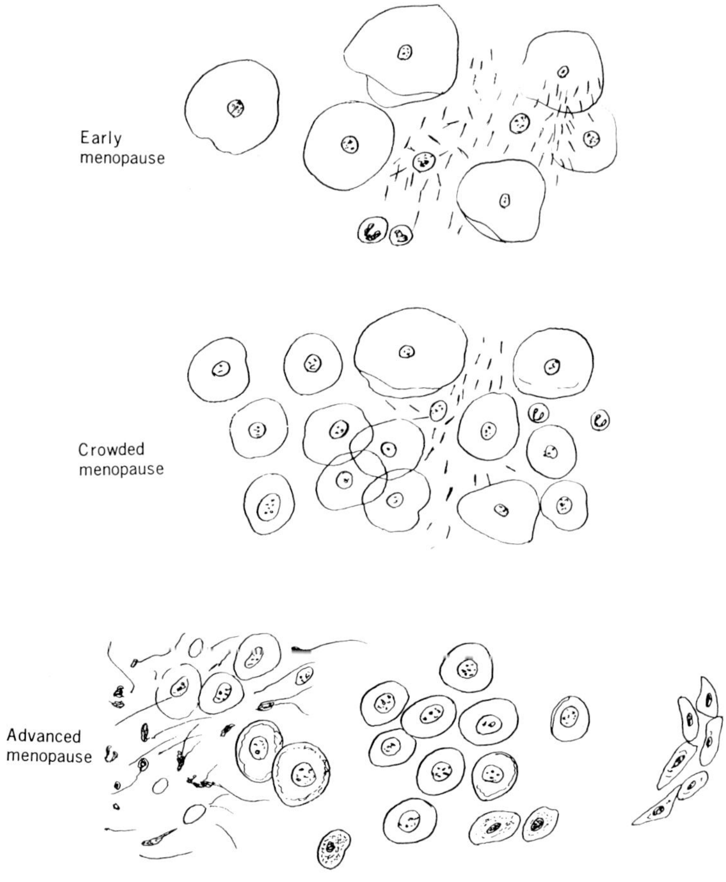

Fig. 27 Vaginal smears during the menopause.

9. Round or elliptical, dark-blue globules are occasionally seen. They are inclined to be confused with the nuclei of malignant cells, but they may be distinguished from them by the round, dark-blue substance which looks like an amorphous nucleus at the center of each globule.

In reality, these three stages appear successively, and it is often difficult to distinguish them from one another.

Furthermore, the degree of change in vaginal cells does not necessarily correspond to the length of time after the menopause.

The vaginal cell picture after bilateral oophorectomy resembles that of the menopause, but the changes appear more rapidly. The changes are even more remarkable in the cases accompanied with hysterectomy. The pattern corresponding to early menopause appears within six weeks, and the atrophic picture after three months.

The vaginal cytology during cessation of menses by irradiation differs according to whether the menopause is ovarian or endometrial. In either case, however, the description of vaginal smears during natural menopause is not applicable, because irradiation also may directly affect vaginal cells.

31. Early menopause

This picture resembles premenstrual smear pattern.

Large numbers of intermediate cells, mostly basophilic and overlapped each other, present, mixed with Döderlein bacilli.

Moderate cytolysis is observed.

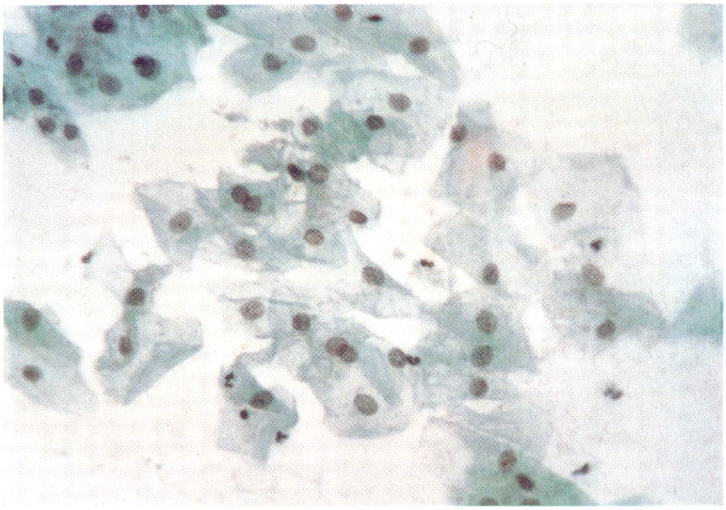

32. Crowded menopause

There are numbers of small intermediate cells showing fairly thick and basophilic cytoplasm and vesicular nuclei.

A few eosinophilic squamous cells, intermediate type, present.

Among those, small numbers of superficial cells as well as parabasal cells are observed.

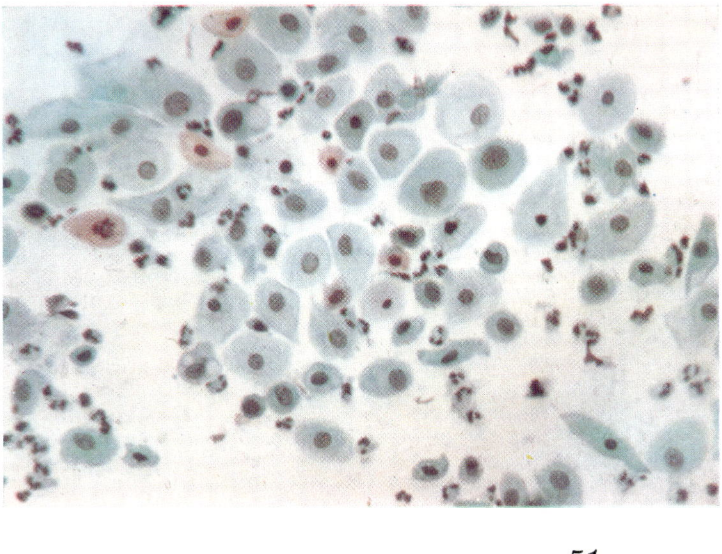

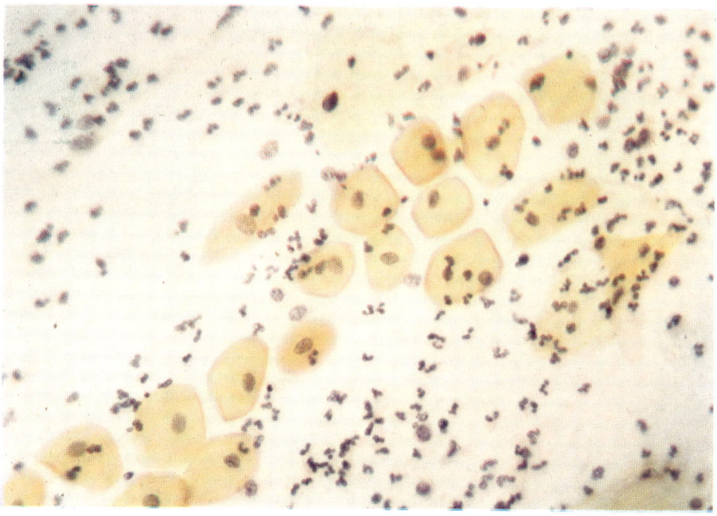

33. Crowded-advanced menopause

a. Low magnification

This is a fairly atrophic vaginal cell pattern almost in the advanced menopause. An intermediate or parabasal cell type.

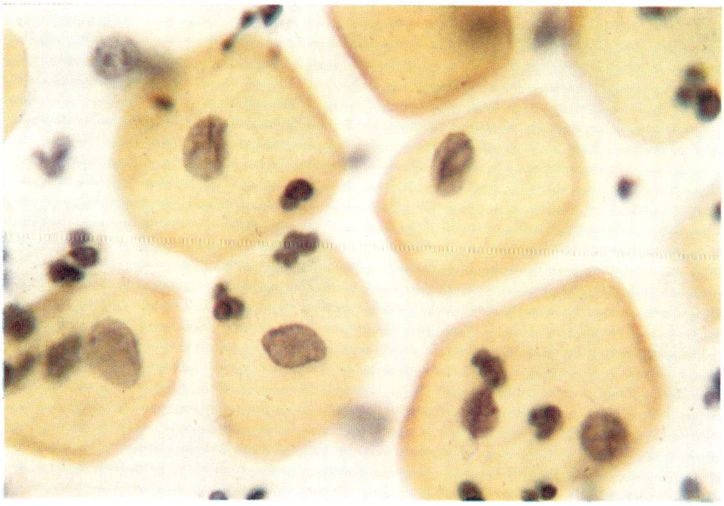

b. High magnification

1. Cells vary in shape and present an angular form.
2. Nuclei are generally vesicular but some are pyknotic. Perinuclear vacuoles are faintly visible.
3. The cytoplasm appears thick. The border is very clear and seems to be thicker than the rest of the cytoplasm.
4. The cytoplasm shows a strong eosinophilic-orange staining reaction.
5. Leukocytes.

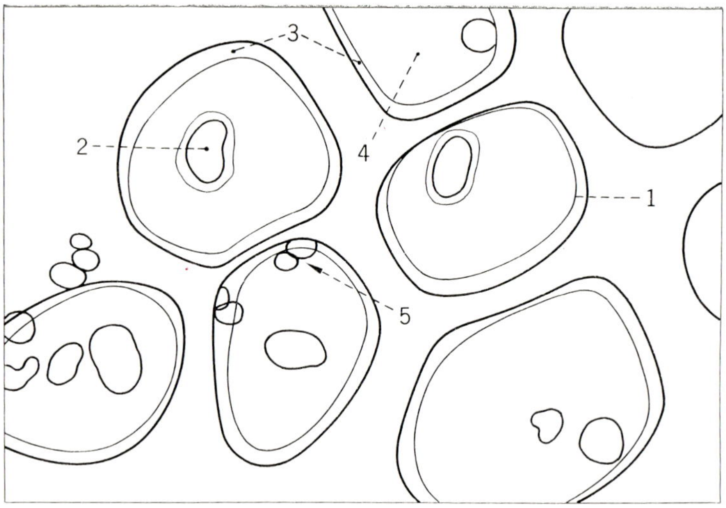

34. Advanced menopause—Atrophic smear

a. Low magnification

Completely atrophic vaginal cells in advanced menopause. The whole picture gives an untidy spotted impression.

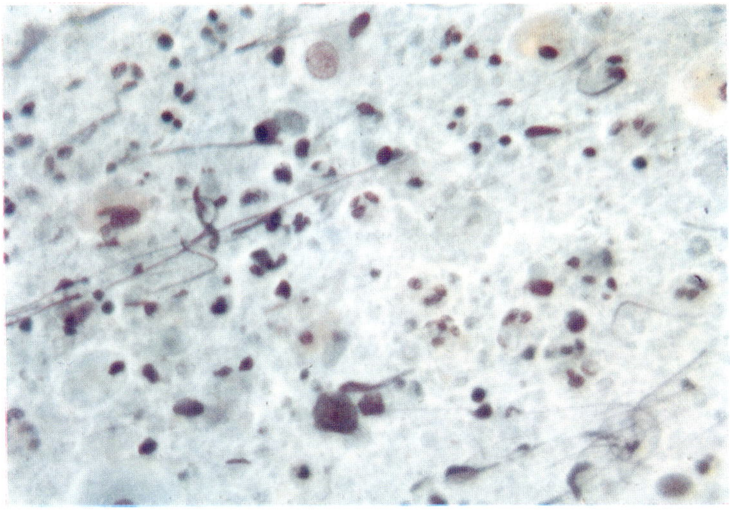

b. High magnification

1. The whole cell is reduced. The cytoplasm stains dark reddish-blue and appears thickened. The nucleus is very pyknotic.
2. This parabasal cell is slightly enlarged by degeneration. Both the nucleus and the cytoplasm have degenerated and are eosinophilically stained. The cytoplasm is partly lost.
3. Ruptured nuclei by karyorrhexis.
4. As a result of karyorrhexis, the nuclear substances are elongated and form thread-like structures.
5. Cell debris produced mainly by autolysis.

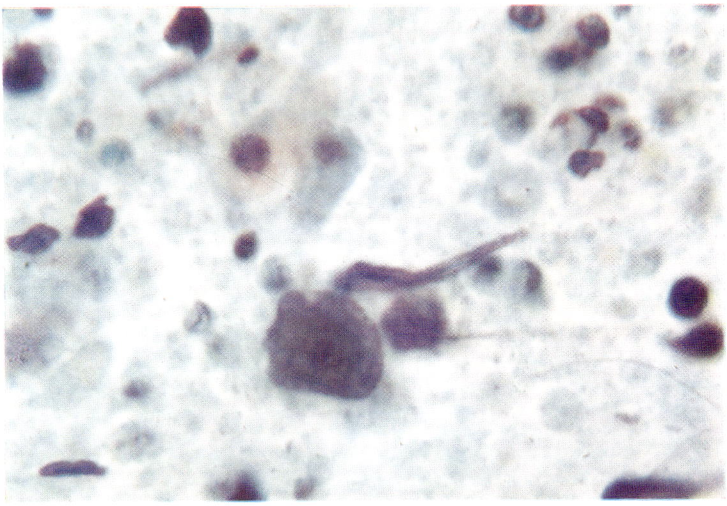

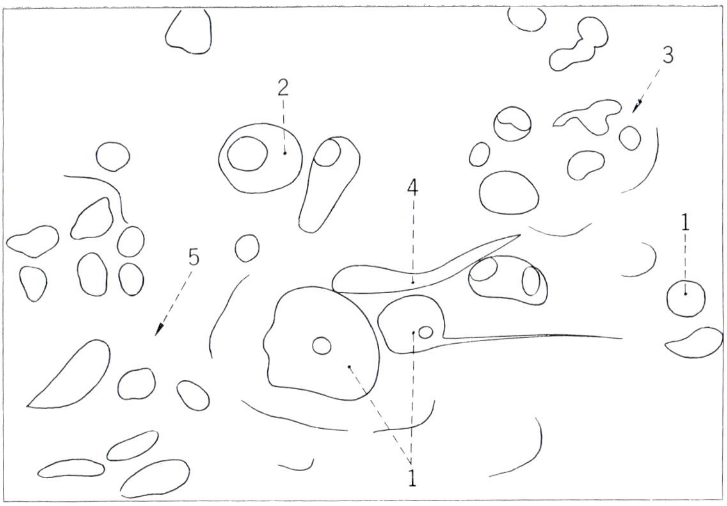

V. Cells Related to Pregnancy

Characteristics of vaginal smears during normal pregnancy

Vaginal cell patterns in normal pregnancy are classified roughly into four types.

1) Navicular type (Fig. 28)

1. Intermediate cells. They often appear in a group.
2. They are boat-shaped or sometimes roundish, and are therefore named "navicular" cells.
3. The borders of these cells are distinct, thickened and sometimes appear doubled.
4. Yellow-brown stained glycogen is definitely observed. It may be considered that the cytoplasm has been pushed to the borders by the glycogen.
5. The nucleus is eccentrically located and round or elliptical. It is often vesiculated, but is sometimes pyknotic.
6. A large number of Döderlein bacilli are observed.

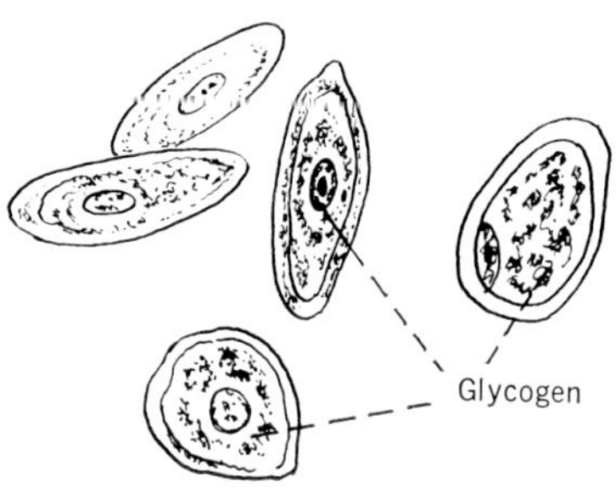

Fig. 28 Navicular cells.

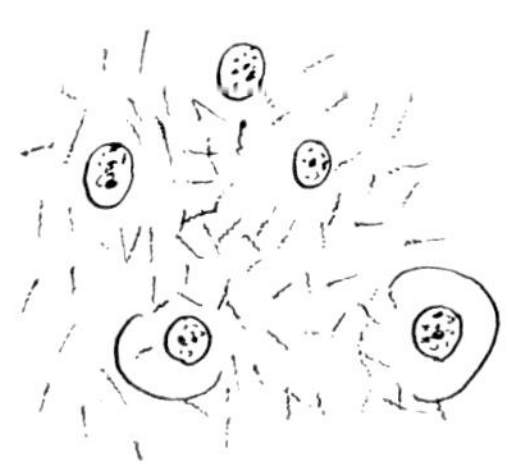

Fig. 29 Cytolysis.

2) Cytolytic type (Fig. 29)

Döderlein bacilli are abundant and cytolysis is marked. Only free-lying nuclei are observed. There are few leukocytes.

3) Inflammatory type

Numerous leukocytes are noticed, so the epithelial cells are sometimes rather atypical. Trichomonads are often observed.

4) Estrogenic type

There are many superficial cells.

Among the four types, the navicular type is considered to be common to pregnancy. It is seen in about 60% of smears during pregnancy. Navicular cells, however, do not appear only during pregnancy. Furthermore, pregnancy cannot be ruled out even when the navicular type or any of the other three types is not seen.

The vaginal smear test has little clinical significance as far as the diagnosis of pregnancy is concerned. The typical smear pattern of pregnancy is not seen before the third month. Besides, a typical picture is not always present.

In addition, pregnancy is easily diagnosed clinically at this stage, and the sero-immunological or biological pregnancy test is far more reliable than the smear test. However, the smear test may have some significance to determine the time of delivery at term of pregnancy.

35. Navicular cells

a. Low magnification

Mostly intermediate cells and in clusters. Boat-shaped cells are observed among them. Yellowish glycogen is clearly seen even under low magnification.

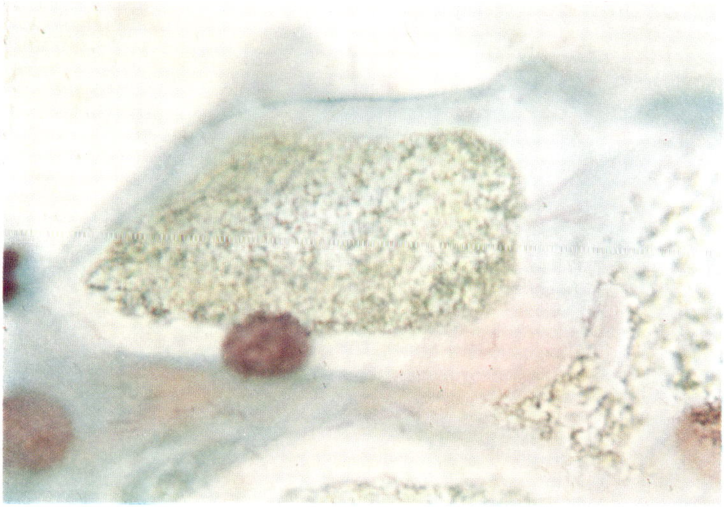

b. High magnification

1. Characteristically filled with glycogen.
2. The cytoplasm appears doubled around the border.
3. The border looks especially thick and distinct.
4. The nucleus is flattened and appears to be pushed to the edge of the glycogen mass.

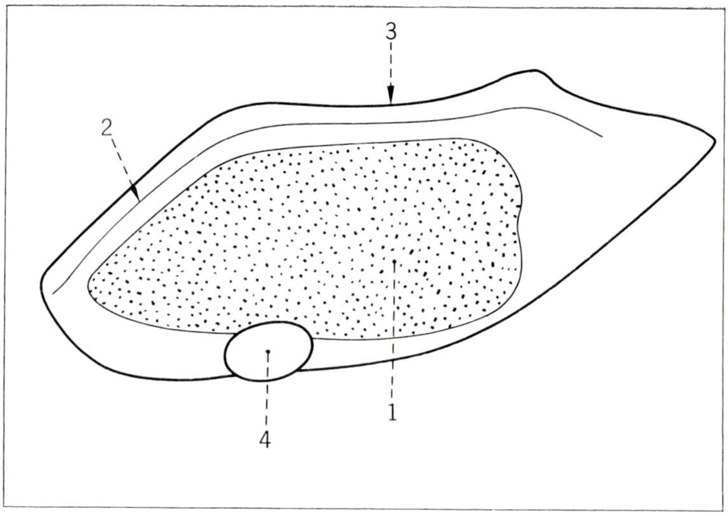

Characteristics of smear patterns during puerperium

The smear pattern changes rapidly during the initial days after delivery.

1. Contains abundant blood, leukocytes, mucus and cell debris within a few days after delivery. The numbers of histiocytes may increase.
2. Navicular type of smear in pregnancy may continue during the first days of puerperium.
3. Smears start to show numerous parabasal cells which give postpartum smears their characteristic appearance, resembling the pattern in menopause, and are named "postpartum cell" by Papanicolaou.
4. Superficial cells are rarely seen in postpartum smears.
5. Early stage of puerperium shows more variation in the proportion and number of epithelial cells.
6. Within several weeks, the postpartum cells become more like normal parabasal cells which remain a constant feature during lactation.
7. When BBT begins to reveal biphasic curve again superficial cells begin to appear in increasing numbers. This change to a mature pattern may complete within 6 weeks after delivery in the nonlactating women.

36. Postpartum smear

a. Low magnification

A cluster of parabasal cells is present among numerous leukocytes and a few intermediate cell.

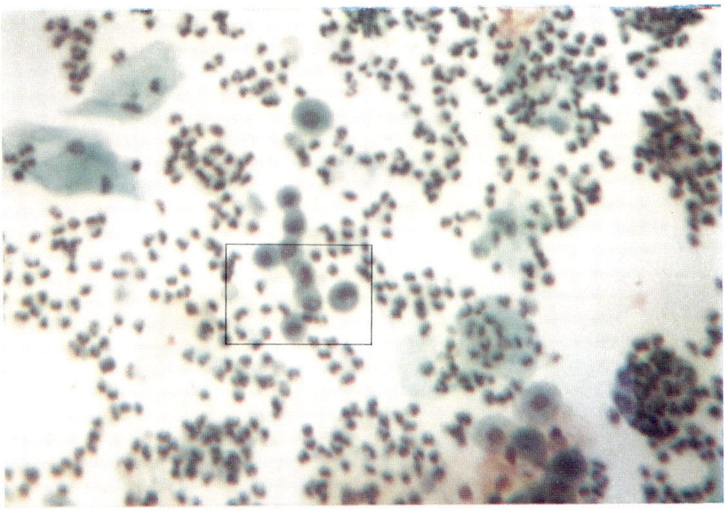

b. High magnification

1. Rounded cells with dense cytoplasm.
2. The cytoplasmic border is fairly thickened.
3. The nuclei are partly eccentric.

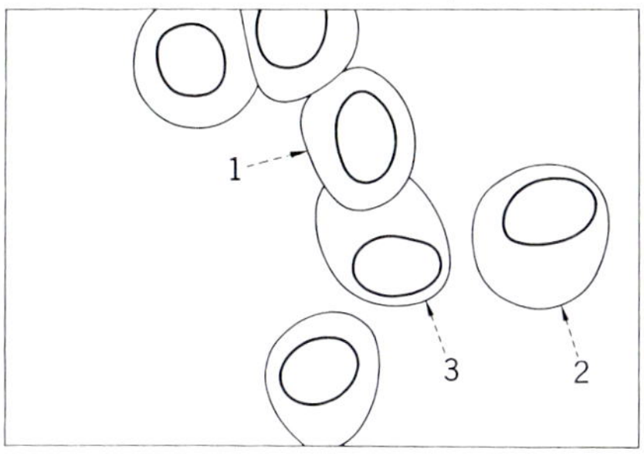

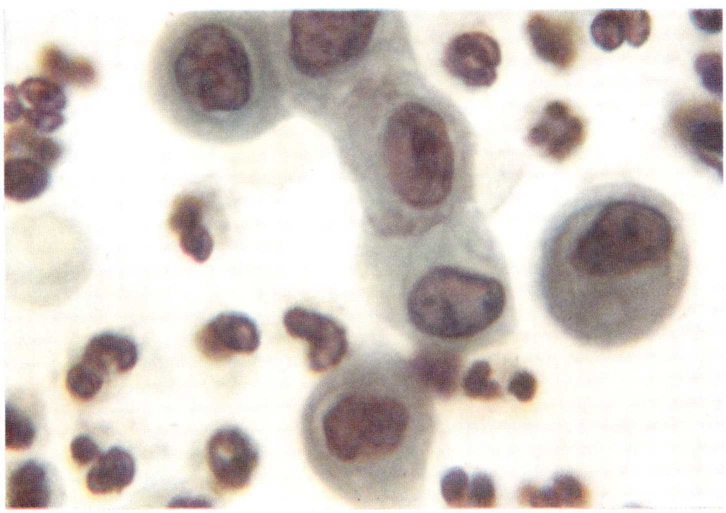

The smear pattern in abortion

If smears of the typical navicular or cytolytic type in pregnancy suddenly start to show more mature cells, this may indicate a tendency towards abortion. This prognostic implication is, however, not reliable during the first trimester, since the typical pregnancy patterns only develop latter as described previously. Roughly speaking, signs of abortion in smears may be as follows:

1. Numerous erythrocytes, leukocytes and cell debris.
2. Presence of eosinophilic superficial cells with pyknotic nuclei.
3. Increasing numbe1 of histiocytes and mucus in large amount.
4. In case of intrauterine death of the fetus, smears may reveal the postpartum pattern.
5. After an incomplete abortion, the vaginal smear may show isolated trophoblastic elements (*cf.* pp. 59–61)
6. One may find muscle fibers in vaginal smear in cases of missed abortion in the later half of gestation.

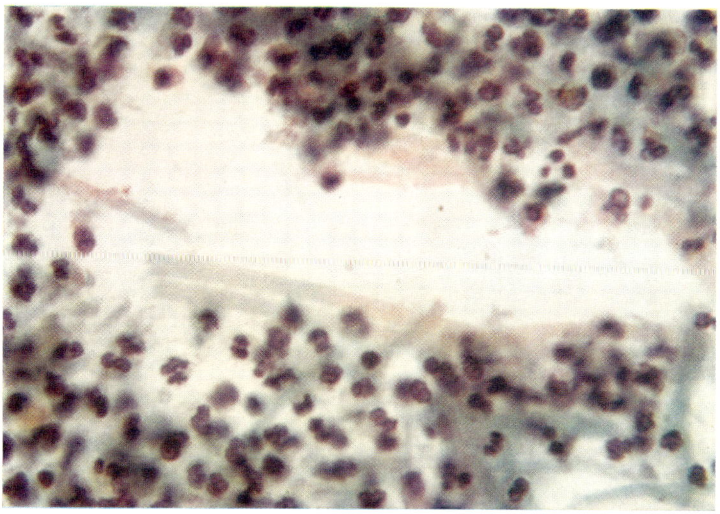

37. Striated muscle fibers of fetal origin in a case of missed abortion

a. Low magnification

Among numerous leukocytes, fragments of muscle-like fibers stained reddish violet are observed.

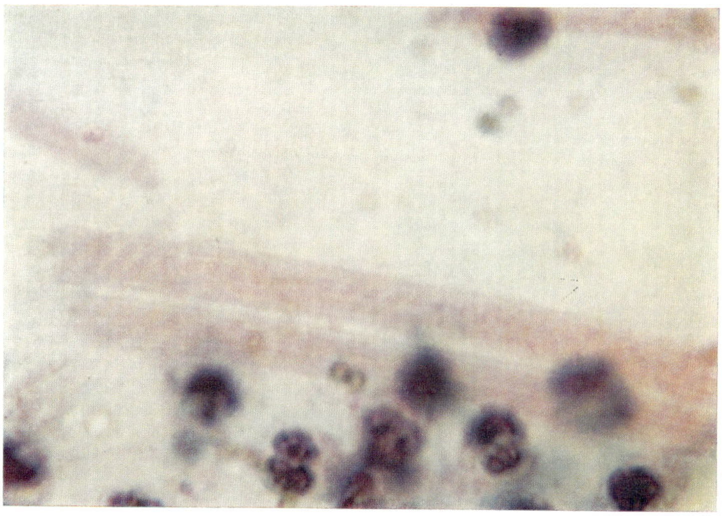

b. High magnification

Fairly clear striae are revealed in those muscle fibers.

Trophoblasts

Trophoblasts are sometimes observed in vaginal smears from such as an incomplete abortion. They are classified into syncytiotrophoblasts and cytotrophoblasts.

Syncytiotrophoblasts (Fig. 30)

1) General

A syncytiotrophoblast is large, round, elliptical or tadpole-shaped and has numerous nuclei.

2) Nuclei

1. Mostly round and uniform in size.
2. Evenly distributed or overlapping.
3. Located at the center of the cell.
4. Chromatin is finely granular and evenly distributed.

3) Cytoplasm

1. Rather dense.
2. Rather deeply basophilically stained.
3. Finely vesiculated, like fine gauze.

4) Background

Usually blood. There are numerous erythrocytes, leukocytes and cell debris.

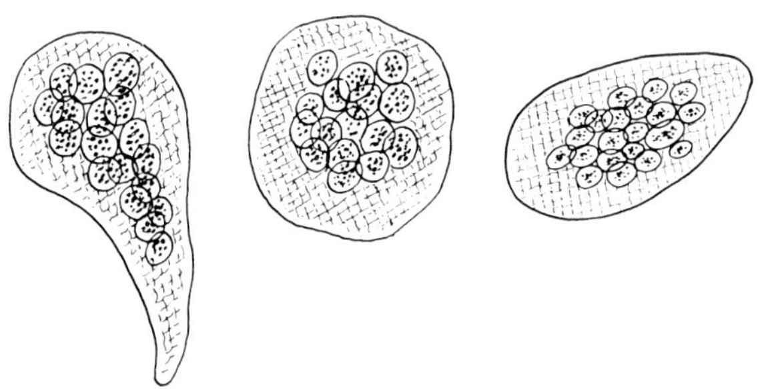

Fig. 30 Syncytiotrophoblasts.

38. Syncytiotrophoblasts

1. Nuclei : Round and uniform in size. Evenly distributed. Mostly overlapping one another. Chromatin is finely granular.
2. Cytoplasm Large, round, elliptical or tadpole-shaped.
3. Blood components in the background — mainly erythrocytes

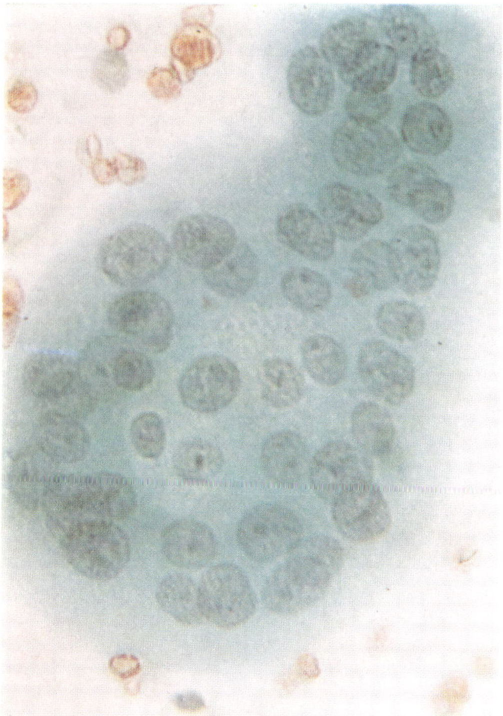

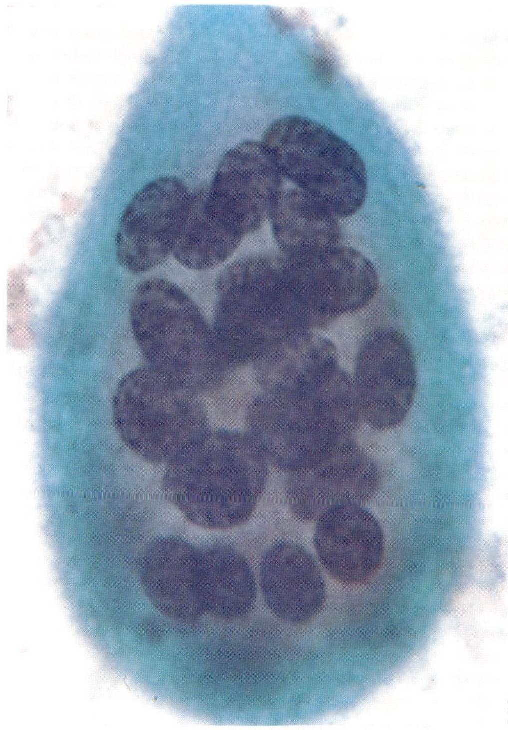

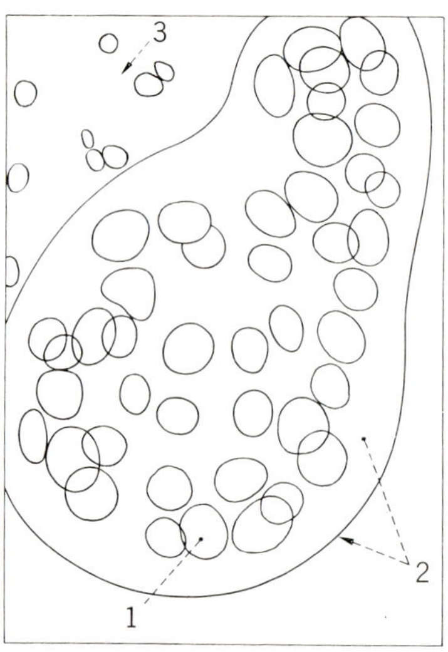

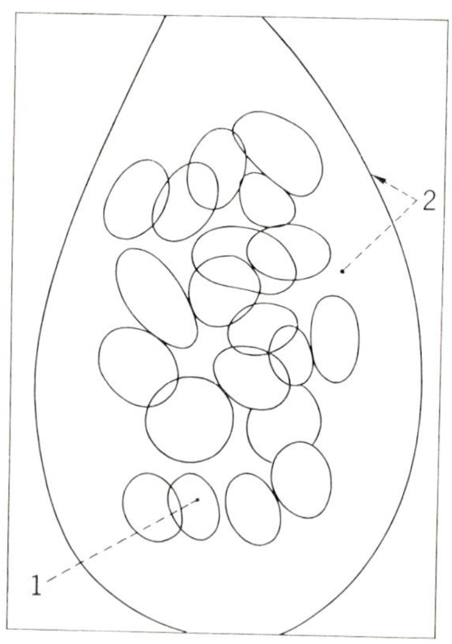

Cytotrophoblasts (Langhans cells) (Fig. 31)

1) General

Unlike giant cell-type syncytiotrophoblasts, cytotrophoblasts appear as a group of individual cells.

2) Nucleus

1. Located near the center of the cell.
2. Not uniform in size. Anisonucleosis. Generally a little larger than the nucleus of a syncytiotrophoblast.
3. Not uniform in shape. Pleomorphism.
4. Clear nuclear border.
5. Less chromatin than a syncytiotrophoblast and finely granular.
6. A nucleolus is sometimes seen.

3) Cytoplasm

1. Not uniform in shape. Often polygonal.
2. The cytoplasm is scanty for the size of the nucleus.
3. The cytoplasmic border can be seen but is not clear.
4. Spongy appearance.
5. Basophilic. Lightly stained.

4) Background

Blood components are often observed.

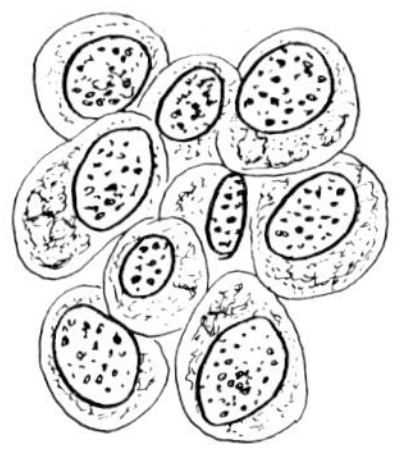

Fig. 31 Cytotrophoblasts (Langhans cells.)

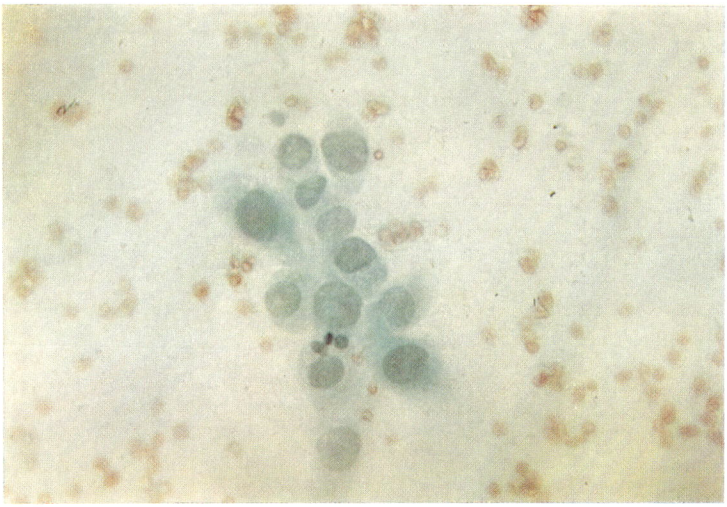

39. Cytotrophoblasts

a. Low magnification

Among numerous erythrocytes a group of large cells is found.

b. High magnification

1. Nuclei: Not uniform in size or shape. Chromatin is finely granular and evenly distributed. The borders are comparatively distinct.
2. A nucleolus.
3. Cytoplasm: Not uniform in shape. The borders are indistinct. Spongy. Stained light-blue.
4. Erythrocytes.

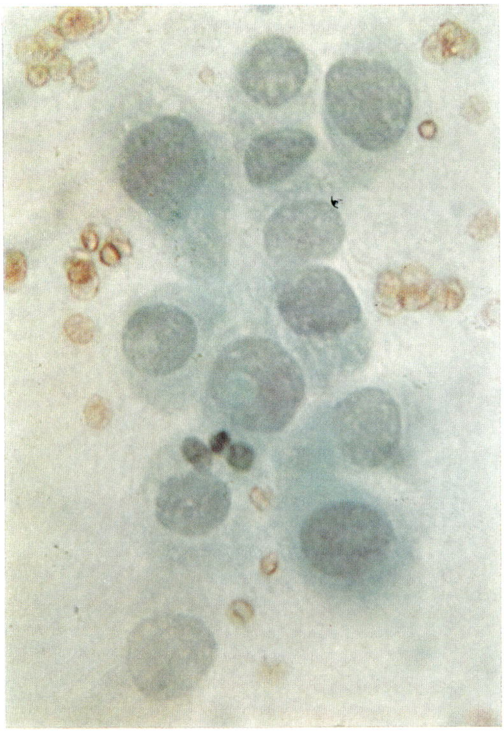

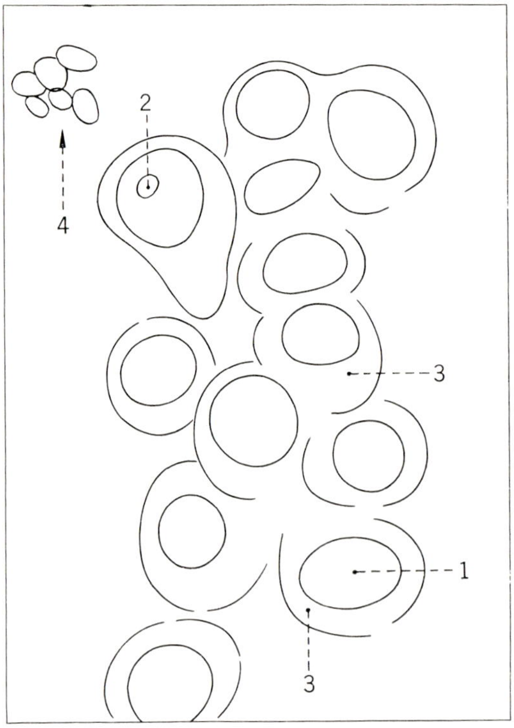

VI. Diagnosis of Carcinoma

Precautions in cancer cytology

Every effort should be made to prevent an incorrect diagnosis. This is especially true when making a diagnosis of malignancy. Therefore, the following should be kept in mind during the smear test.

1. The object of the smear test for carcinoma is to help to detect a malignant tumor in the early stages. However, it is not only meaningless, but sometimes leads to an incorrect diagnosis to make only smear tests on the patient in whom carcinoma can be obviously diagnosed, either clinically or macroscopically. For instance, in very advanced invasive carcinoma of the squamous epithelium, it is sometimes impossible to obtain representative cells to make the diagnosis, because the surface may be markedly necrotic. In such cases, clinical or histological diagnosis should take precedence.
2. A smear should be examined first under low magnification (about $\times 100$). This is very important to grasp the general impression of the whole cell pattern. A rough diagnosis could be made under low magnification.
3. Even when a questionable cell is found, the diagnosis should not be made before the whole smear has been screened. Also, if a questionable cell is observed under low magnification, the details of the nucleus and the cytoplasm should always be examined under higher magnification ($\times 400$ — oil immersion).
4. A diagnosis of malignancy should not be made even if only one questionable or suspicious cell has been observed in the smear. Smear tests should be repeated.
5. First, determine whether the tumor is malignant or benign. Even for the experienced examiner, it is not always easy to decide the type or the location of a malignant tumor from the smear test. Therefore, such decisions should not be made hastily.
6. The degree of staining reaction and the size of nucleus should be compared with leukocytes and other cells scattered in the same smear. It is inadequate to make an evaluation of hyperchromatism from an isolated cell.
7. When a cell puzzles the examiner, he should ask to repeat the smear test or have enough courage to ask the opinion of someone else with more experience.
8. A negative diagnosis does not deny the existence of a malignant tumor. One should explain to the clinician that it only means that a malignant cell has not been found on that particular smear. Both false-positives and false-negatives often present a problem. However, the examiner should always remember that a false-negative evaluation may particularly delay the discovery of carcinoma, often with tragic results.
9. The diagnosis of Class III should be made cautiously and only after all other possibilities have been eliminated. When no other diagnosis but Class III can be made, one should suggest to the clinician to collect more cellular

material, to perform a biopsy, conization, endometrial curettage or colposcopy, and to make further clinical investigations.

10. Clinical histories, observations and examinations yield valuable information, so it is foolhardy to disregard them and to make a diagnosis only from smear tests. If clinical data are not available, the examiner should note in his report that his findings are based "on the smear observations alone . . . .".
11. An incorrect diagnosis is sometimes made when the smear is either improperly fixed, dried, or stained. The experienced examiner, however, can make a relatively accurate diagnosis even from poorly prepared smears.
12. In a laboratory handling a large number of smears, accidents of mislabeling and contamination by cells from other smears can occur. Therefore, a check should be made when a cell which does not agree with the clinical symptoms and/or observations.

Diagnostic criteria for malignant cells

There is no yardstick which can be used for every malignant cell. Therefore, when making a diagnosis of malignancy, the examiner should investigate every questionable cell in the smear with all the knowledges he has in his power. Experience is invaluable. Just as a textbook knowledge of surgery cannot substitute for a surgeon's skill, a knowledge of cytology is incomplete without practical appearance with the microscope.

1) Abnormalities of the nucleus
    1. Size : Not uniform. Generally enlarged.
    2. Shape : Varied and irregular. Elongated, spindly or bizarre-shaped.
    3. The nucleus is large compared to the cytoplasm. In extreme cases, the nucleus is stripped of cytoplasm. Accordingly, the nuclear-cytoplasmic ratio is high.
    4. Hyperchromatism : This is usually seen under low magnification. The chromatin is increased, the distribution is uneven, and small clumps of chromatin are found in the nucleus. Chromatin tends to accumulate around the nuclear border. On the other hand, some nuclei become hypochromatic and chromatin accumulates near the borders or at various places in the nuclei.
    5. A nucleolus may appear, especially in a normally exfoliated adenocarcinoma cell.
    6. Mitosis and multinucleated cells. Mitosis of a squamous cell often indicates malignancy, but it has no definite relationship to the stage of carcinoma. Mitosis of an endocervical cell has little diagnostic significance.
2) Abnormalities of the cytoplasm
    1. Size : Not uniform. Caution is required, especially when the amount of cytoplasm is small in relation to the size of the nucleus.
    2. Shape : Varied. Look out for bizarre (elongated, spindly or tadpole-shaped) cells. Indistinct borders.
    3. Vacuolation. This is a sign of degeneration and one of the characteristics of cells from adenocarcinoma in vaginal smears. It is, however, sometimes seen during a benign inflammatory change.

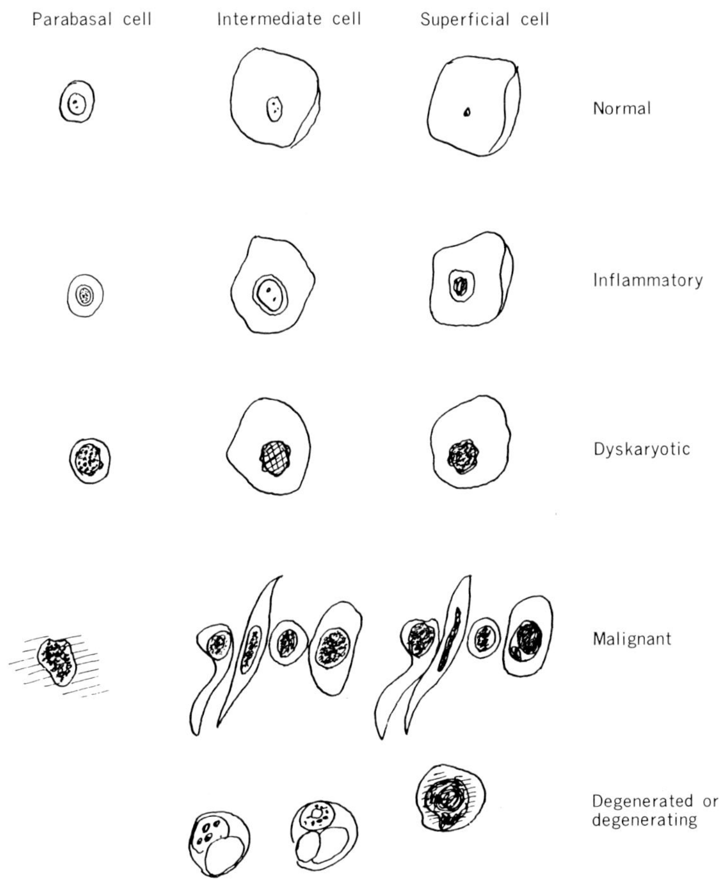

Fig. 32 From normal squamous cells to malignant cells.

4. Changes in staining reaction. The cytoplasm is stained somewhat deeper. The cytoplasm of a keratinized cell of squamous carcinoma is often stained orange or yellowish-pink. This is one of the typical characteristics of pulmonary carcinoma of the bronchogenic-type.

3) Other non-characteristic abnormalities

1. Signs of inflammation or hemorrhage : There is an increase in polymorphonuclear leukocytes, lymphocytes, erythrocytes and histiocytes, and the smear has a dirty appearance.
2. Maturation of the squamous epithelium is marked. There are numerous superficial cells, although the patient is quite old. This may be seen in adenocarcinoma.

VII. Epithelial Abnormalities of the Uterine Cervix

Nuclear abnormalities of squamous cells — Dyskaryosis

Dyskaryosis (Fig. 33) is a cytologic change in which the nucleus is abnormal while the cytoplasm is normally differentiated.

Classification of dyskaryosis

Dyskaryosis	Mild (Early stage)	Squamous type	Superficial
			Intermediate
	Advanced (Late stage)	Endocervical type	Parabasal

The squamous type is the most important.

Superficial and intermediate dyskaryotic cells

1) Size of the cell

Approximately as large as a normal superficial or intermediate cell.

2) Nucleus
   1. Moderately larger than normal.
   2. Not uniform in size.
   3. Indistinct border.
   4. Moderately hyperchromatic. Chromatin is coarsely granular and unevenly distributed.
   5. A perinuclear vacuole is sometimes observed.

3) Cytoplasm

Not much different from that of a normal squamous cell. Thin and delicate. The border is distinct.

Parabasal dyskaryotic cells

1) Size and shape

Nearly the same as those of the normal parabasal cell.

2) Nucleus
   1. Markedly enlarged.

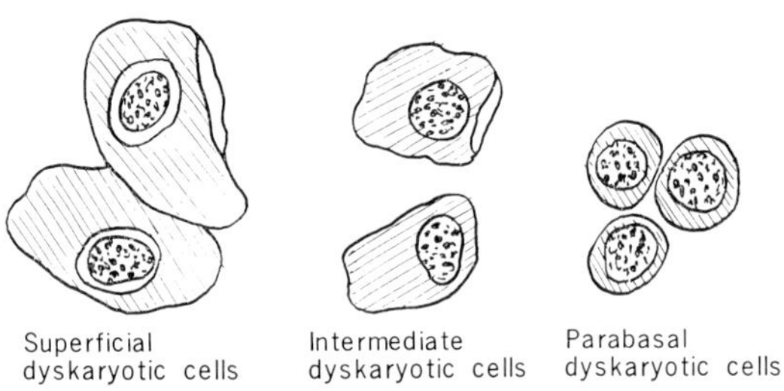

Fig. 33 Dyskaryotic cells.

2. Various and irregular in shape.
3. Distinct border.
4. Hyperchromatic. Chromatin is unevenly distributed and clumps are observed.
5. A nucleolus may be noted.

At first sight it resembles the nucleus of a malignant cell.

3) Cytoplasm

The same as that of a normal parabasal cell. The border is distinct. However, the nuclear-cytoplasmic ratio is definitely higher than a normal parabasal cell.

Dysplasia

Dysplasia is a series of atypical epithelial lesions that are sometimes called borderline lesions, atypical hyperplasia or koilocytotic atypia. The epithelial atypia may be ranked between carcinoma in situ and mere atypia, *e. g.* inflammatory atypia whatever it is called.

Superficial or intermediate dyskaryotic cells are mainly observed, but parabasal dyskaryosis are also often noticed in this lesion. It is a characteristic of dysplasia, however, that definitely malignant cells which appear in carcinoma in situ are never present.

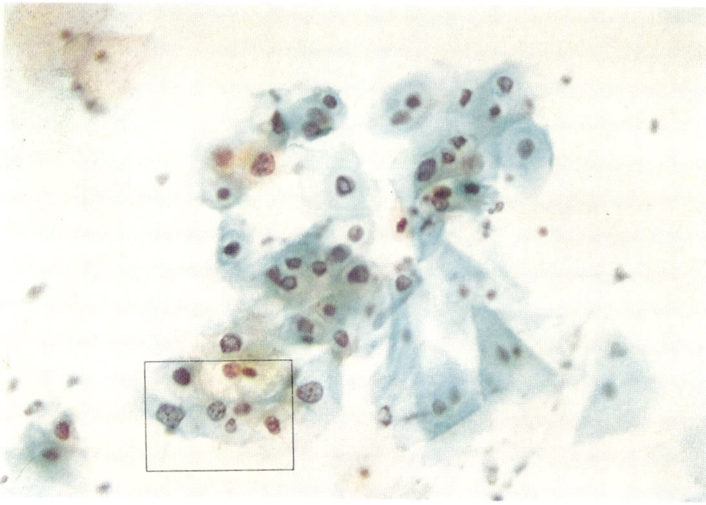

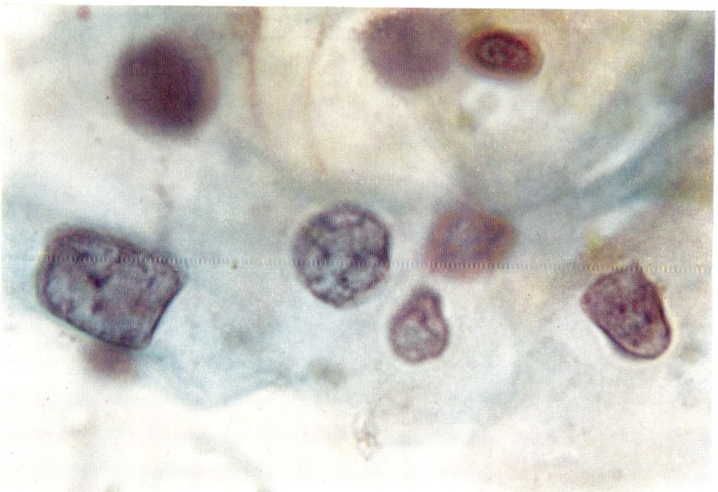

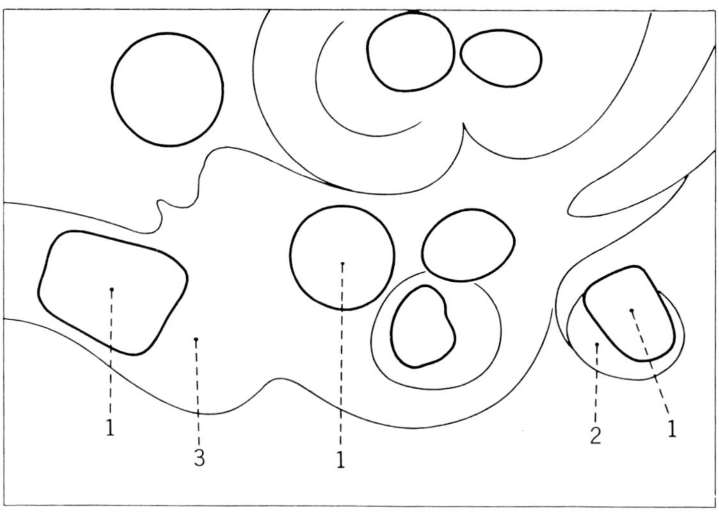

40. Dysplasia of the uterine cervix

a. Low magnification

This is a typical cell pattern of this lesion. Intermediate dyskaryosis. Moderate numbers of dyskaryotic cells are noticed among normal superficial and intermediate cells. There are comparatively few inflammatory cells.

b. High magnification

1. Nucleus: Note the irregular shape, anisonucleosis, enlarged size and hyperchromatism due to an increase in the amount of chromatin. The chromatin is rather evenly distributed, and there are few coarse granules.
2. A perinuclear vacuole.
3. Cytoplasm: Almost as abundant as that of the normal intermediate cell, but the shape is more irregular and the border is indistinct.

41. Dysplasia of the uterine cervix

a. Low magnification

Groups of atypical parabasal dyskaryotic cells are observed. Their nuclei are more or less abnormal. A large, multinucleated atypical cell is present in the left to them.

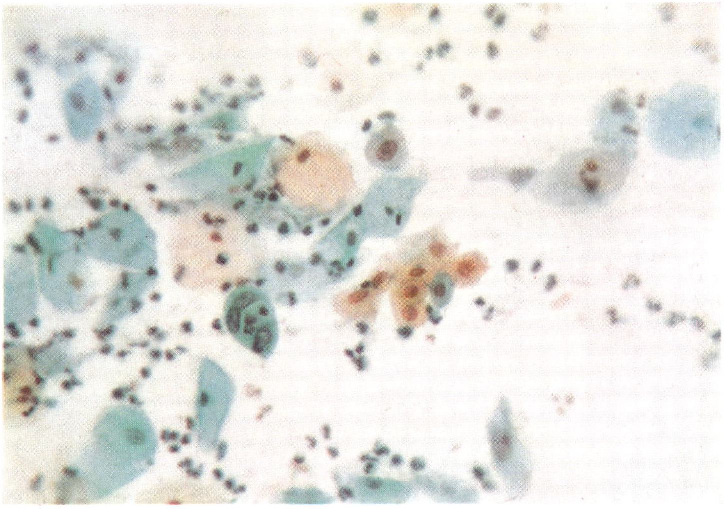

b. High magnification

1. Nuclei: Three large nuclei are noticed. Every nucleus is enlarged and rather irregularly shaped. Moderate degree of hyperchromatism is observed. Chromatin granules are distinct.
2. A nucleolus is faintly noticeable.
3. Intracellular vacuole.
4. Leukocytes included in the vacuole.

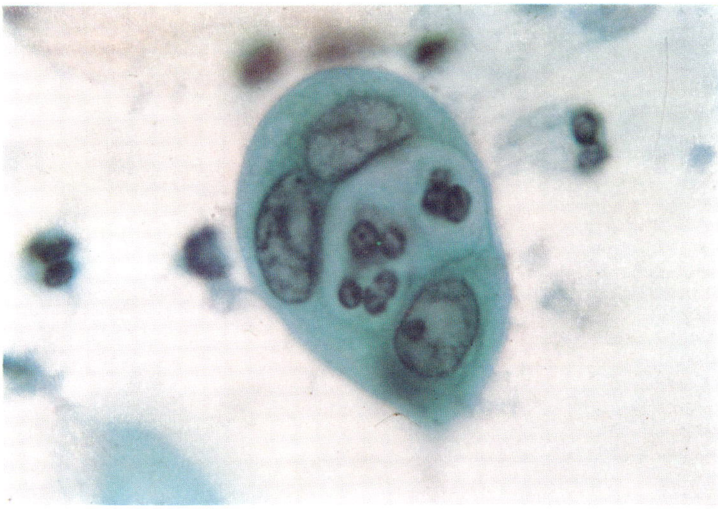

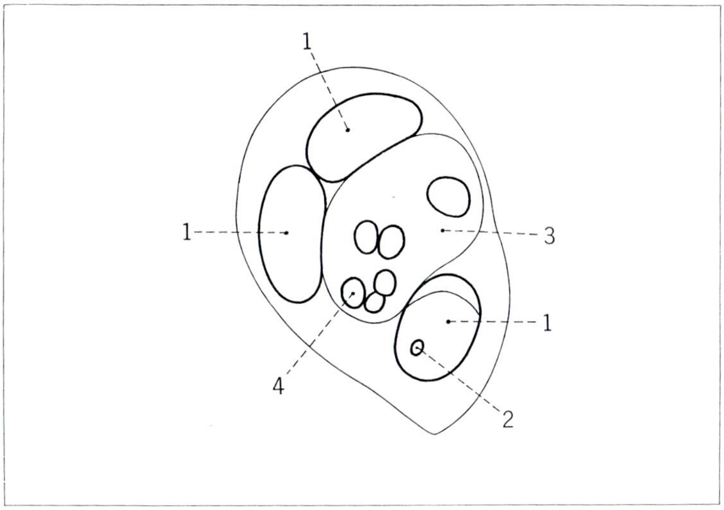

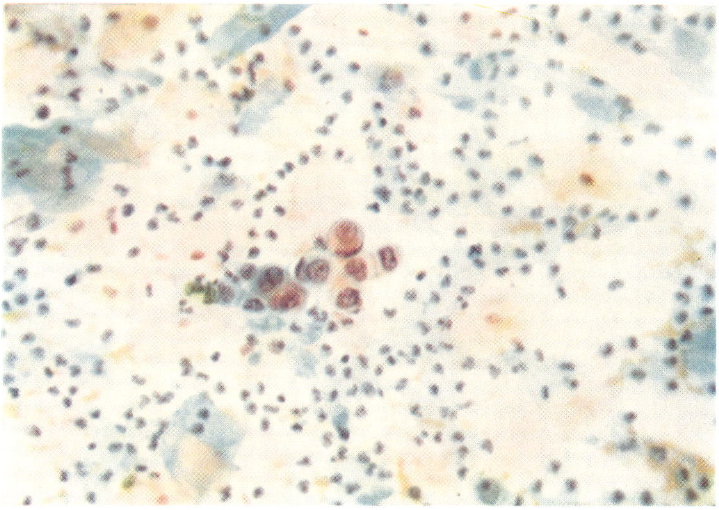

42. Dysplasia of the uterine cervix

a. Low magnification

A group of atypical cells is seen in the center. Background reveals numerous leukocytes and scattered benign-looking squamous cells.

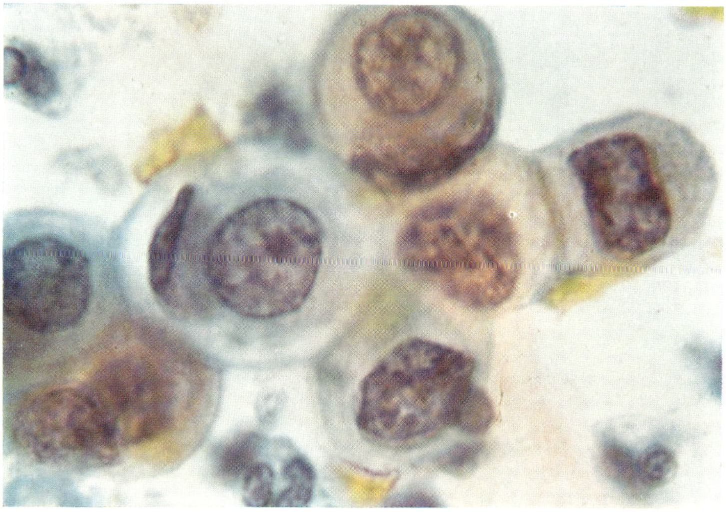

b. High magnification

1. Irregularly shaped nucleus located in the center of the cell.
2. Rounded nucleus showing increased chromatin content along the border.
3. A nucleolus is observed.
4. Overlapping nuclei.
5. Cytoplasm: Quite abundant, and the border appears thickened and distinct.

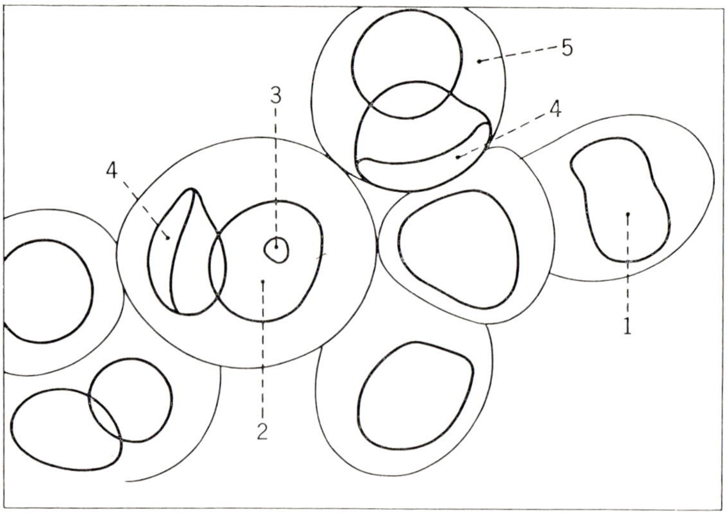

43. Dysplasia of the uterine cervix —advanced

a. Low magnification

Several definitely abnormal cells are observed on a comparatively dirty background with numerous erythrocytes. All are parabasal dyskaryotic cells.

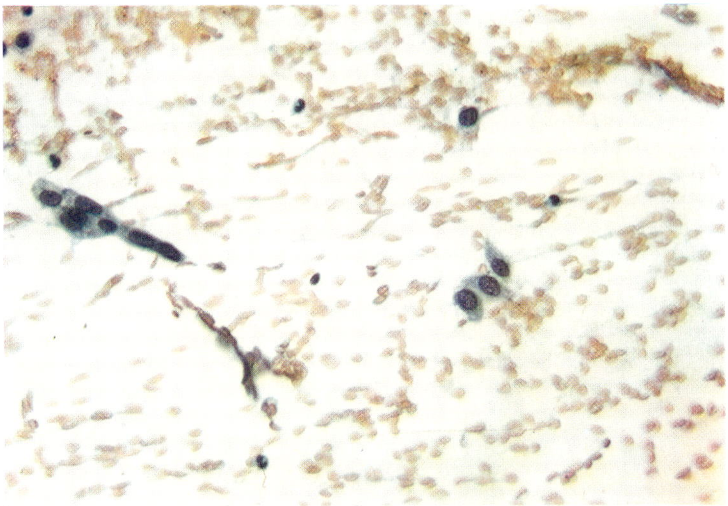

b. High magnification

1. Nuclei: Remarkably enlarged and not quite uniform in size. Hyperchromatism and coarse granulations are noticed.
2. Cytoplasm. Fairly abundant and lacy in appearance. The boder is rather distinct.
3. Numerous erythrocytes.

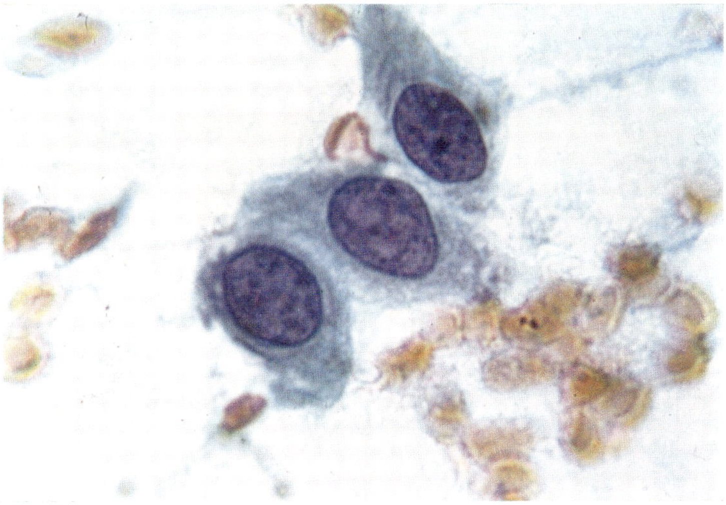

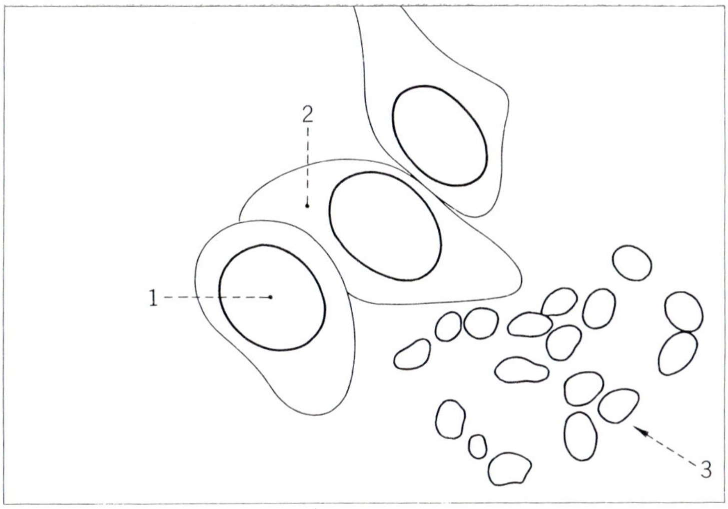

VIII. Cytodiagnosis of Carcinoma of the Female Genital Tract

Carcinoma in situ of the uterine cervix

Carcinoma in situ (Fig. 34)

"Carcinoma in situ is a carcinomatous lesion limited within the mucosal epithelium. It never invades the stroma beneath the basement membrane."

Cytologic characteristics of vaginal smears

1. These cells are more often observed mainly in cervical smears, because they infrequently exfoliate naturally and appear in vaginal smears.
2. Signs of necrosis and inflammation are only sometimes observed in the background. Therefore, the smear has a clean appearance and seldom includes dirty cellular fragments, polymorphonuclear leukocytes and erythrocytes. This does not apply when cervicitis exists independent of carcinoma in situ.
3. These cells are mainly parabasal dyskaryotic cells with a few malignant cells. Numerous benign squamous cells are also seen.

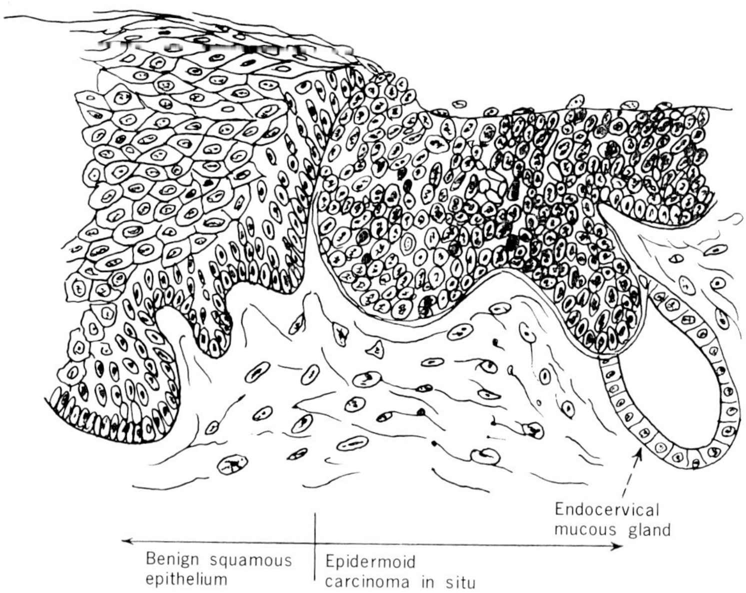

Fig. 34 Tissue of carcinoma in situ of the uterine cervix.

Characteristics of cells of carcinoma in situ

1) For the characteristics of parabasal dyskaryotic cells see pp. 66–67.

2) Characteristics of the malignant cells appearing in carcinoma in situ

Nucleus :

1. Size : Slightly enlarged, but often not very large. Smaller than the nucleus of a dyskaryotic cell.
2. Shape : Varies enormously. Round, elliptical, polygonal and other bizarre shapes are common.
3. Border : Sometimes distinct. Becomes indistinct in degeneration.
4. Chromatin : Extremely abundant and hyperchromatic. Sometimes coarse chromatinic granules are distributed unevenly, but usually the nucleus is concentrated and deeply stained, and loses the normal configuration. The nucleus in degeneration is hypochromatic and sometimes enlarged.
5. Nucleoli and mitosis may be observed in smears.
6. Occasionally multinucleated.
7. Generally, the nuclei of malignant cells of carcinoma in situ are more uniform in size and shape than those of invasive carcinoma.

Cytoplasm :

1. Amount : Very scanty. Therefore, the nuclear-cytoplasmic ratio is much higher than in dyskaryotic cells. The ratio becomes higher as the degree of differentiation decreases.
2. Shape : Varied. It is sometimes spindly in accordance with the shape of the nucleus, or tadpole-shaped. Generally, however, the shape of cells of carcinoma in situ varies less than those of invasive carcinoma.
3. The border is often indistinct.
4. Either thin or thick.
5. Staining reaction : Relatively strong basophilic-blue. However, it often stains eosinophilic-orange when keratinization is marked.
6. In degeneration : Vacuolation is noticed.
7. Sometimes a malignant cell is present included in another malignant cell.

44. Carcinoma in situ of the uterine cervix

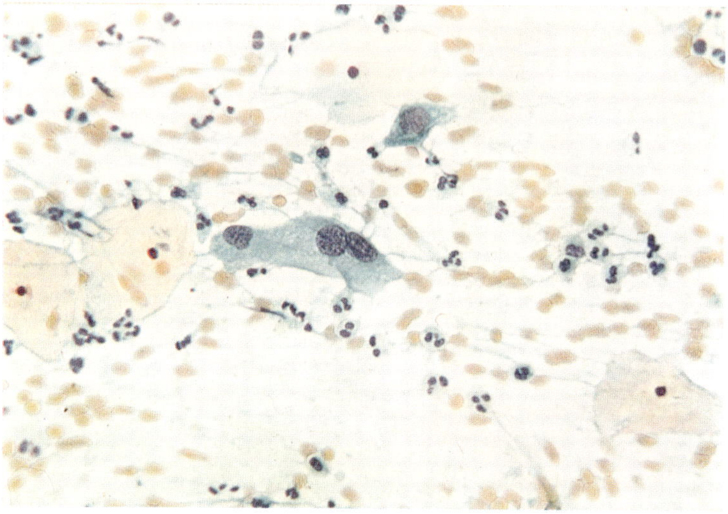

a. Low magnification

Parabasal dyskaryotic cells are seen among numerous erythrocytes as well as squamous cells.

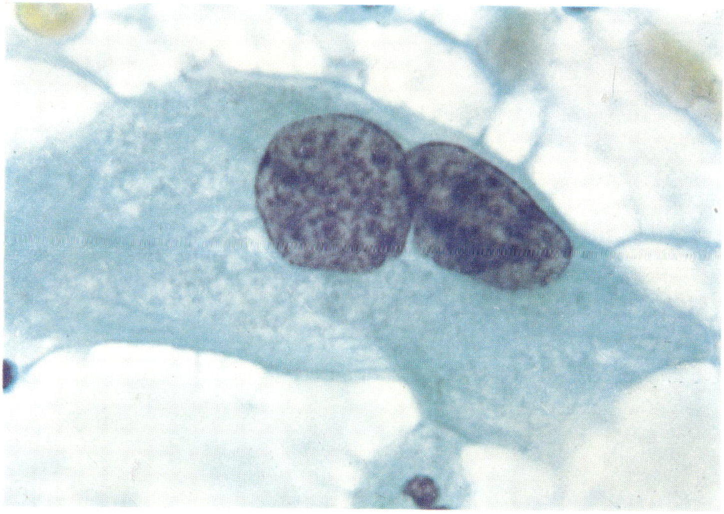

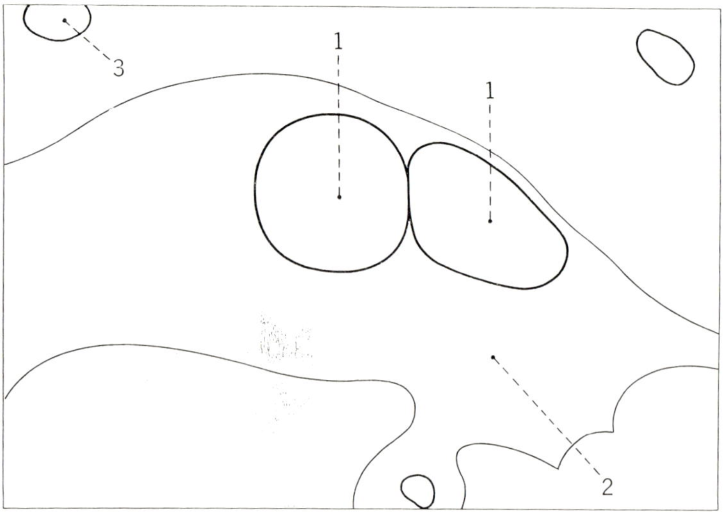

b. High magnification

1. An abnormally large and angular nucleus. Chromatin is increased and coarsely granular.
2. The cytoplasm is comparatively distinct. This is a parabasal dyskaryotic cell.
3. Erythrocytes.

45. Carcinoma in situ of the uterine cervix

a. Low magnification

Parabasal dyskaryotic cells are scattered on a clean background.

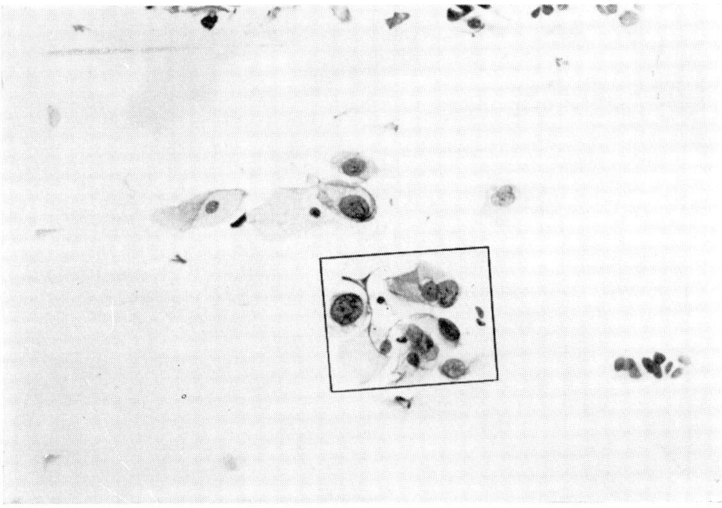

b. High magnification

1. A parabasal dyskaryotic cell of moderate degree. Chromocenters are distinctly observed. Chromatin has accumulated around the nuclear border, and this part is more hyperchromatic. The cytoplasm is overlapping and looks thickened.
2. A more advanced parabasal dyskaryotic cell. The nucleus is hyperchromatic and the nuclear membrane is thickened. Coarse chromatin granules are noticed. The cytoplasm is scanty and the border is irregular and indistinct.
3. This cell approximates a malignant cell. The amount of nuclear chromatin is even more increased, and the nucleus looks almost structureless. The cytoplasm is scanty and the border is indistinct.
4. An almost benign-looking intermediate cell. The nucleus, however, is rather dyskaryotic.
5. A parabasal dyskaryotic cell of moderate degree. The amount of nuclear chromatin has slightly decreased. The nuclear membrane is noticeably thickened.
6. Polymorphonuclear leukocytes. A comparison of the staining of the nuclei of polymorphonuclear leukocytes and dyskaryotic cells shows how hyperchromatic the latter are.

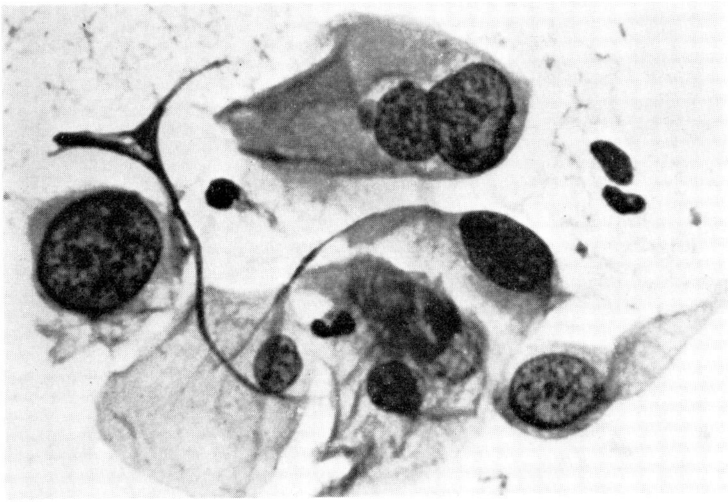

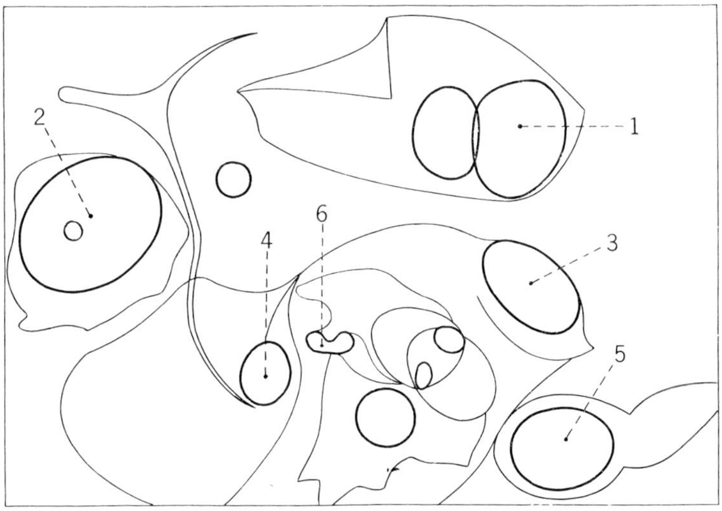

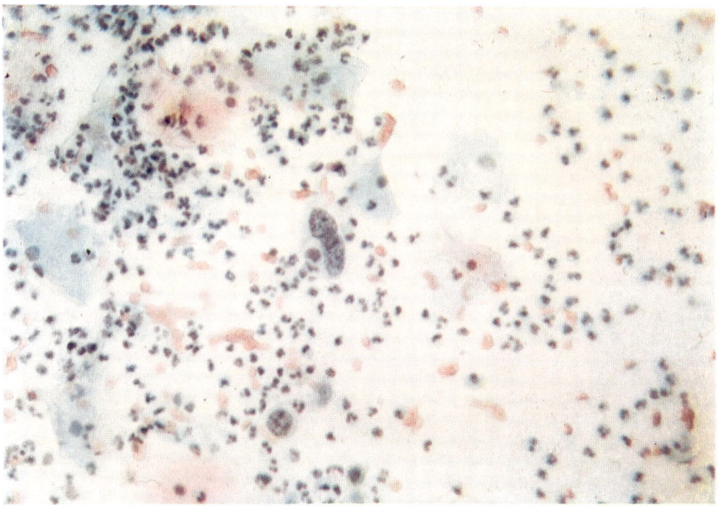

46. Carcinoma in situ of the uterine cervix

a. Low magnification

Dyskaryotic and malignant cells are observed among numerous leukocytes and scattered benign squamous cells.

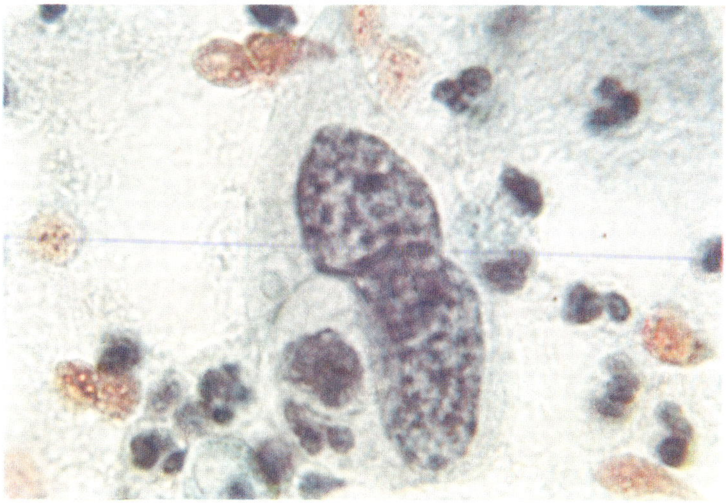

b. High magnification

1. A large malignant cell. As already noticed under low magnification, the nuclear chromatin is more abundant than in the other cells, and the nucleus is remarkably hyperchromatic. Coarse chromatinic granules are observed.
2. There is a small malignant cell within the cytoplasm of the large malignant cell.
3. Polymorphonuclear leukocytes.
4. Fairly well-preserved cytoplasm.

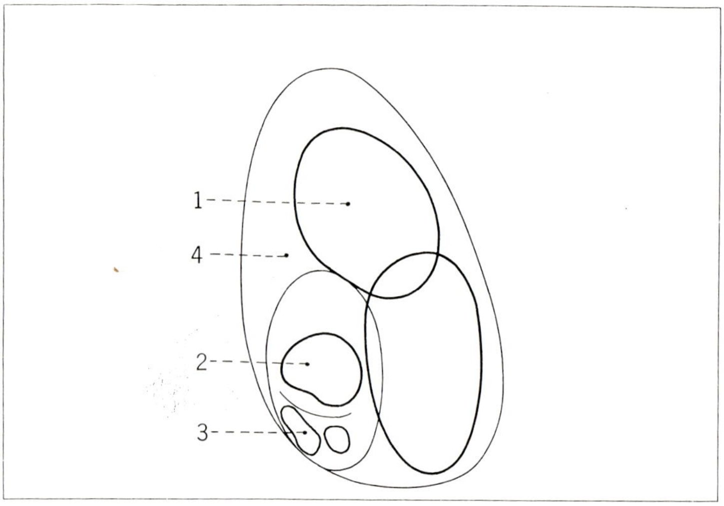

47. Carcinoma in situ of the uterine cervix

a. Low magnification

Several dyskaryotic cells are observed among a fairly large number of leukocytes and some benign cells.

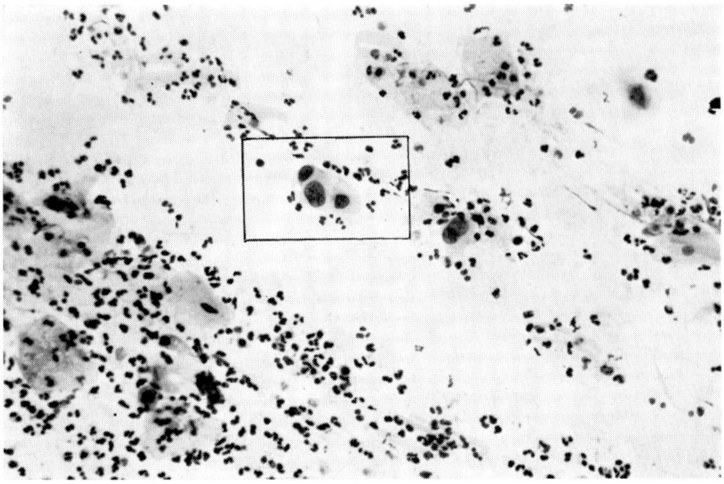

b. High magnification

1. Relatively advanced parabasal dyskaryotic cells. Note the enlarged and hyperchromatic nuclei and the comparatively well preserved cytoplasm.
2. An advanced parabasal dyskaryotic cell. It closely approximates a malignant cell. The size of the nucleus is reduced and it is more hyperchromatic. Note the scanty cytoplasm.
3. Polymorphonuclear leukocytes.

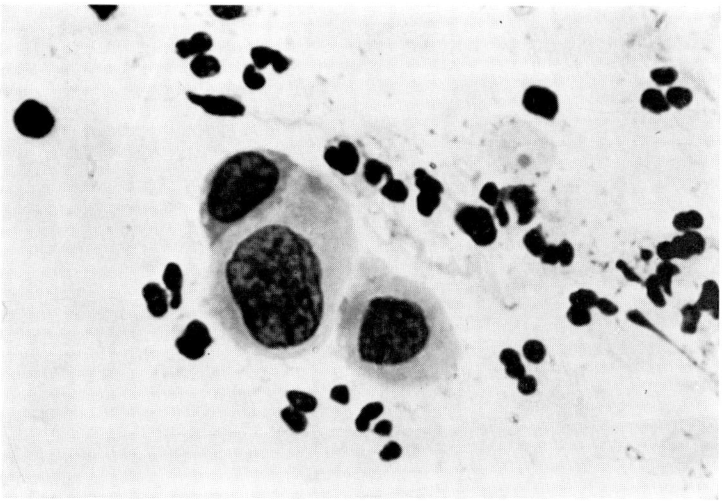

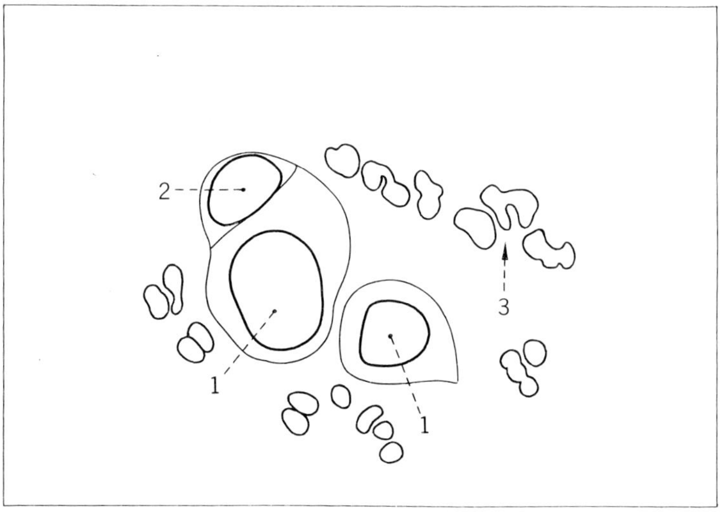

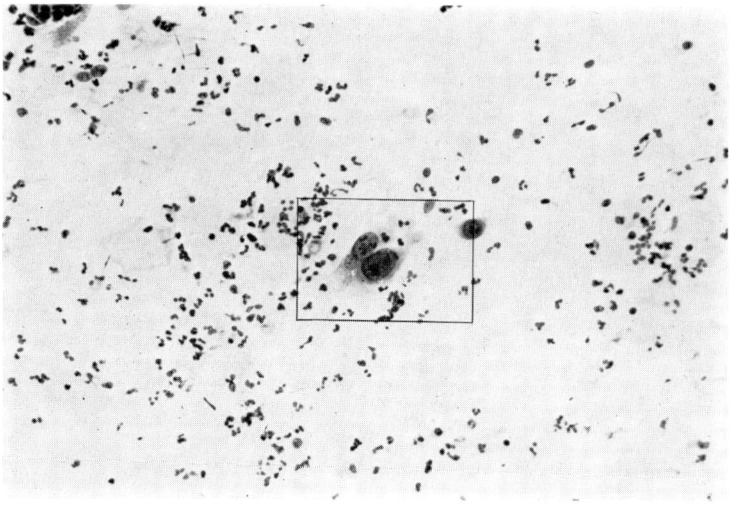

48. Carcinoma in situ of the uterine cervix

a. Low magnification

Three advanced dyskaryotic cells are seen. Normal superficial cells and leukocytes are scattered. Note the size and hyperchromatism of the nuclei of dyskaryotic cells compared with those of normal cells.

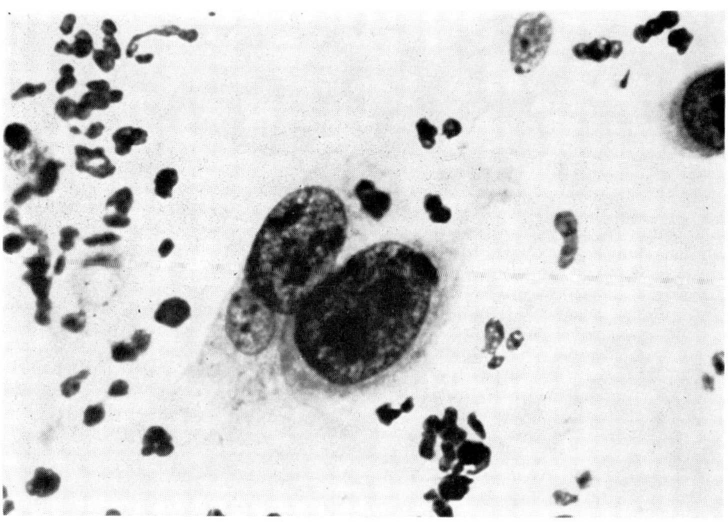

b. High magnification

1. Nuclear chromatin is coarsely granular.
2. Marked clumps of chromatin are observed.
3. The cytoplasm is scanty and the borders are indistinct.
4. Leukocytes included among the dyskaryotic cells.
5. The nucleus of a non-dyskaryotic parabasal squamous cell.
6. This is a benign cell, but the nucleus is naked due to cytolysis.
7. Polymorphonuclear leukocytes.

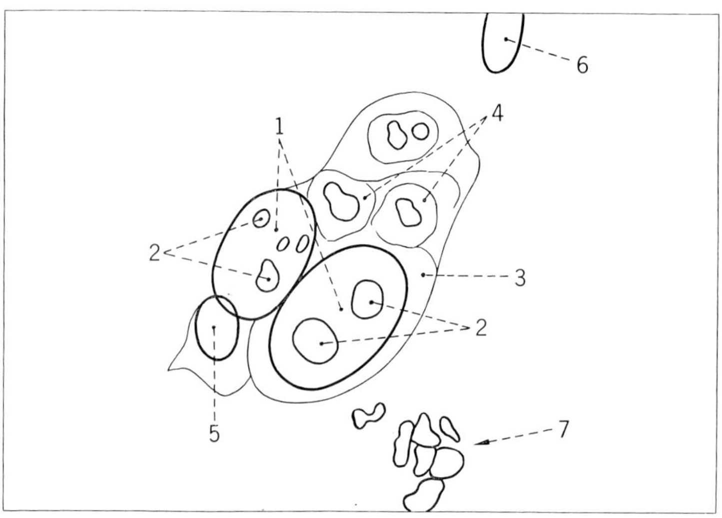

49. Carcinoma in situ of the uterine cervix

a. Low magnification

At least two dyskaryotic cells are seen in the center, one of which has keratinized cytoplasm shows multinucleation. Several benign squamous cells are present above those.

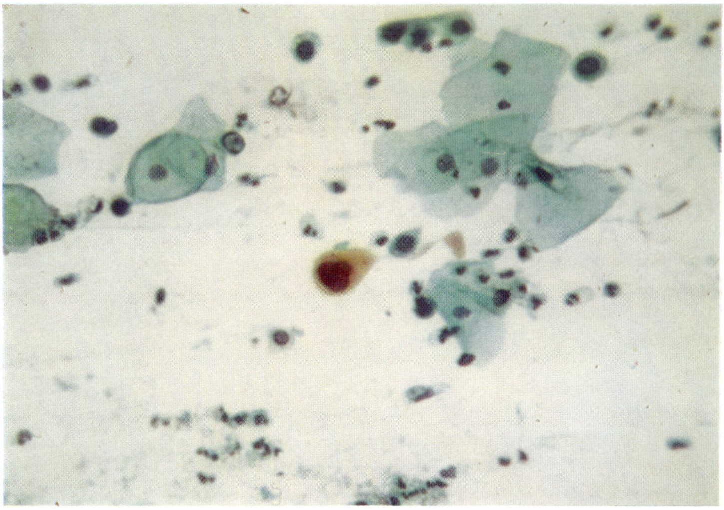

b. High magnification

1. Nuclei: Large in size, round or elliptical in shape and multinucleated. They are markedly hyperchromatic.
2. Abundant and remarkably orange-colored cytoplasm which appears thickened. The border is clear and distinct.
3. Smaller parabasal dyskaryotic cell.
4. The nucleus is round and hyperchromatic.

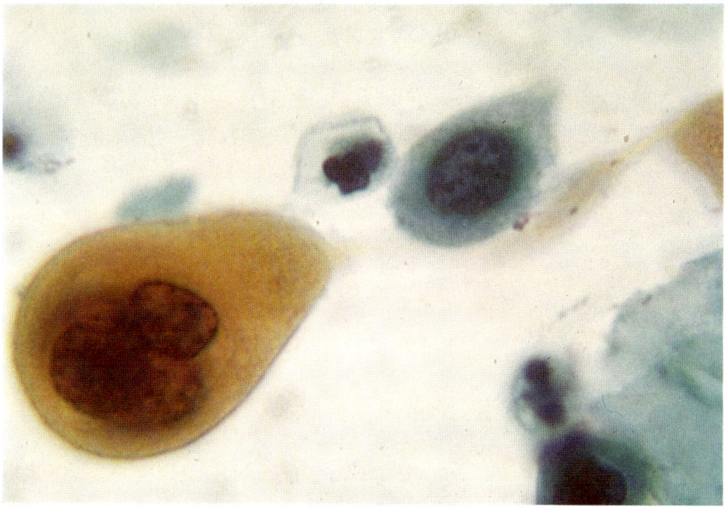

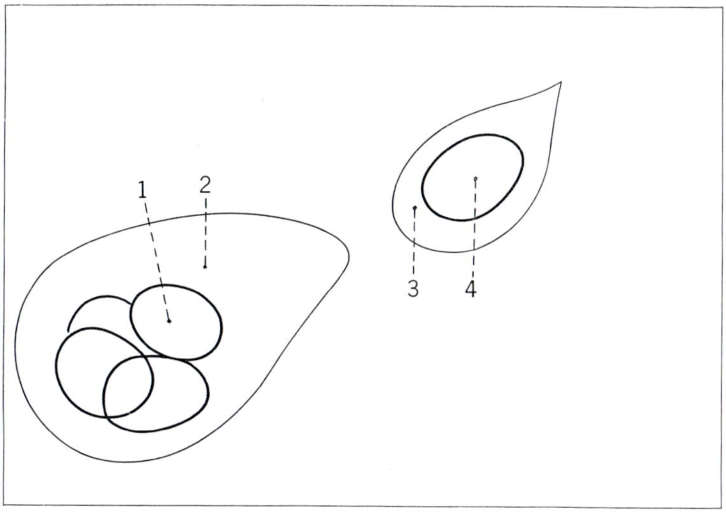

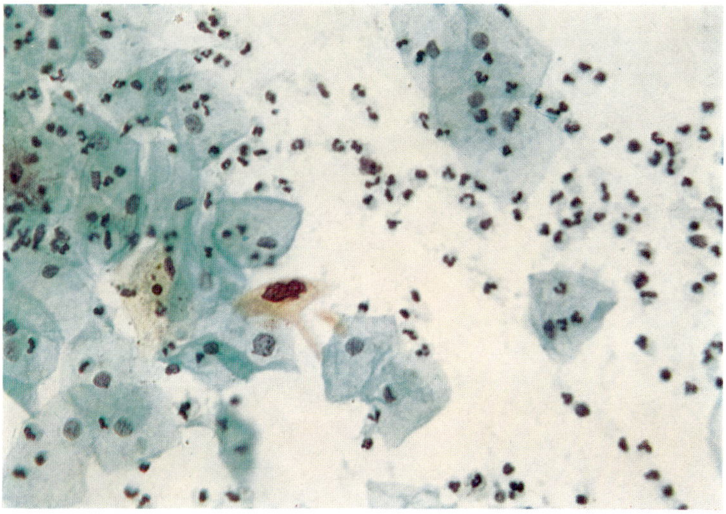

50. Carcinoma in situ of the uterine cervix

a. Low magnification

A single bizarre dyskaryotic cell is observed, showing marked keratinization.

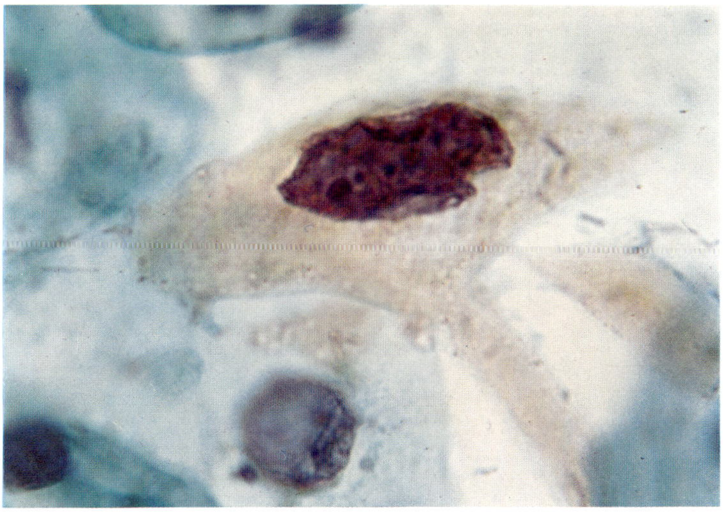

b. High magnification

1. The nucleus reveals an irregular shape and hyperchromatism.
2. Bizarrely shaped cytoplasm showing marked keratinization.
3. Perinuclear vacuole.

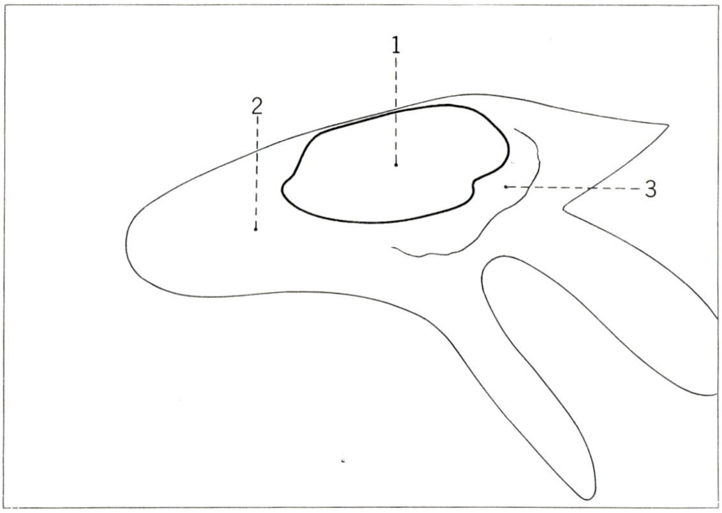

51. Carcinoma in situ of the uterine cervix

a. Low magnification

Numerous dyskaryotic cells are observed. The background is comparatively clean with small numbers of inflammatory cells. This is a typical picture of carcinoma in situ.

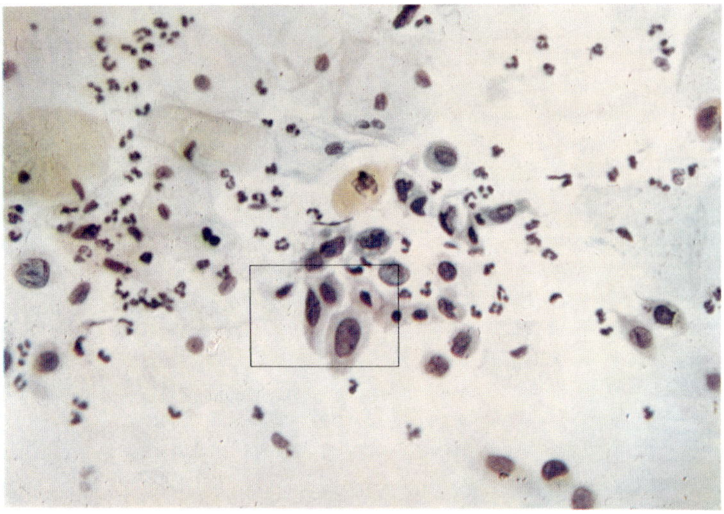

b. High magnification

1. Note the regularly elliptical dyskaryotic nucleus and coarse chromatin granules.
2. Cytoplasm : Abundant and the border is distinct. Lacy and thin.
3. Elongated nucleus.

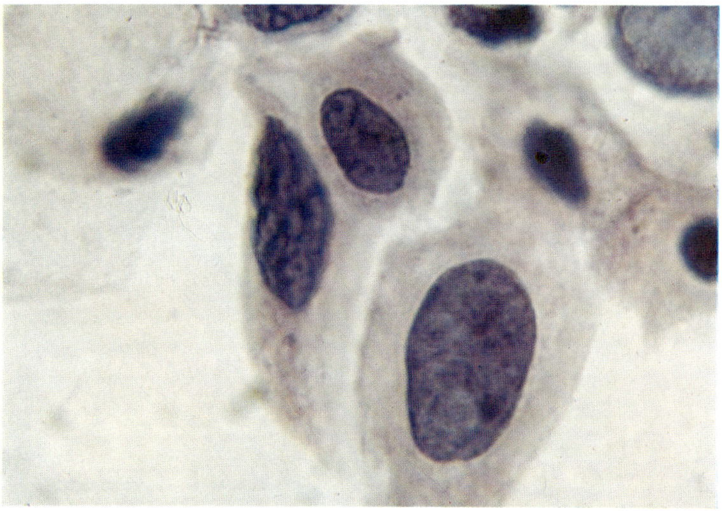

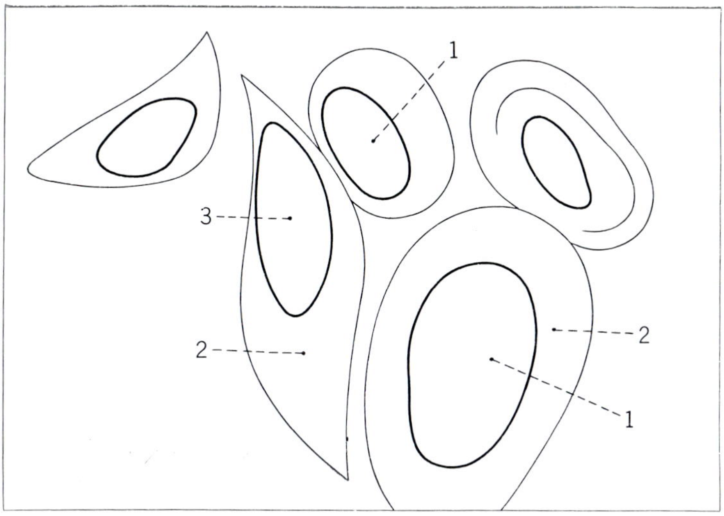

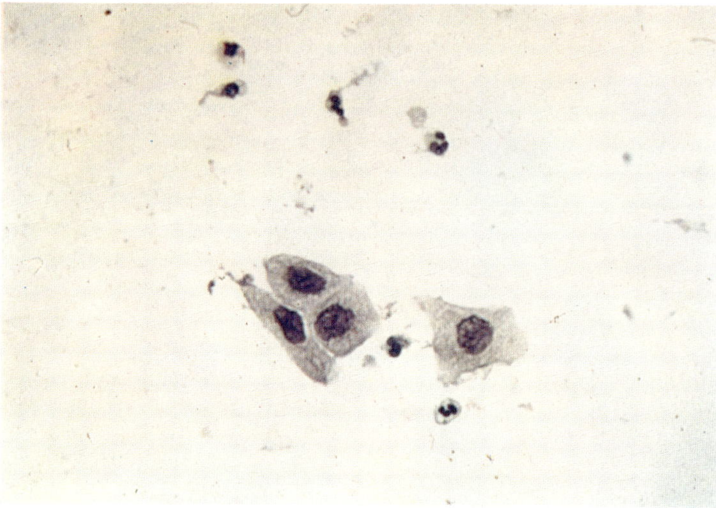

52. Carcinoma in situ of the uterine cervix

a. Low magnification

Four markedly dyskaryotic cells are seen at the center.

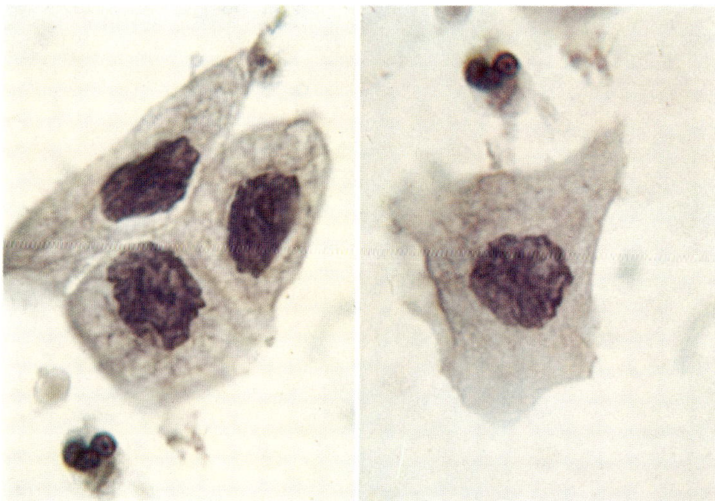

b. High magnification

1. Nuclei: Have irregular borders. They are markedly hyperchromatic. Chromatin accumulation is especially marked around the borders.
2. Cytoplasm: Thick and appears like wire netting. The border is very distinct.
3. Polymorphonuclear leukocytes.

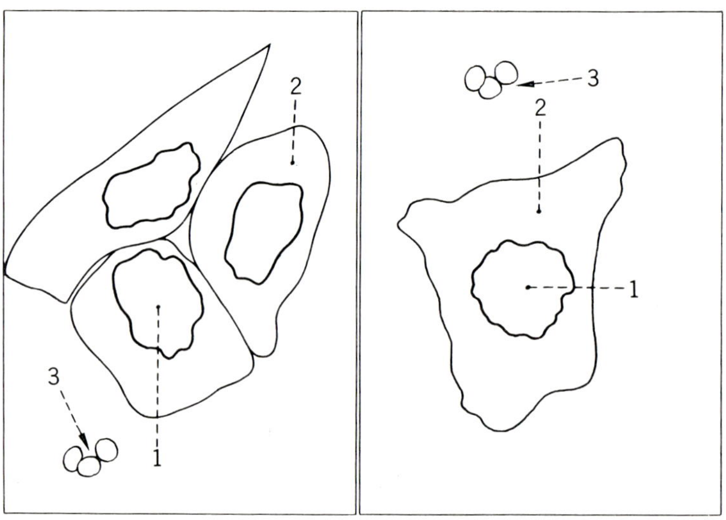

53. Carcinoma in situ of the uterine cervix

a. Low magnification

A group of parabasal dyskaryotic cells among numerous leukocytes and cell debris.

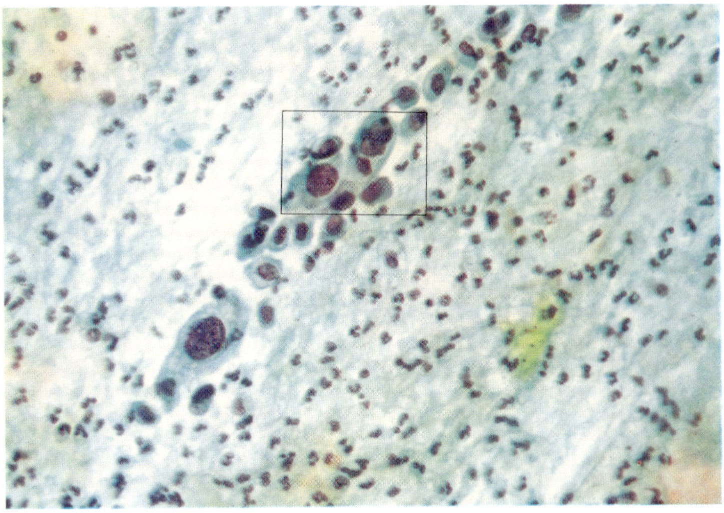

b. High magnification

1. Multinucleation of parabasal dyskaryotic cells.
2. Perinuclear vacuole.
3. Large and round dyskaryotic nucleus.
4. Leukocytes.

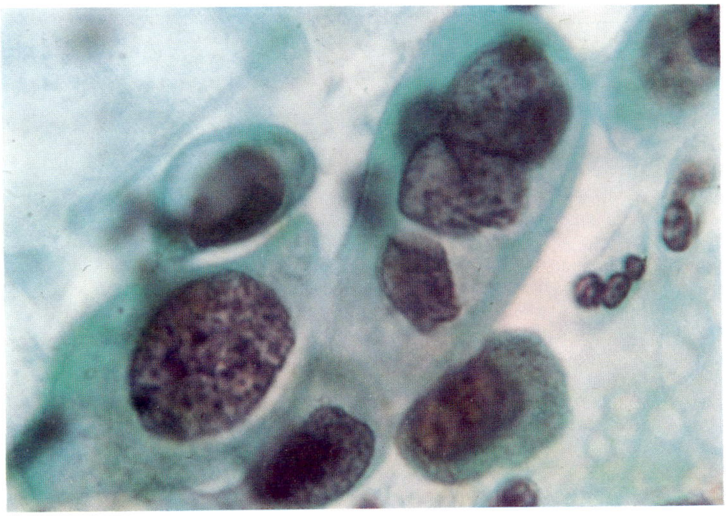

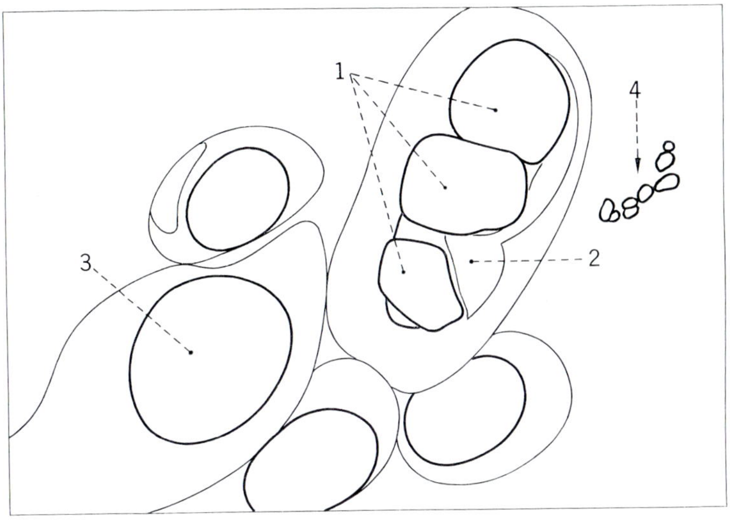

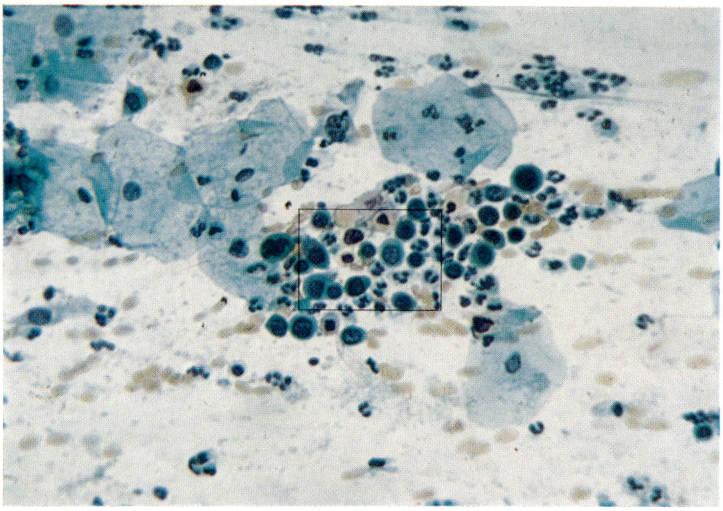

54. Carcinoma in situ of the uterine cervix

a. Low magnification

There are numerous dyskaryotic cells mixed with a small numbers of malignant cells. Compare these dyskaryotic nuclei with those of benign squamous cells located above.

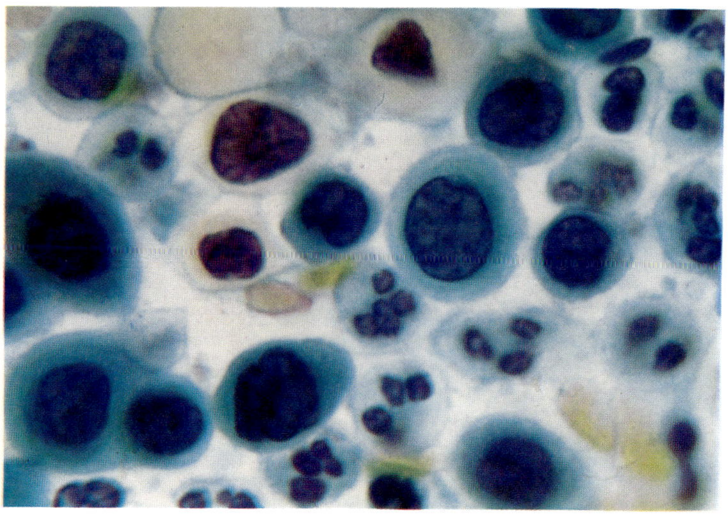

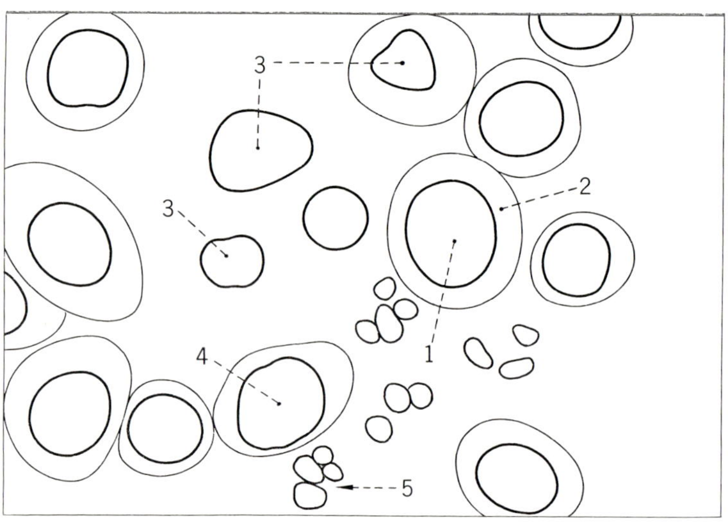

b. High magnification

1. Typical parabasal dyskaryotic cell showing round and hyperchromatic nucleus.
2. Cytoplasm is sparse in amount compared with the size of the nucleus. Consequently N/C ratio becomes small. The cellular border is distinct.
3. Malignant cells showing irregular and hyperchromatic nuclei and vague cytoplasm.
4. Parabasal dyskaryotic cell with irregularly shaped nucleus.
5. Polymorphonuclear leukocytes.

55. Carcinoma in situ of the uterine cervix

a. Low magnification

A group of malignant cells is observed in the center. A few leukocytes are in the background.

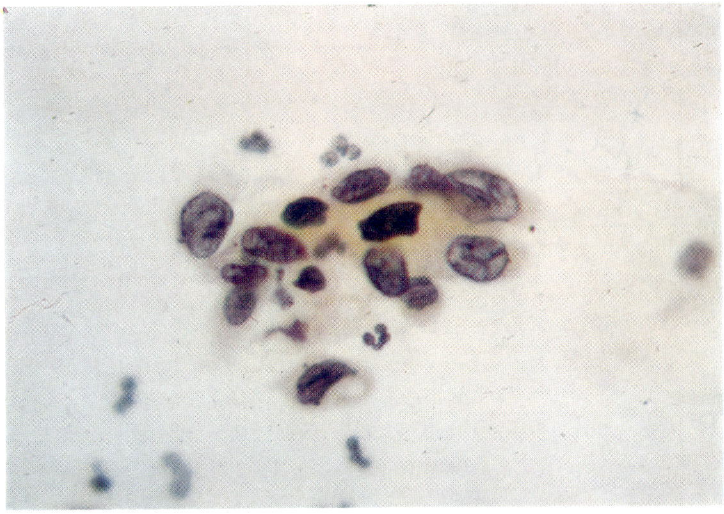

b. High magnification

1. Nuclei: Various in shape: round, elliptical, kidney-shaped and irregular. Not uniform in size. A marked characteristic is that chromatin is distributed more abundantly around the border, and thus the nuclear membrane appears thickened. On the contrary, chromatin is rather scanty in the middle Small chromocenters are noticed.
2. The cytoplasm is extremely scanty. No cellular border can be recognized.
3. The nucleus is stripped of cytoplasm.
4. Leukocytes.

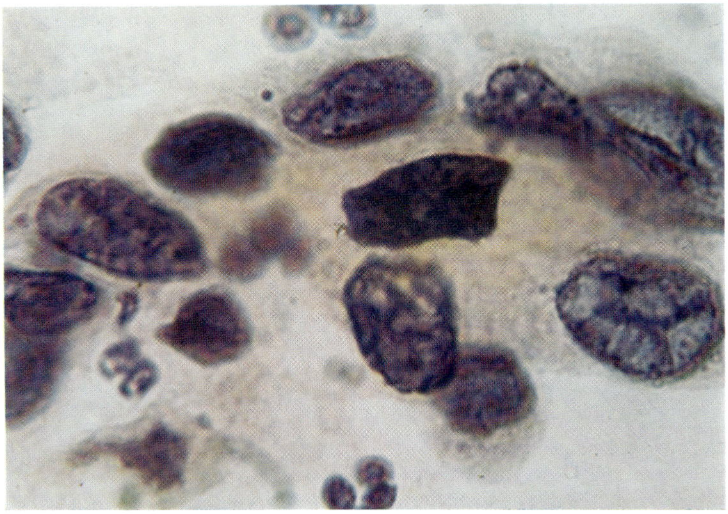

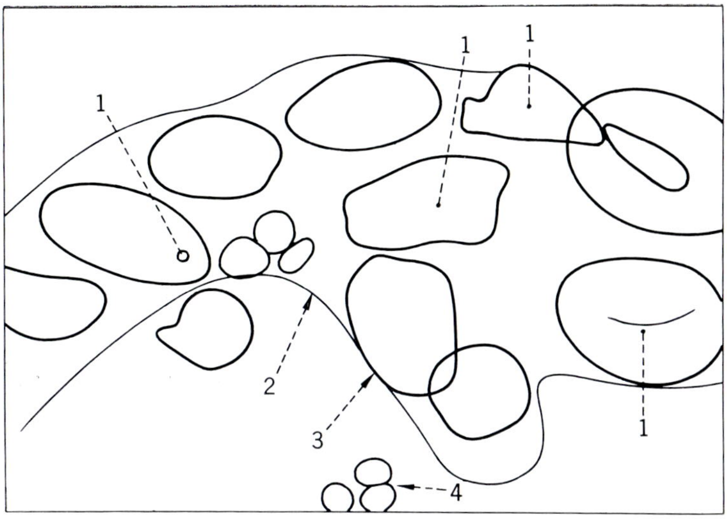

Invasive (epidermoid) carcinoma of the uterine cervix

General characteristics (Fig. 35)

1) Characteristics of the smear
   1. The whole cell is smaller than normal squamous cell. The undifferentiated malignant cell is especially smaller than the keratinized differentiated one.
   2. It is said that the number of malignant cells appearing in vaginal smears may have a relation to the stage of the carcinoma.
   3. It is one of the characteristics of malignant cells that they appear separately, but sometimes they appear in clusters.
   4. Numerous cellular fragments due to necrosis and polymorphonuclear leukocytes due to inflammation are also seen in the smear.
   5. Malignant cells of all types are also a characteristic of invasive carcinoma. Of course, dyskaryotic cells are occasionally observed.
   6. Sometimes malignant cells form pearls.

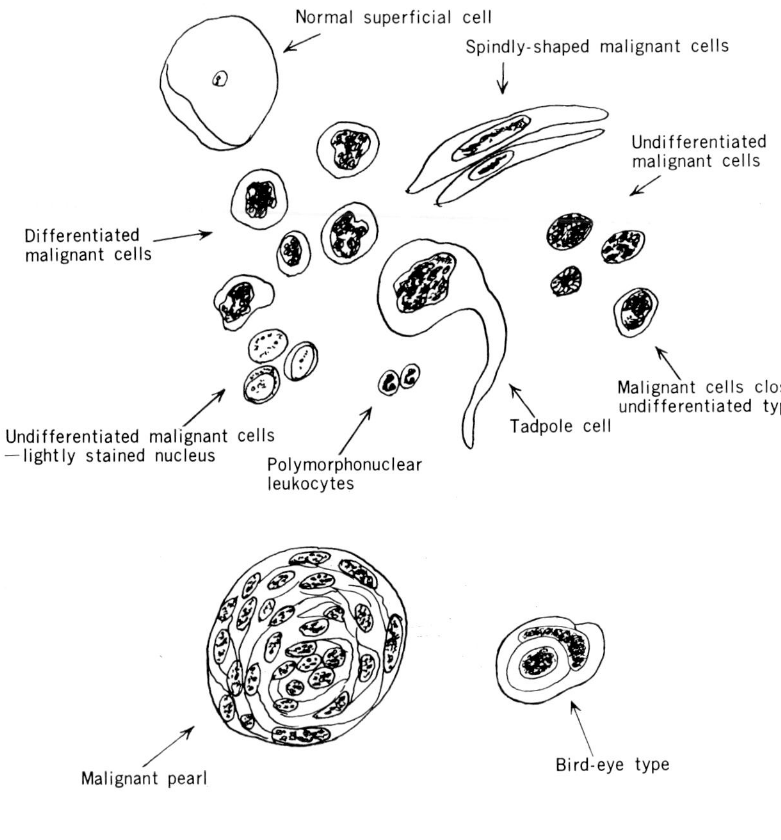

Fig. 35 Various types of malignant cells.

2) Nucleus

1. Size : Twice as large as the nucleus of a superficial cell, or even larger. While the nuclear-cytoplasmic ratio of a normal superficial cell is about 2%, the ratio of a malignant cell is on the average higher than 35%. The N/C ratio of an undifferentiated carcinoma cell is even higher, and the cytoplasm is often almost completely lost. Thus the N/C ratio is generally very high.
2. Shape : Variety in shape is one of the characteristics; round, elliptical, polygonal, spindly, etc.
3. The border is distinct, but becomes indistinct in degeneration.
4. Chromatin : Increases greatly. Remarkable hyperchromatism is one of the striking features.
   It is coarsely granular in 50% of smears and finely granular in 25%. In degeneration, the hyperchromatism disappears and the nucleus becomes rather translucent and structureless. Occasionally, chromatin accumulates around the border, and this part becomes hyperchromatic, while the inside is rather hypochromatic.
5. Nucleolus : An eosinophilic nucleolus is clearly seen in more than 10% of smears. Sometimes more than two nucleoli are observed.
6. Nuclear vacuolation : Rarely observed.

3) Cytoplasm

1. Varies in size and shape.
2. Round, elliptical, elongated, spindly, tadpole and other irregular shapes are noticed. Elliptical and spindly cells are often observed in comparatively undifferentiated tumor, and bizarre cells in the keratinized differentiated type.
3. The border is generally indistinct.
4. The cytoplasm is sometimes almost lost in the undifferentiated type.
5. Staining reaction : Basophilic blue in 50% of smears. This staining is often observed in undifferentiated carcinoma. Twenty-five per cent are stained eosinophilic-orange, red and yellow. This is often noticed in keratinized differentiated carcinoma. The other twenty-five per cent show no definite staining reaction.
6. A cytoplasmic vacuole may be observed in the degenerated cell. This is more often observed in adenocarcinoma, but is sometimes seen in epidermoid carcinoma, too. A leukocyte is occasionally included in the vacuole.

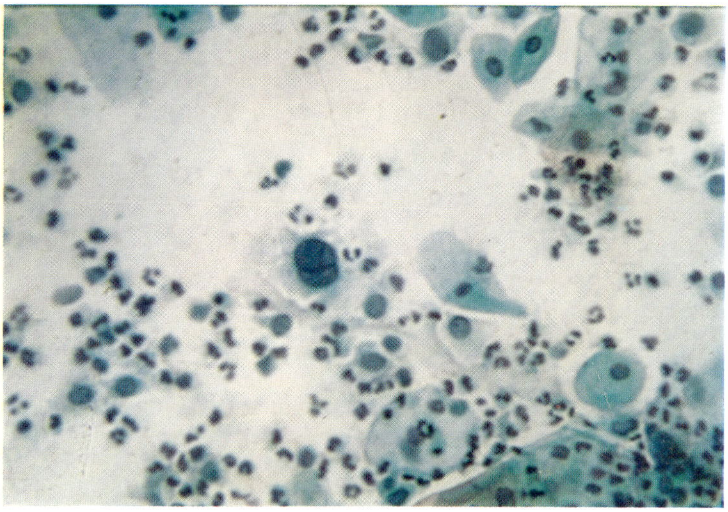

56. Invasive carcinoma of the uterine cervix

a. Low magnification

A large cell is observed among numerous leukocytes and scattered squamous cells, some of which are dyskaryotic.

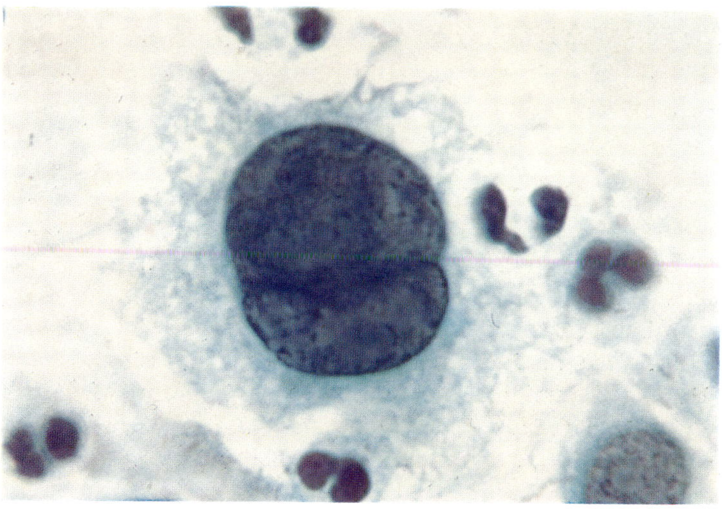

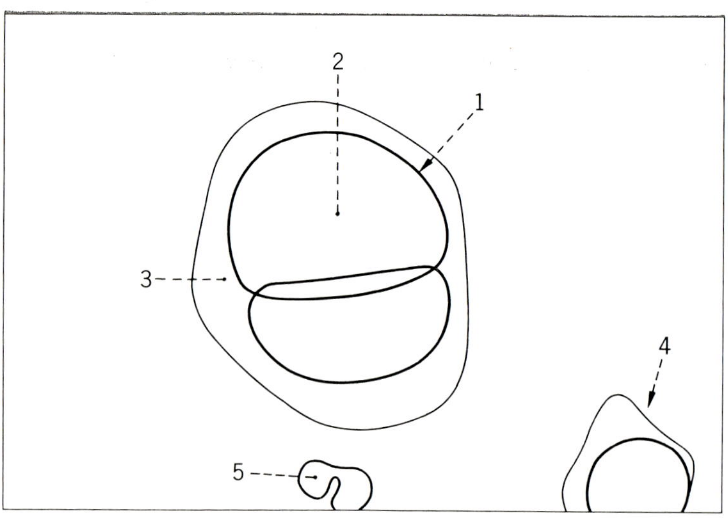

b. High magnification

1. The nuclear border is very distinct and thickened. Very hyperchromatic, and may contain two nuclei.
2. No nucleolus is observed.
3. The cytoplasm is scanty and the borders are generally vague.
4. Dyskaryotic cell.
5. Leukocyte.

57. Invasive carcinoma of the uterine cervix — Snake cell

a. Low magnification

Long, snake-shaped malignant cells are observed in the center. Spindly, soybean-shaped and round malignant cells are also scattered in the smear.

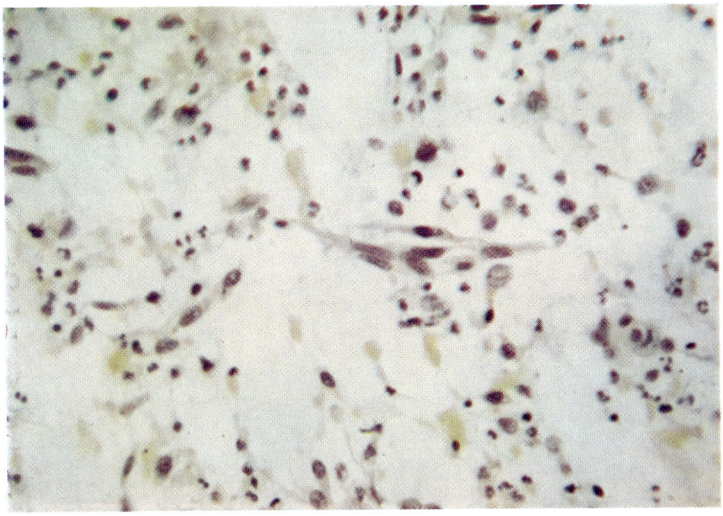

b. High magnification

1. A snake cell. The nucleus is enlarged, and the chromatin is coarsely granular.
2. The cytoplasm is elongated, and the border is comparatively distinct.
3. A malignant cell with an irregularly shaped and hyperchromatic nucleus. The cytoplasm is indistinct.

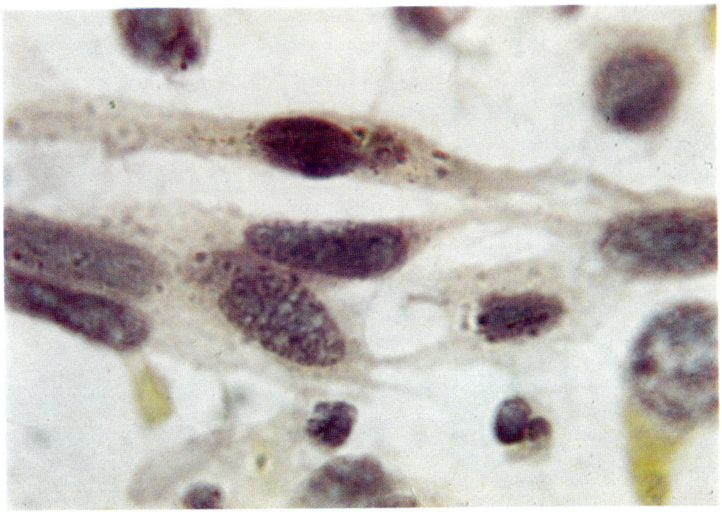

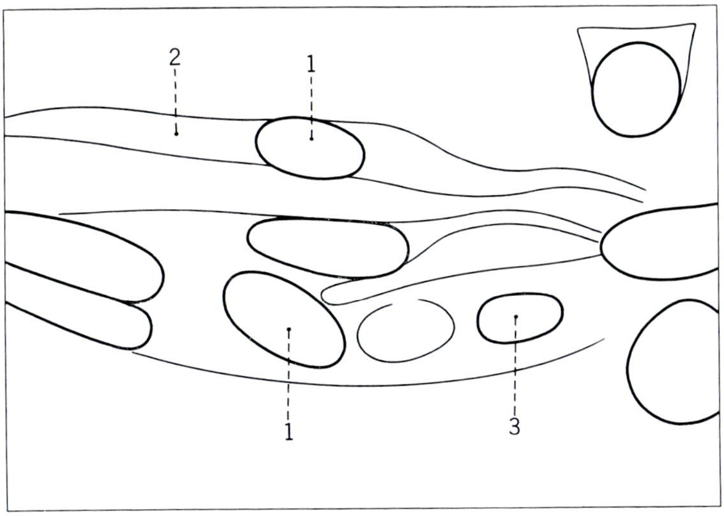

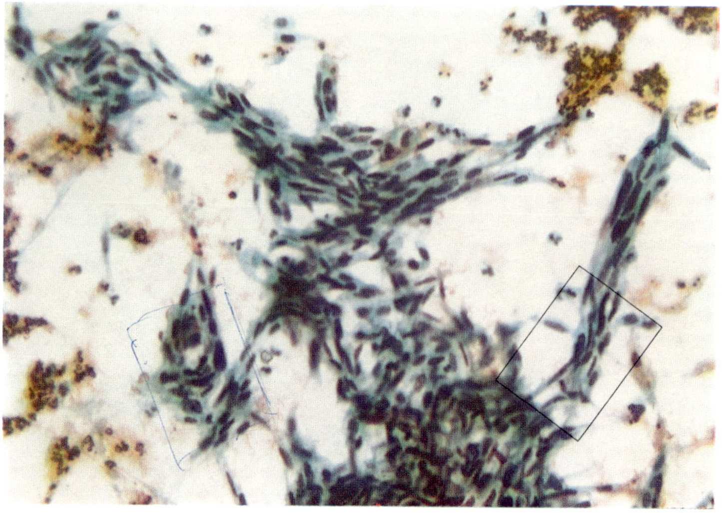

58. Invasive carcinoma of the uterine cervix — Fiber cell

a. Low magnification

A group of spindly malignant cells resembling fiber cells is observed. Many leukocytes are seen in the background.

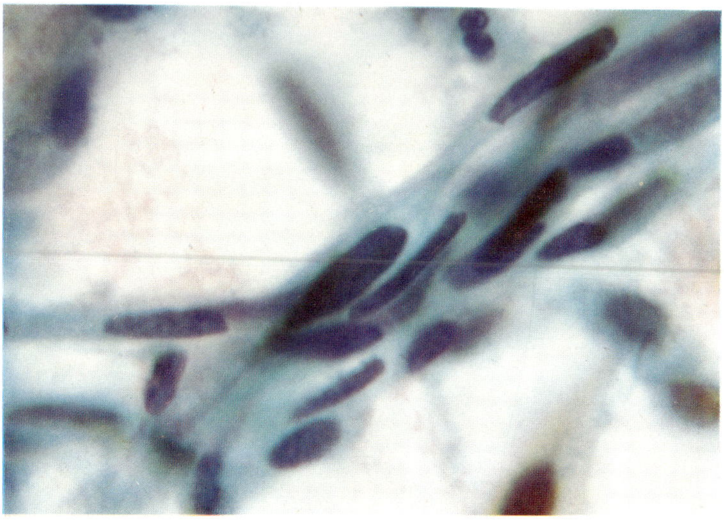

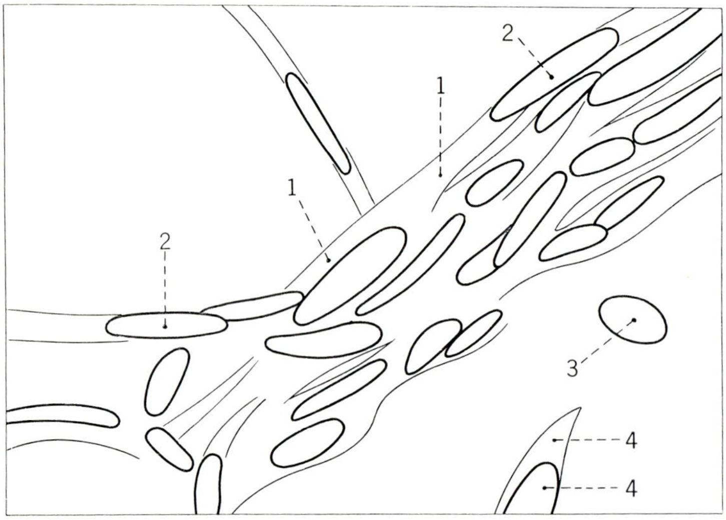

b. High magnification

1. Differentiated malignant cells resembling typical fiber cells. The nuclei are spindly and hyperchromatic. The elongated cytoplasm is comparatively distinct.
2. The nuclei are markedly hyperchromatic.
3. A rather elliptical nucleus is also observed.
4. A differentiated malignant cell with an enlarged and round nucleus is also seen.

59. Invasive carcinoma of the uterine cervix — Tadpole cell

a. Low magnification

Many definitely malignant cells are observed. They are variously shaped: tadpole, elongated and others.

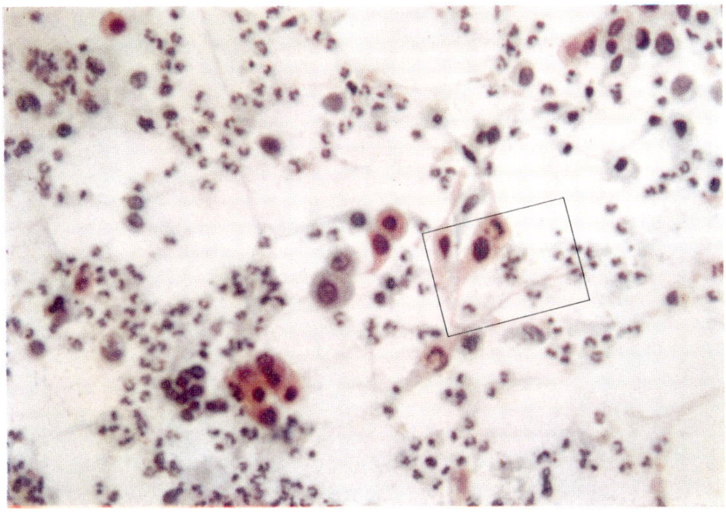

b. High magnification

1. Nuclei: Very large and irregularly shaped. The chromatin pattern is not clear due to hyperchromatism.
2. Cytoplasm: The border is comparatively distinct, but it is very scanty for the size of the nucleus. It is peculiarly tadpole-shaped.
3. Another malignant cell appears to be included in the head of the tadpole-shaped cell.

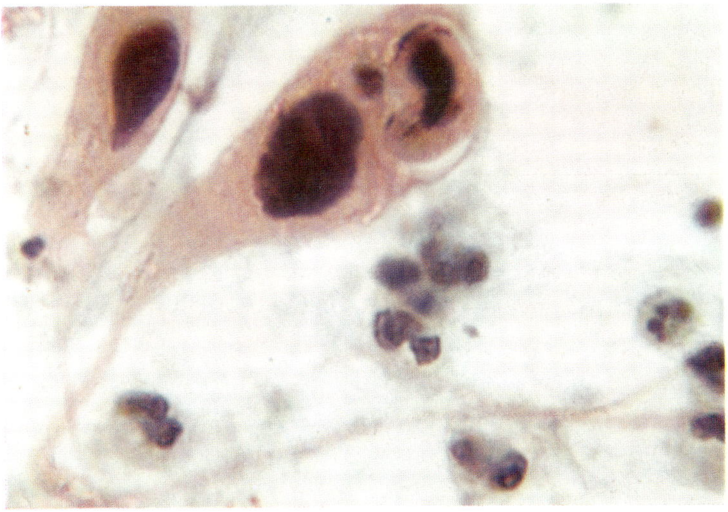

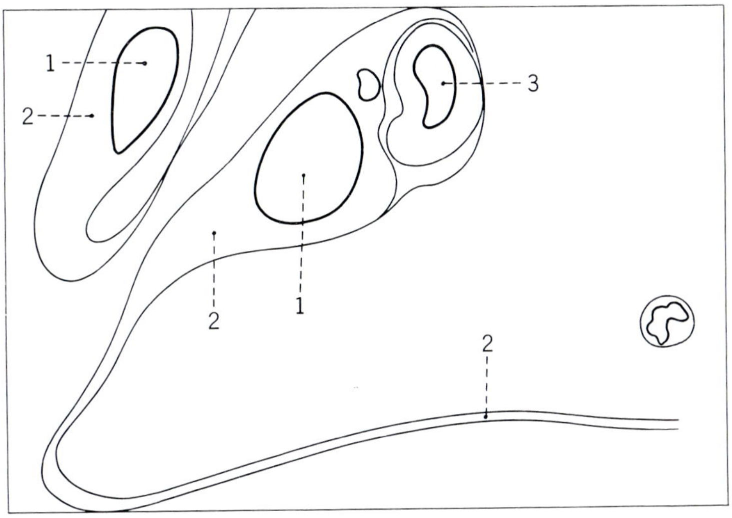

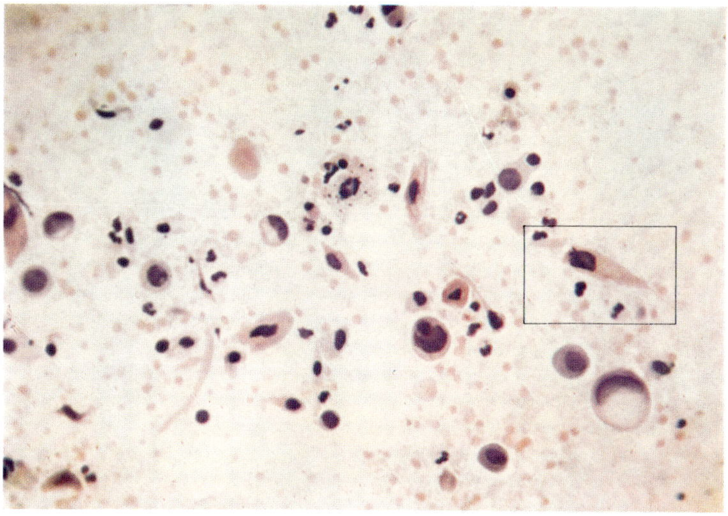

60. Invasive carcinoma of the uterine cervix — Tadpole cell

a. Low magnification

Scattered malignant cells are seen. Their shapes are various: signet-ring like, round, elliptical, tadpole-shaped or irregular.

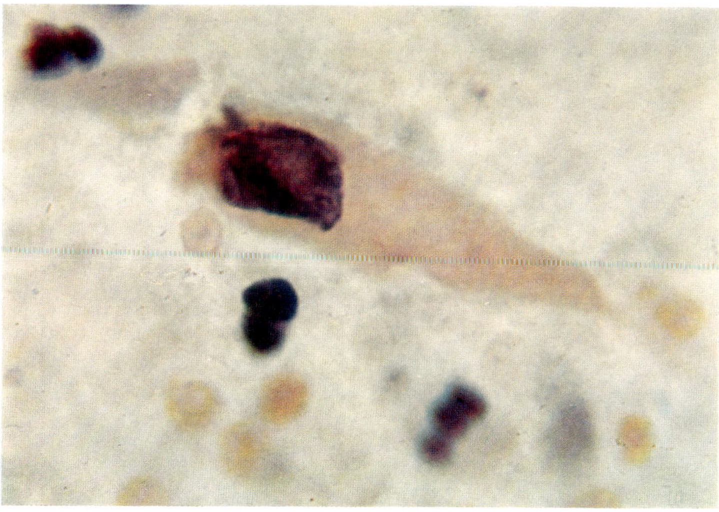

b. High magnification

1. The nucleus of the tadpole cell is irregularly elliptical and hyperchromatic.
2. The cytoplasm has a long tail and is comparatively distinct. Eosinophilically stained.

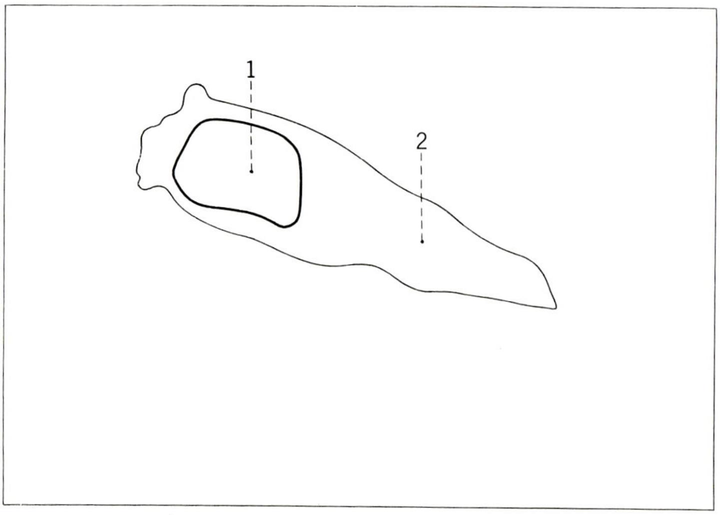

61. Invasive carcinoma of the uterine cervix — Malignant pearl

a. Low magnification

Many hyperchromatic cells in a laminated form are observed in the middle of numerous inflammatory cells. This is called a malignant pearl. The pearl is sometimes formed by benign squamous cells, but this is a pearl formation of malignant cells.

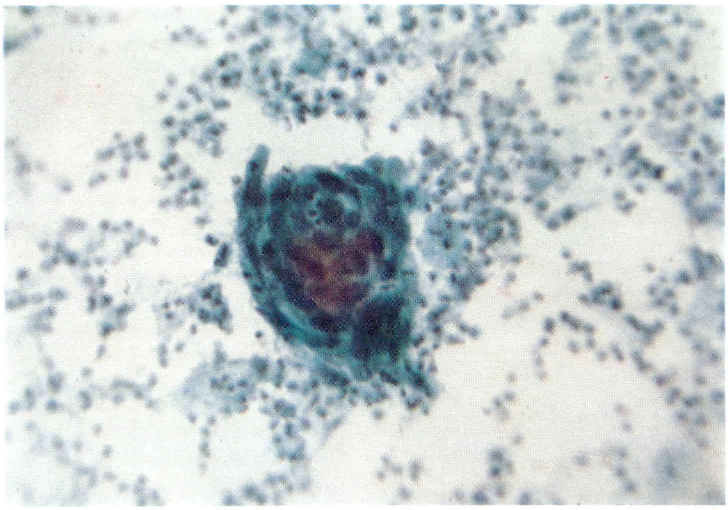

b. High magnification

1. The boundaries of individual cells are obscure. The cytoplasm of the cells is undergoing involution, and numerous nuclei are arranged irregularly overlapping one another. The nuclei are not uniform in size, but are generally much larger than the leukocytes scattered around the pearl. They vary in shape and chromatin content.
2. Leukocyte.

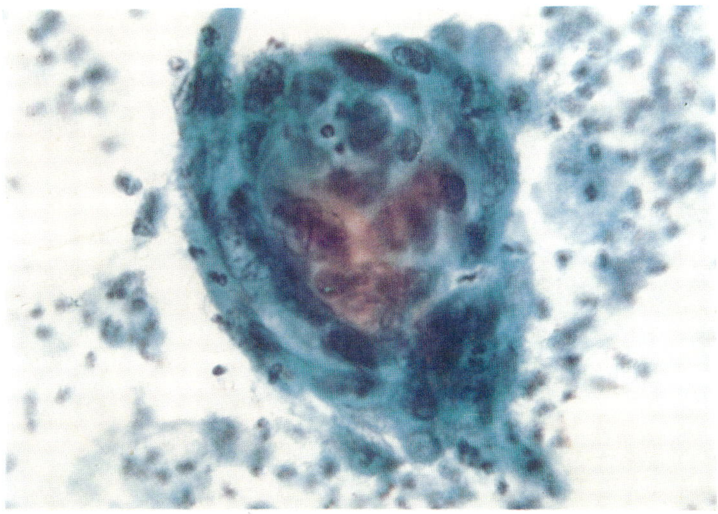

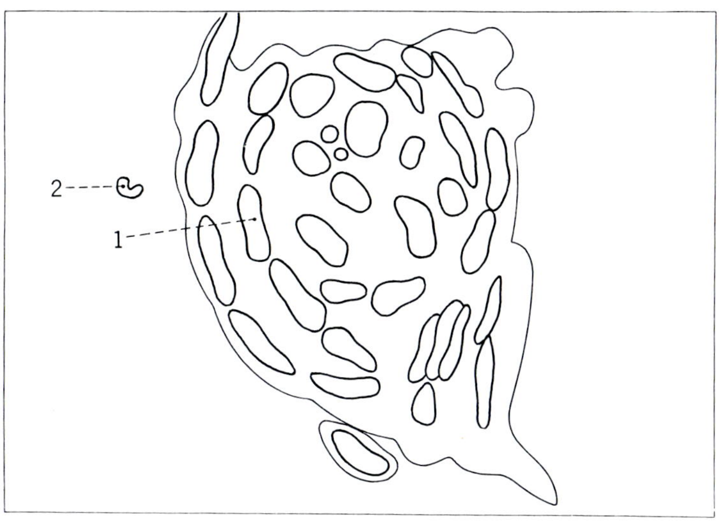

62. Invasive carcinoma of the uterine cervix

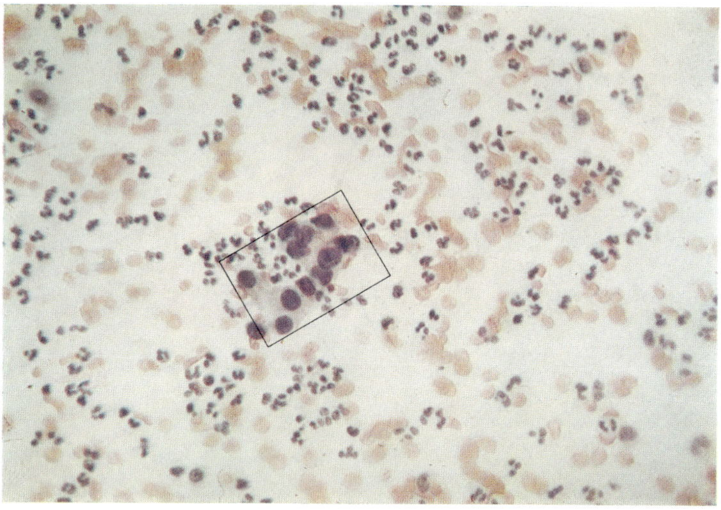

a. Low magnification

There is a group of more than ten almost undifferentiated malignant cells observed among numerous erythrocytes and leukocytes.

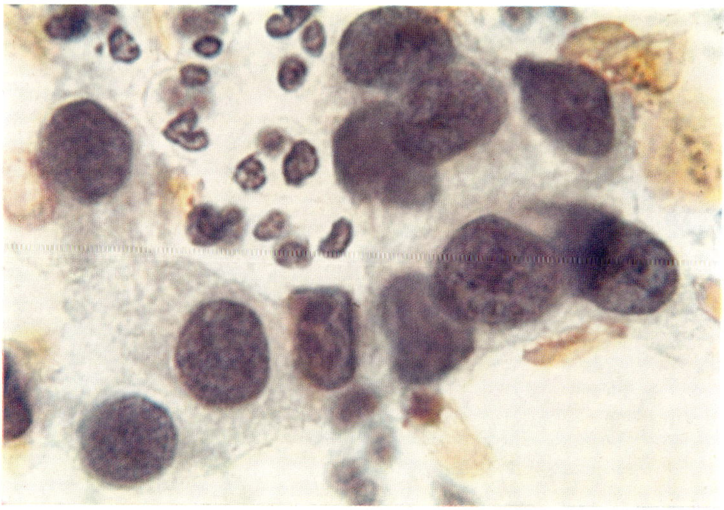

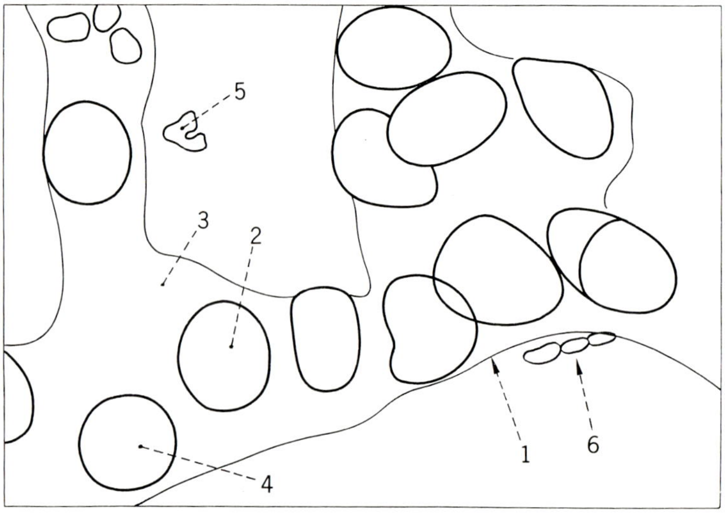

b. High magnification

1. The nuclei are enlarged but comparatively uniform in size. Not uniform in shape. Hyperchromatic.
2. Chromatin is comparatively coarsely granular.
3. The border of each cell is indistinct.
4. A malignant cell with very little cytoplasm.
5. Polymorphonuclear leukocyte.
6. Erythrocytes.

63. Invasive carcinoma of the uterine cervix

a. Low magnification

A group of malignant cells is observed among numerous leukocytes. They are differentiated.

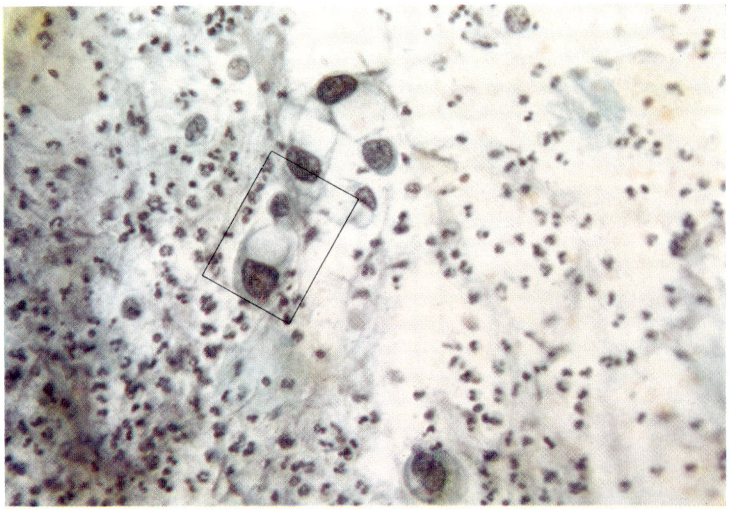

b. High magnification

1. Nuclei irregularly vary in size and shape. They show large amounts of chromatin content and are consequently hyperchromatic.
2. Large cytoplasmic vacuole.
3. Cytoplasm is generally distinct in the border and appears dense.
4. Some are clearly defined.
5. Leukocytes.

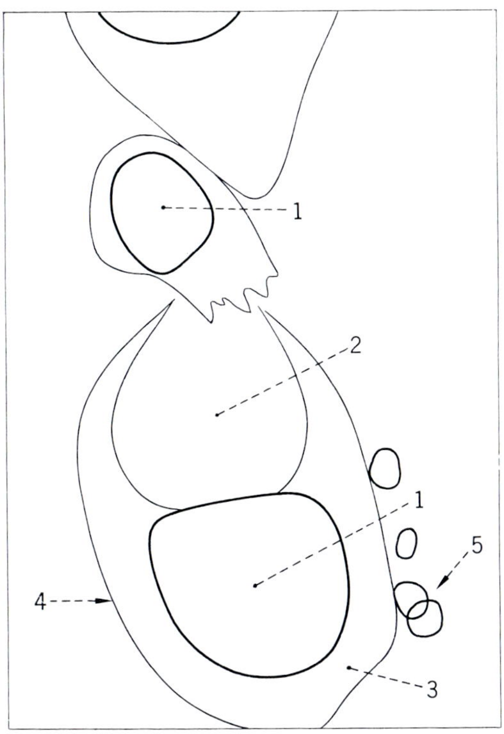

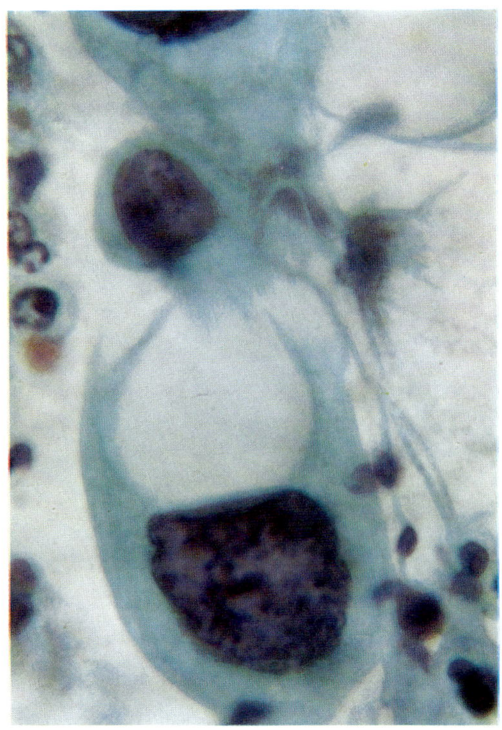

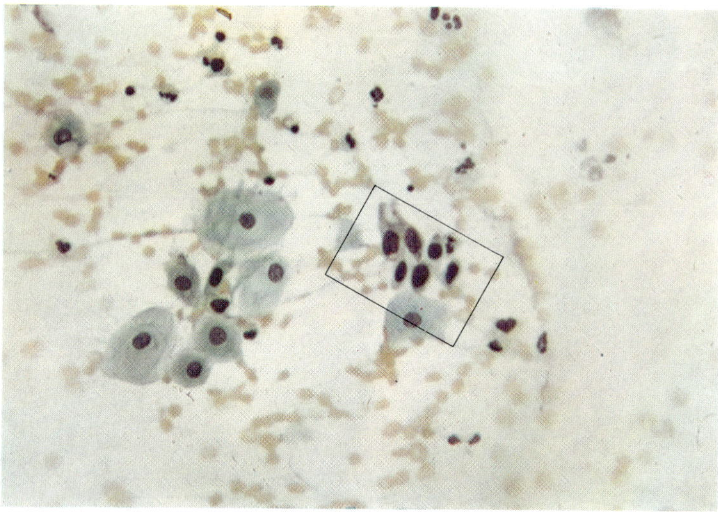

64. Invasive carcinoma of the uterine cervix

a. Low magnification

There are groups of malignant cells in the background showing numerous erythrocytes or cell debris due to necrosis.

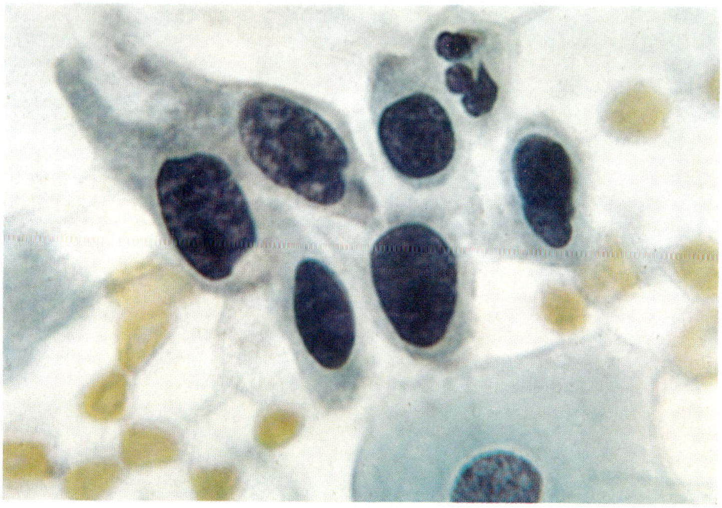

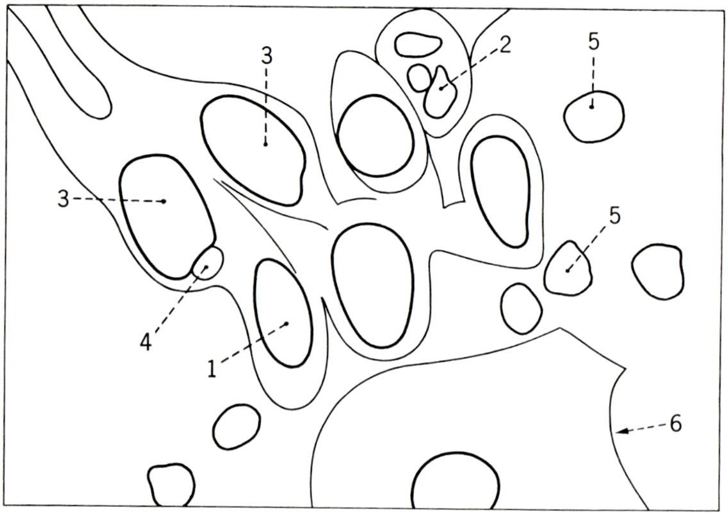

b. High magnification

1. Hyperchromatic nuclei.
2. Pyknotic nuclei. No distinct cytoplasm can be observed.
3. Coarse chromatin granules are seen in the elliptical nucleus. The cytoplasm is comparatively distinct.
4. Cytoplasmic vacuole.
5. Erythrocytes.
6. Slightly dyskaryotic squamous cell.

65. Invasive carcinoma of the uterine cervix — Undifferentiated type

a. Low magnification

Numerous stripped nuclei are scattered in the background having relatively few inflammatory changes. These are undifferentiated malignant cells.

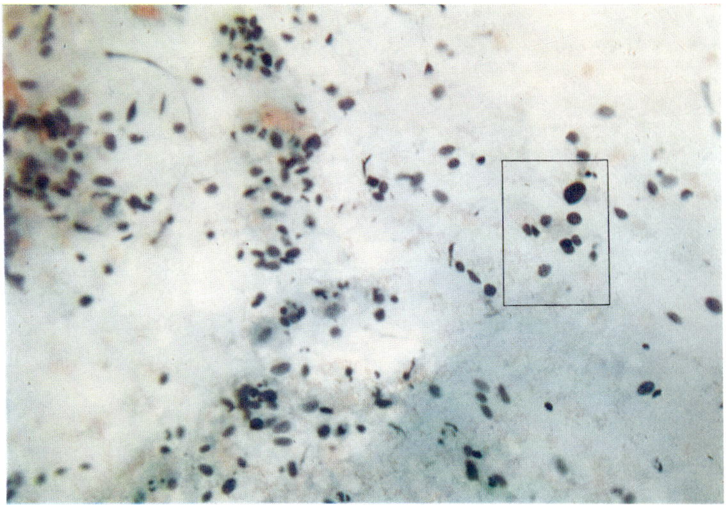

b. High magnification

1. The scattered cells show loss of cytoplasm and the nuclei appear stripped and extremely hyperchromatic.
2. Not uniform in size.
3. Mostly round or elliptical.
4. Hyperchromatic. Chromatin is increased but is distributed comparatively evenly and is coarsely granular

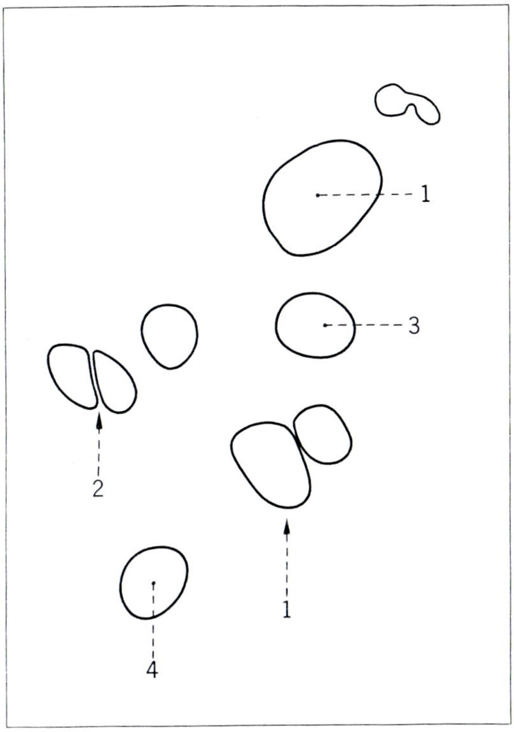

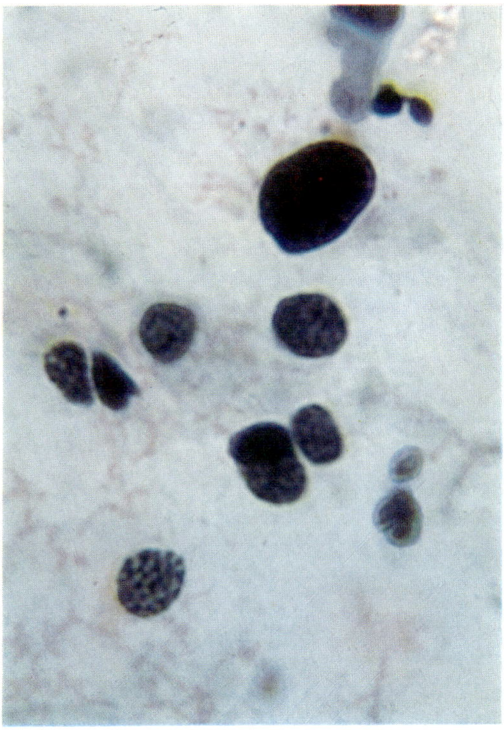

66. Invasive carcinoma of the uterine cervix — Undifferentiated type

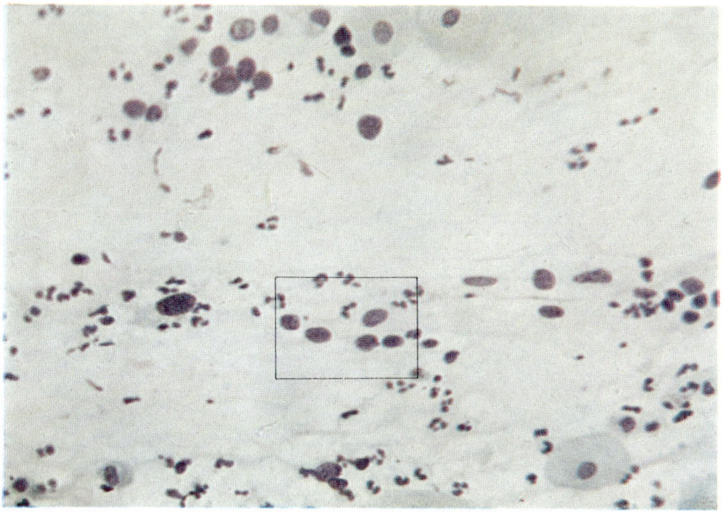

a. Low magnification

Small, round, stripped nuclei are scattered like sesame seeds among some leukocytes, cellular debris and mucus. These are compatible with highly undifferentiated malignant cells.

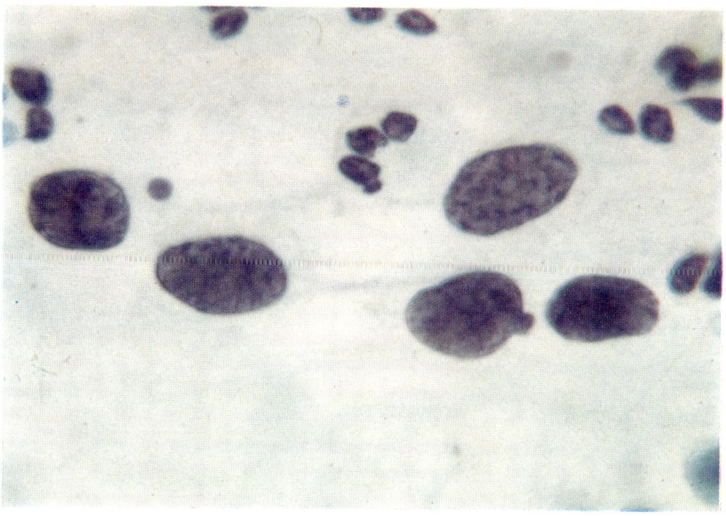

b. High magnification

1. Nuclei: All are hyperchromatic in appearance. The nuclear structures are mostly indistinct. Close observation reveals that they vary in size and shape.
2. Cytoplasm: Entirely or almost completely missing.
3. Cellular debris due to necrosis are observed in the background along with leukocytes and mucus.

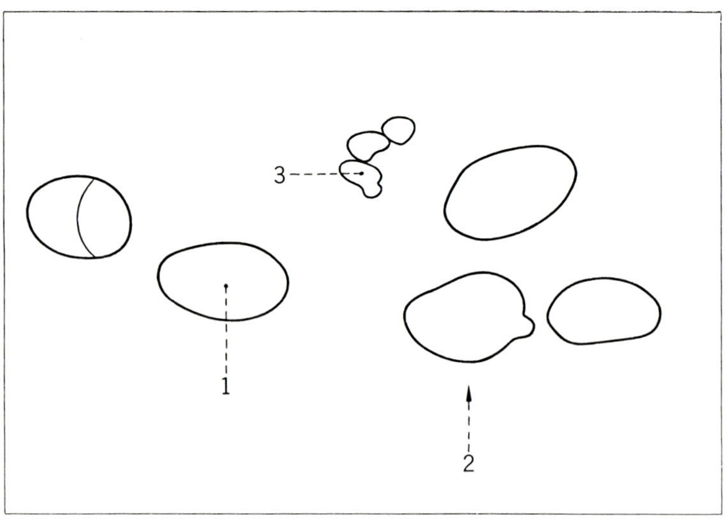

Endocervical and endometrial adenocarcinoma

General characteristics (Fig. 36)

1) General

The most striking features are that degenerated malignant cells appear in clusters in the smear. The cluster is sometimes very dense with cells overlapping one another, or it may be comparatively sparse and the borders of individual cells are distinguished.

2) Nucleus

1. Size : Twice or three times as large as the nucleus of a normal endocervical or endometrial cell. The size generally increases with the lack of differentiation.
2. Shape : Round or elliptical. The degenerated nucleus is sometimes confetti-shaped. It may be flattened when a vacuole exists in the cytoplasm.
3. Staining reaction : Generally stained more lightly than the nuclei of cells from an epidermoid carcinoma. When the nucleus has degenerated, chromatin accumulated around the nuclear border and the nuclear membrane looks thickened, while the interior is hypochromatic.

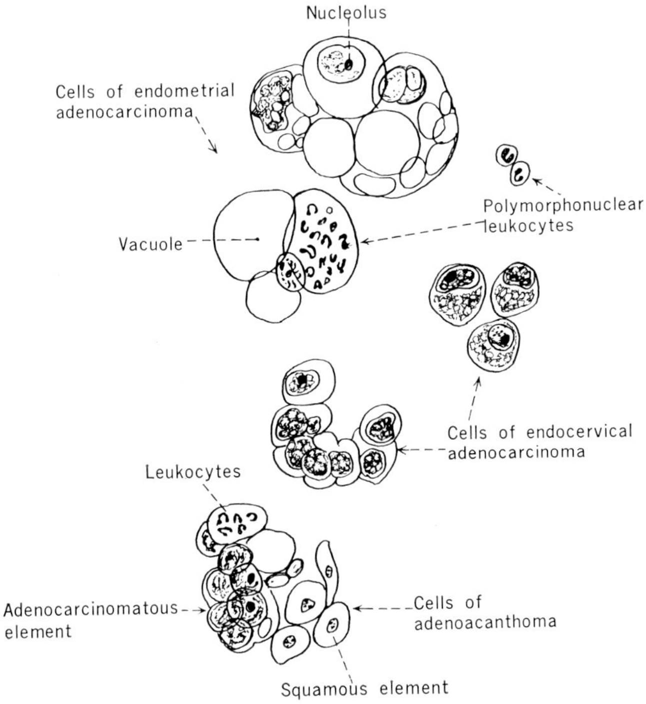

Fig. 36

4. Chromatin pattern : Mostly finely granular, but sometimes hyperchromatic or hypochromatic. Coarsely granular on other occasions. The percentages of these patterns are shown in Fig. 37.

5. Nucleolus : Distinctly observed in most cases, and especially marked in the undifferentiated type. The more degenerated the malignant cells are, the more prominent it may be. Therefore, the malignant cells found in the vaginal smear may reveal more distinct nucleoli than in cells from aspiration smears directly taken from malignant tumors of the endometrium.

3) Cytoplasm

1. The cells observed in vaginal smears are often degenerated and lightly stained. A marked vacuole may exist in the cytoplasm, and consequently the nucleus is pushed near the border. Such vacuolated cells may be sometimes observed in endometrial aspiration smears as well. It is also one of the characteristics that many polymorphonuclear leukocytes are included in the vacuole.
2. When malignant cells appear in a cluster, the border of each cell is indistinct partly due to vacuolation, and the cluster gives the impression that large foams are overlapping one another, while it appears that degenerated nuclei are located here and there.
3. Cytoplasmic vacuoles are often not observed in endometrial aspiration smears.
4. Staining reaction : Cells from an endometrial adenocarcinoma may frequently stain basophilic and endocervical adenocarcinoma cells eosinophilic, but this is not always true.
5. Endocervical adenocarcinoma : These cells may contain mucus-droplets in the cytoplasm, and sometimes look like golfballs. Such cells occasionally form a grape-like cluster.
6. Endometrial adenoacanthoma : Cells apparently originating from squamous epithelium are observed besides adenocarcinomatous cells. At first sight, these cells do not look malignant.

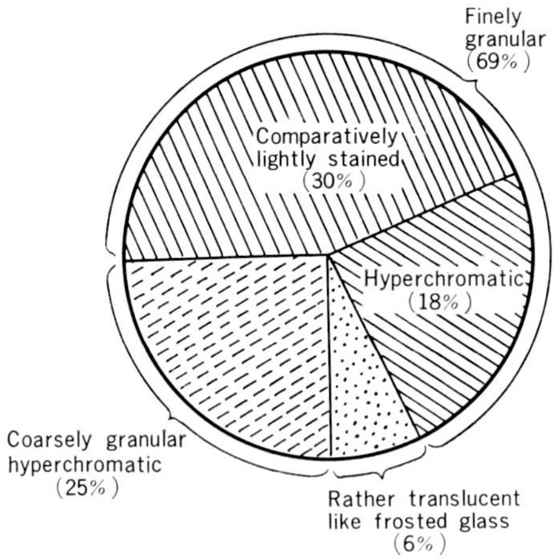

Fig. 37 Nuclear chromatinic patterns.

67. Endocervical adenocarcinoma — Grape-like cluster of malignant cells

a. Low magnification

A papillary cluster of malignant cells is observed.

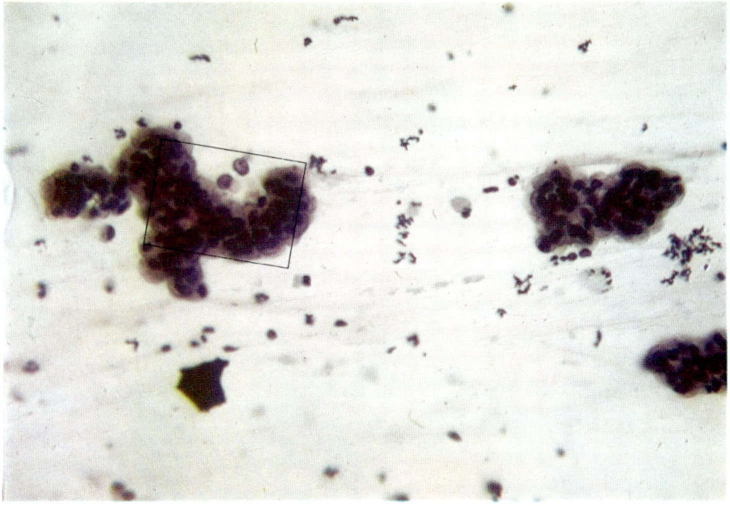

b. High magnification

1. The cluster appears grape-like.
2. Nuclei : Not uniform in size or shape. Fairly hyperchromatic.
3. Chromatin is increased around the nuclear border.
4. Cytoplasm : The borders of individual cells are indistinct, but rim of the cluster is clearly seen.
5. The cytoplasm is vesiculated and appears apparently mucus-producing.

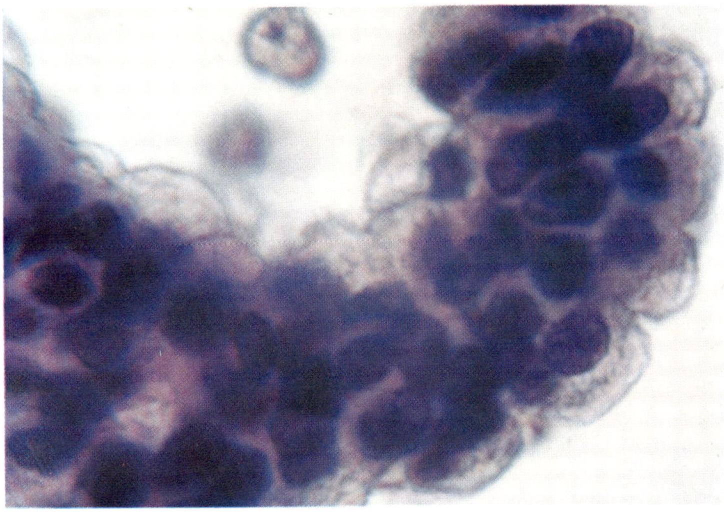

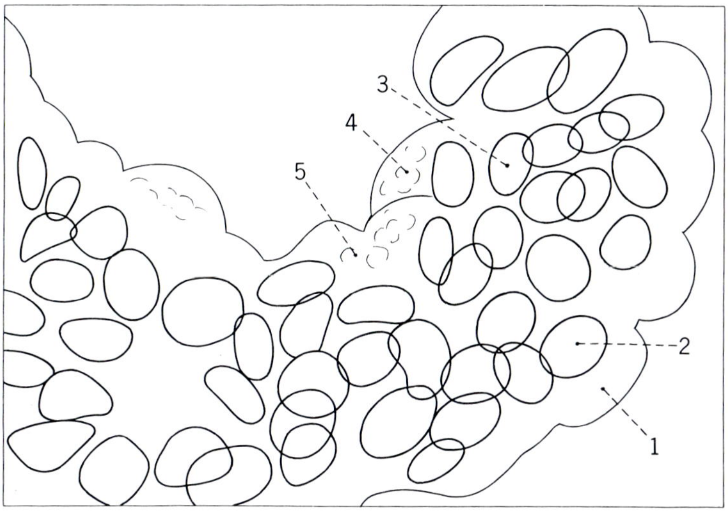

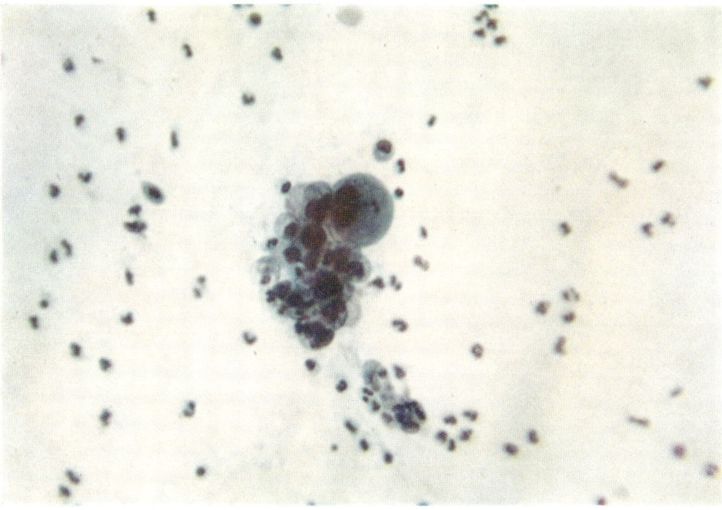

68. Endocervical adenocarcinoma

a. Low magnification

A dense mass of cells is observed. These cells seem to have a round and abundant cytoplasm. Leukocytes are scattered around the cell cluster.

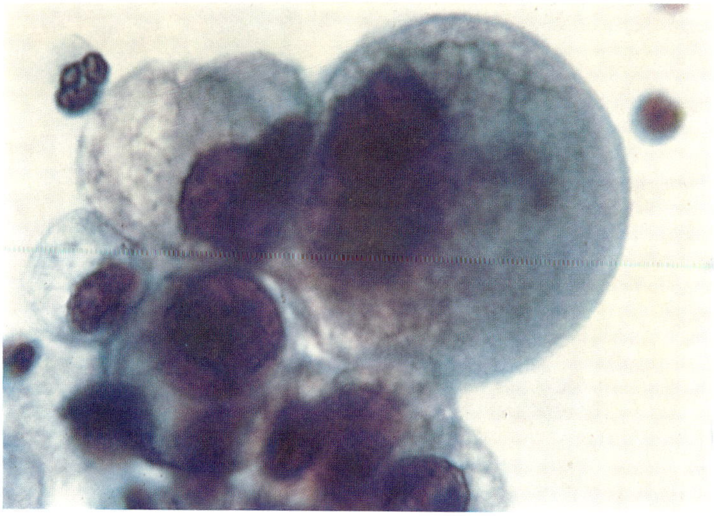

b. High magnification

1. The nuclei are enlarged, but they are various in size and shape, and are mostly located eccentrically. Chromatinic pattern is not too distinct. It is particularly abundant around the border. Hyperchromatic. Nucleoli are poorly visible.
2. The borders of individual cells are fairly distinct. The cytoplasm is foamy in appearance and the cells overlap each other. Large vacuolation is infrequently observed.

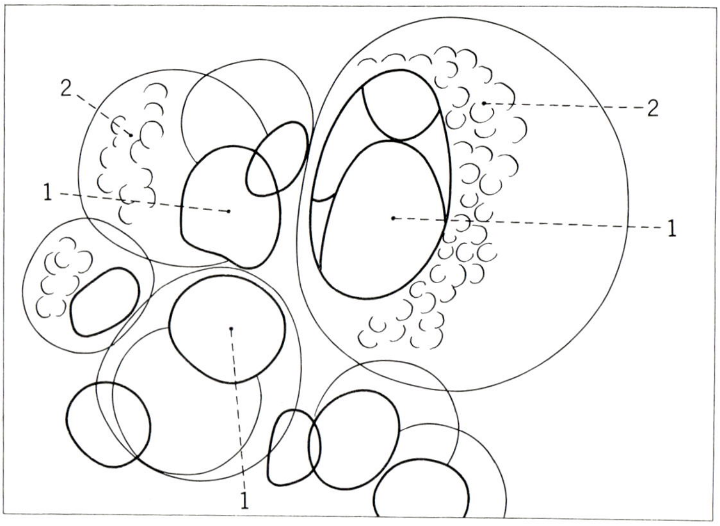

69. Endocervical adenocarcinoma — markedly mucus-producing

a. Low magnification

Malignant cells appear in groups on the smear. In this case, individual cells are separate from one another. The cells are generally much smaller than benign squamous cells (on the upper left), but the nuclei are much larger.

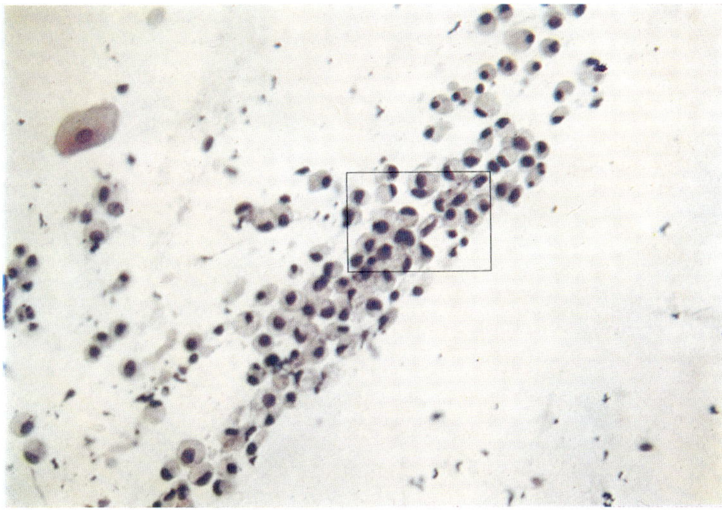

b. High magnification

1. Nuclei: Often located near the borders.
2. They are definitely enlarged and vary in shape: round, elliptical, polygonal and pea-shaped.
3. The borders are distinct.
4. Chromain is increased. Hyperchromatic. No nucleolus, nuclear vacuole or mitosis can be observed.
5. Cytoplasm: Mostly round or elliptical and abundant.
6. The cytoplasmic border is distinct and the cytoplasm is net-like, lacy and vesicular. Thus it looks as though they may be mucus-producing. These are the main characteristics of the cells.

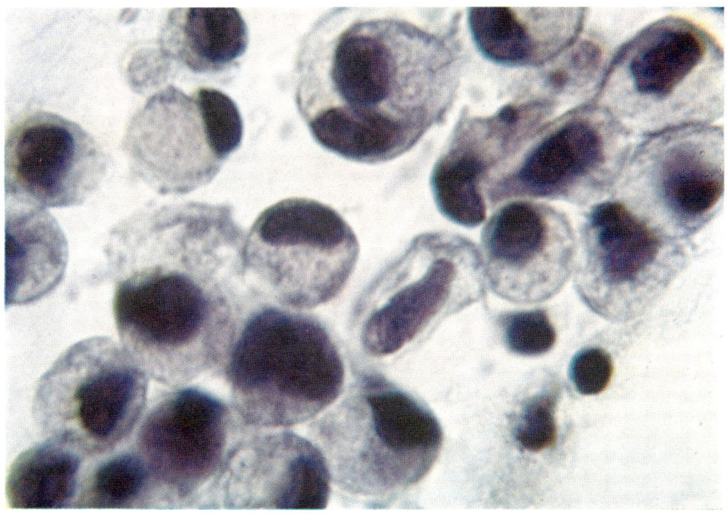

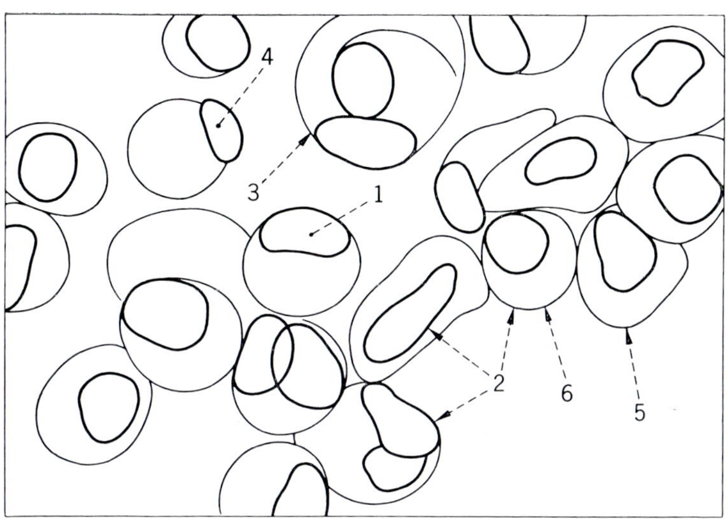

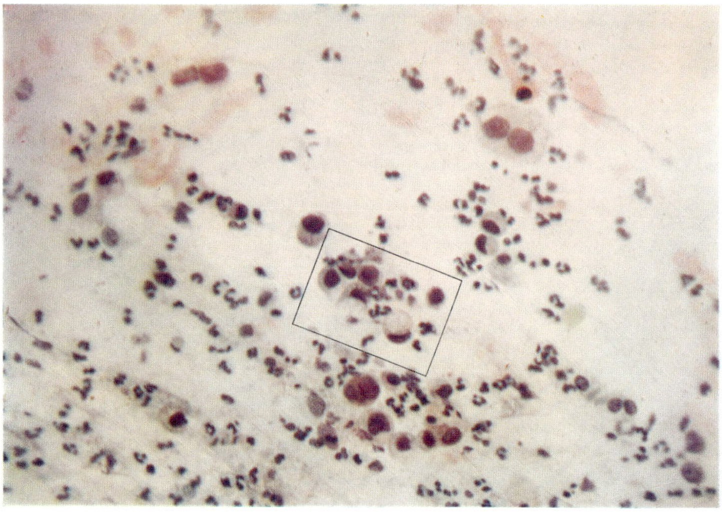

70. Endocervical adenocarcinoma

a. Low magnification

A group of malignant cells is observed on a messy background including numerous erythrocytes and leukocytes.

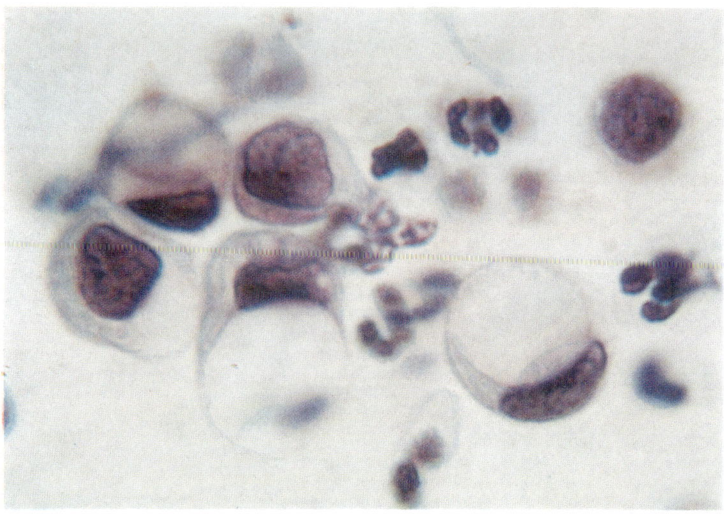

b. High magnification

1. Cell that shows large vacuole in the cytoplasm and crescent-shaped nucleus pushed towards the cellular border.
2. Nuclei : Round, elliptical or polygonal and moderately enlarged. Compare them with leukocytes. Chromatin; coarsely granular and generally rather hyperchromatic.
3. The nuclear membrane is denser.
4. Nucleoli are also present, although they are not very distinct in the picture.
5. The borders of individual cells can be vaguely distinguished. The cytoplasm is thin and lacy.

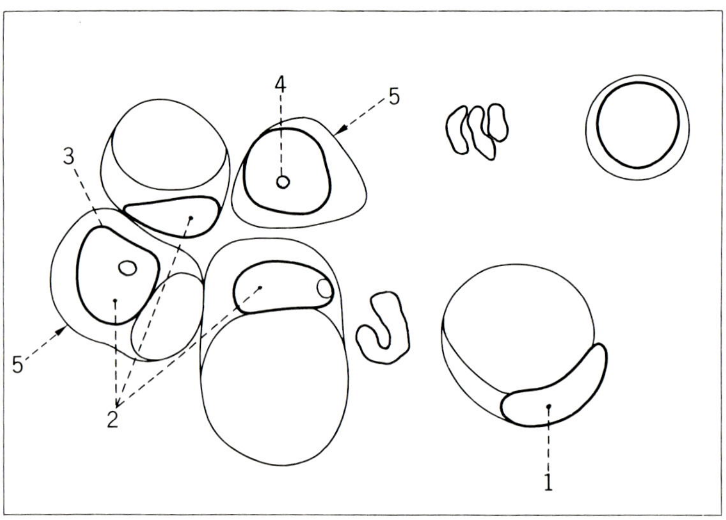

71. Endometrial adenocarcinoma

a. Low magnification

There are two large nuclei showing prominent nucleoli.

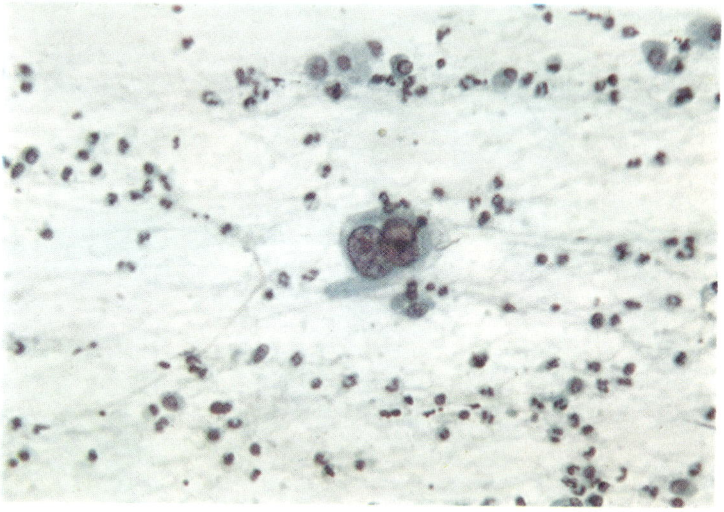

b. High magnification

1. Large and round nucleus. Chromatinic accumulation is marked in the border.
2. Prominent nucleoli.
3. Lacy cytoplasm the border of which is fairly distinct.

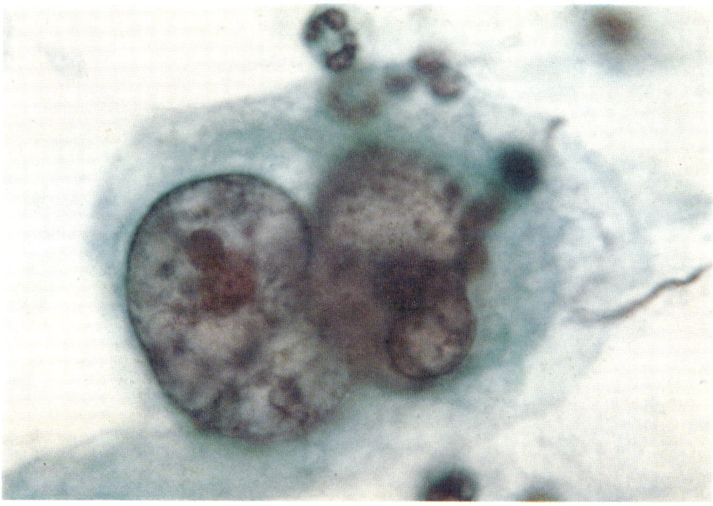

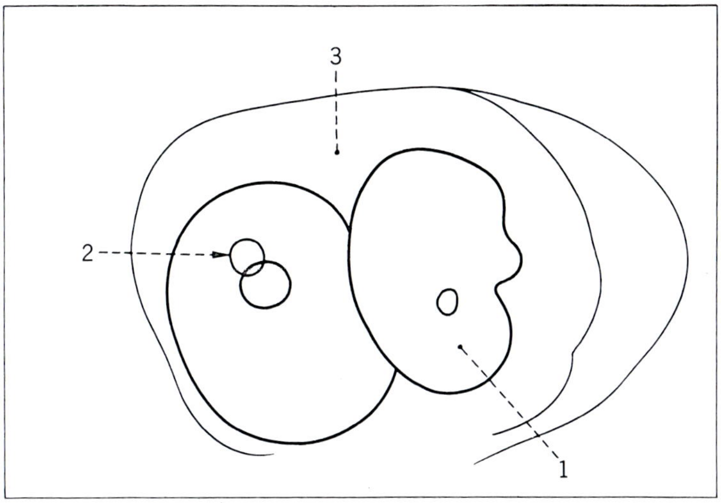

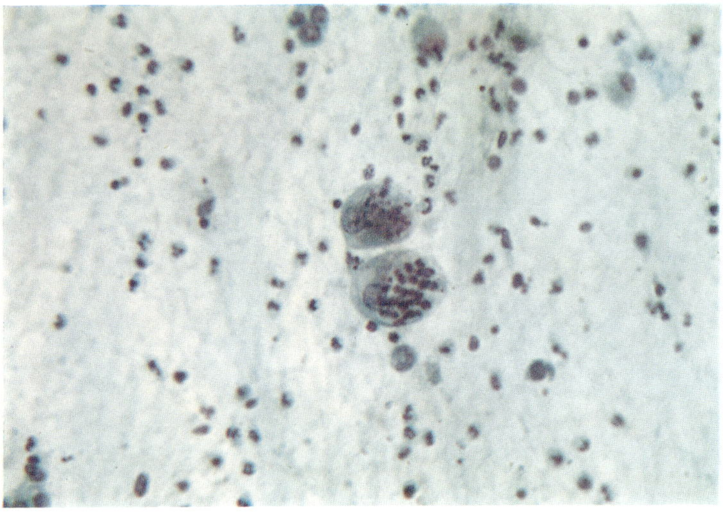

72. Endometrial adenocarcinoma
a. Low magnification
There are two large cells showing large vacuoles in the cytoplasm. Many leukocytes are included in the vacuoles.

b. High magnification
1. Large nucleus: Not too hyperchromatic but a dense chromatinic concentration appears near the border.
2. A nucleolus is faintly observed.
3. Cytoplasmic vacuole containing numerous leukocytes.
4. No leukocytes are seen in the vacuole.

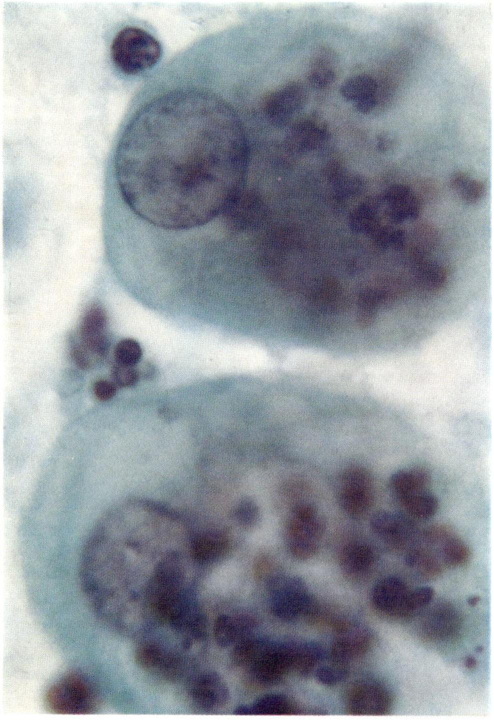

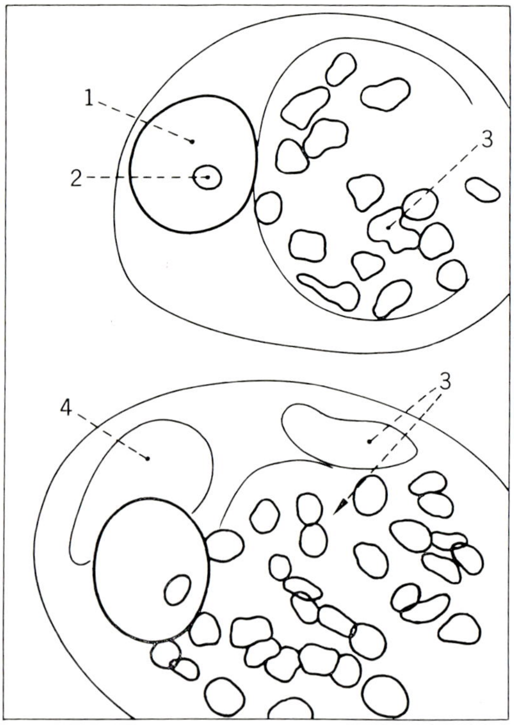

73. Endometrial adenocarcinoma

a. Low magnification

A group of vacuolated glandular malignant cells is seen in the middle of the smear.

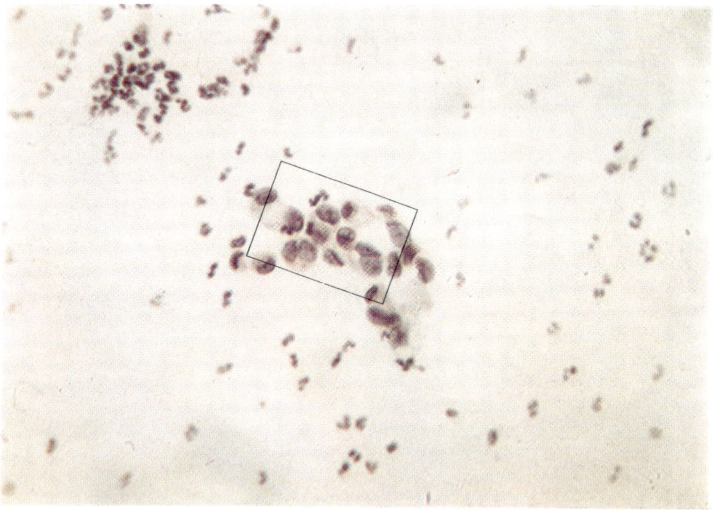

b. High magnification

1. Nucleus: Large and elliptical, and rather hyperchromatic. The chromatin pattern is reticulated.
2. A prominent nucleolus.
3. Cytoplasm: Characteristic cytoplasmic vacuoles.
4. The nucleus is sometimes deformed into a crescent or triangular shape.

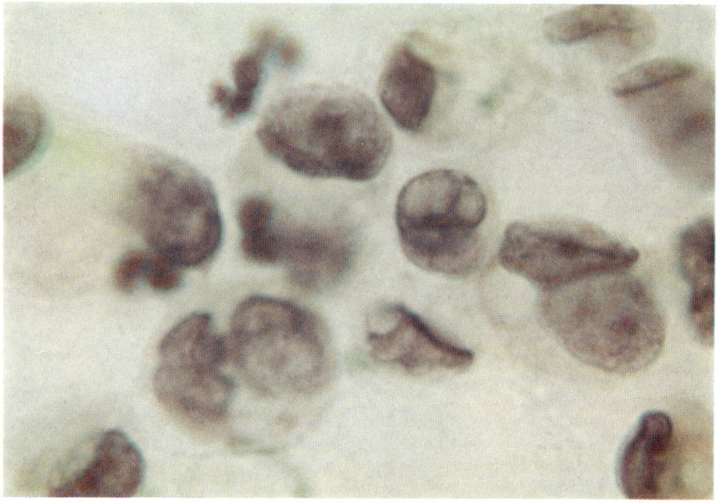

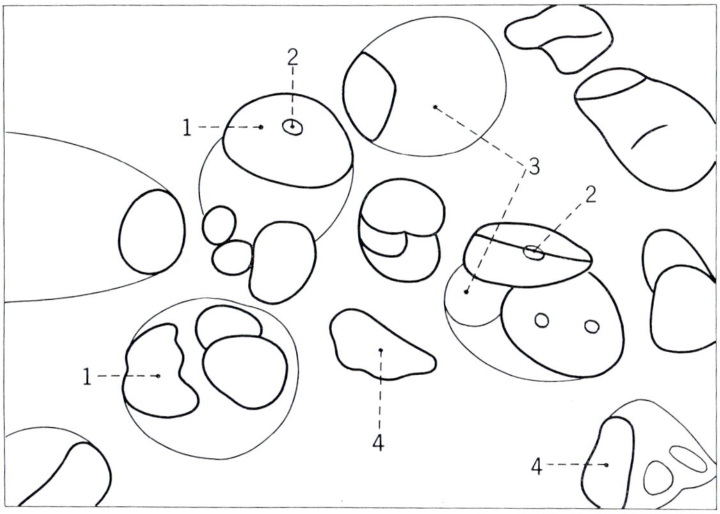

74. Endometrial adenocarcinoma (Endometrial aspiration smear)

a. Low magnification

Groups of hyperchromatic cells are located in the center and at the lower right. Malignant glandular cells are scattered at random.

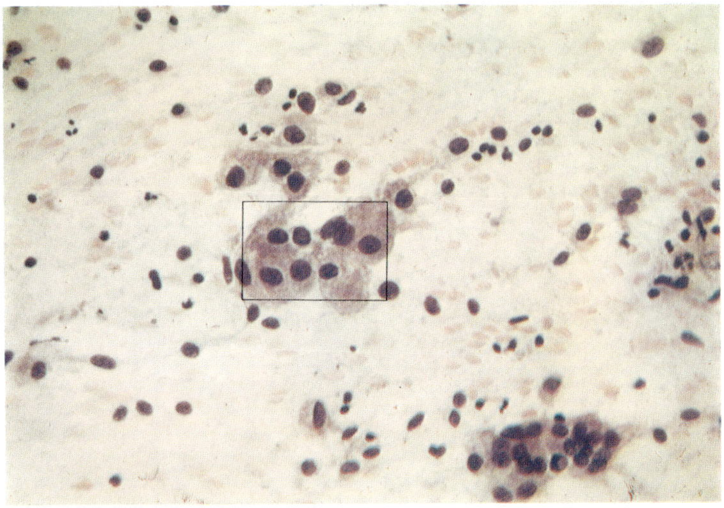

b. High magnification

1. Nuclei: Round or elliptical and fairly large. Very hyperchromatic. Coarse chromatin granules are densely distributed. No particular chromatinic condensation is noticed around the rims.
2. The boundaries of individual cells are indistinct, but the border of the group is moderately clear. Note the characteristically reticulated and lacy cytoplasm. Cells in the aspiration smear are well preserved; therefore, marked vacuolation mostly due to degeneration is usually not encountered, and prominent nucleoli are poorly visible in aspiration smears.

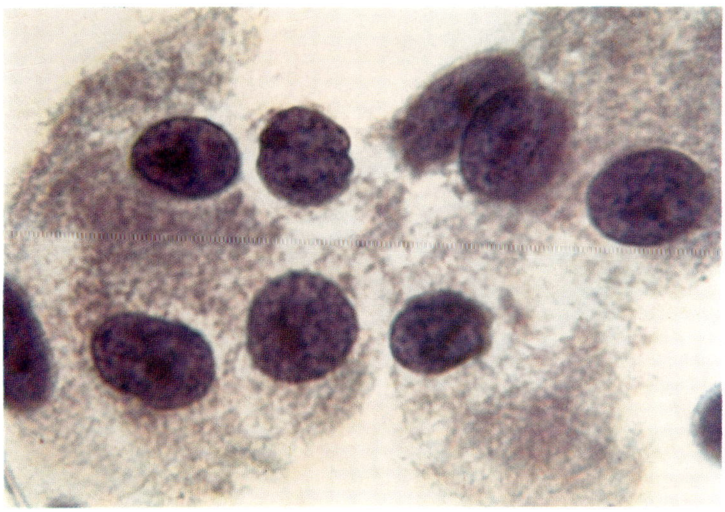

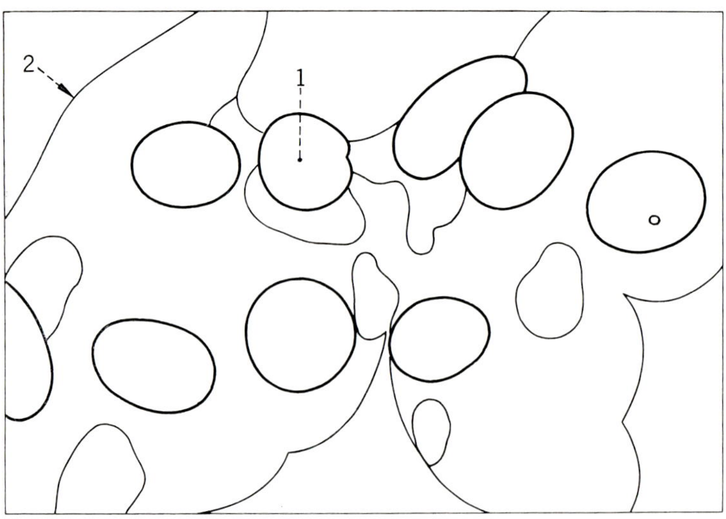

75. Endometrial adenocarcinoma (Endometrial aspiration smear)

a. Low magnification

A tight clumping of malignant glandular cells is observed on a background of bleeding.

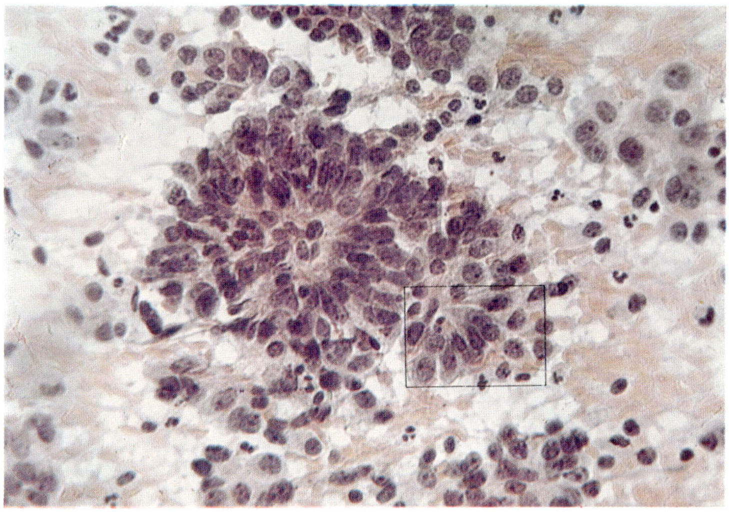

b. High magnification

1. Nuclei appear concentrically arranged and the chromatinic pattern is coarsely granular.
2. Some of the nuclei reveal more hyperchromatism and irregularity in shape.
3. Prominent nucleoli are noted.
4. Cytoplasm is triangular in shape and cellular borders are quite definitely recognized.
5. Lacy cytoplasm.

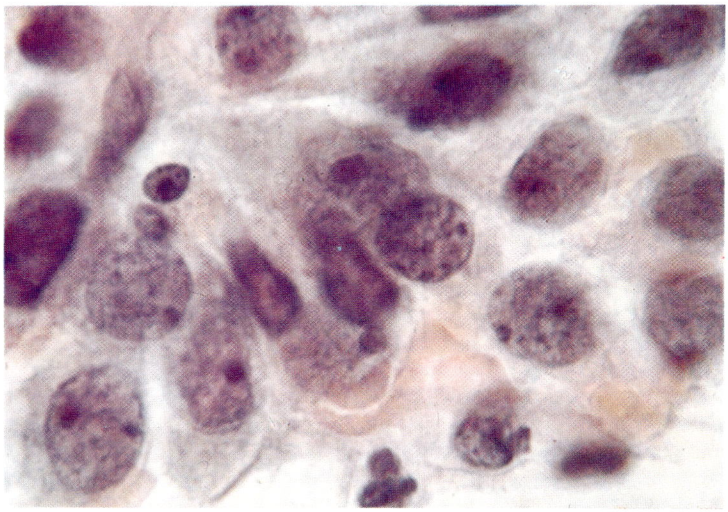

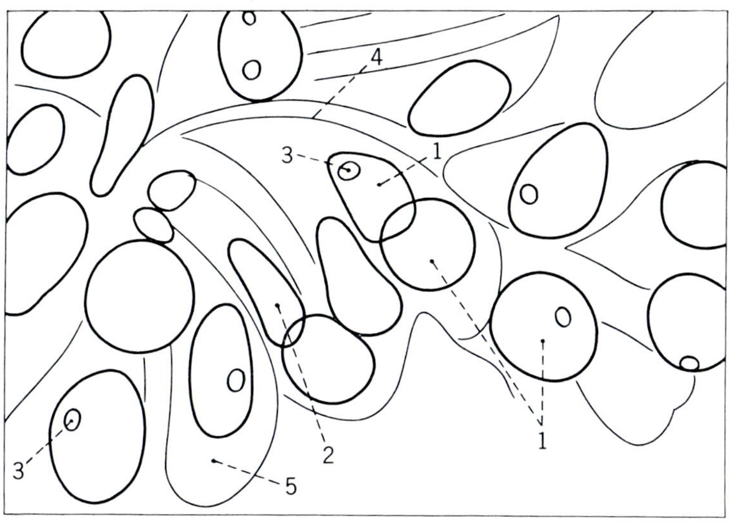

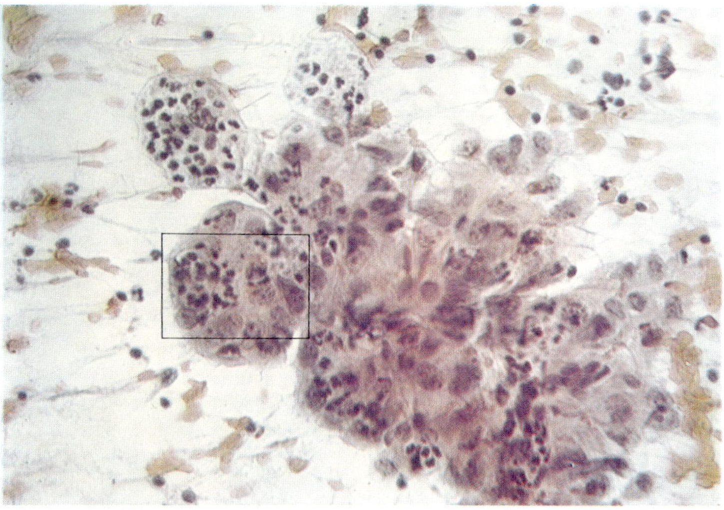

76. Endometrial adenoacanthoma — Adenocarcinomatous constituents (Endometrial aspiration smear)

a. Low magnification

A large clump of malignant glandular cells is observed. The nuclei are comparatively uniform in size and various in shape. Note the marked intrusion of leukocytes into the vacuoles in partially degenerated cells.

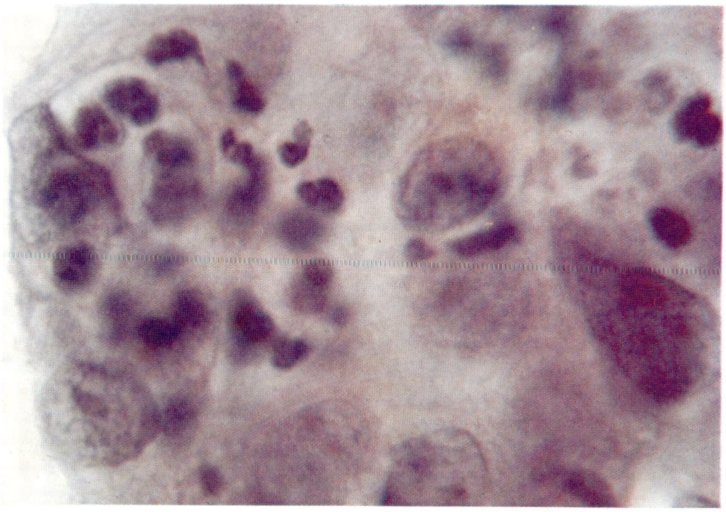

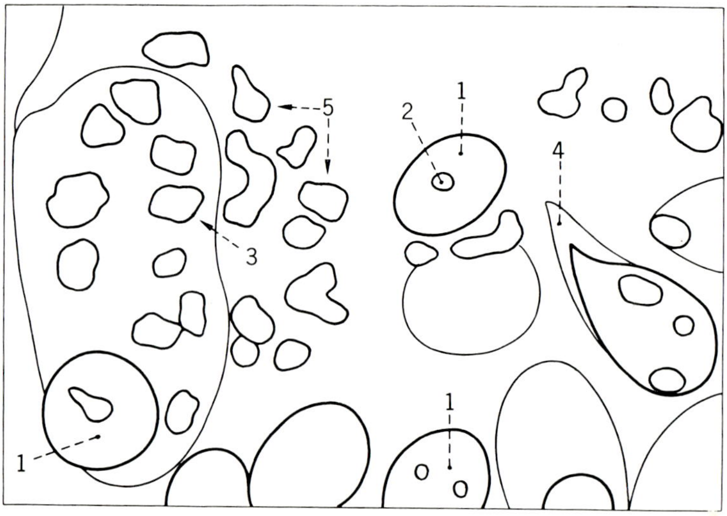

b. High magnification

1. Nuclei : Mostly elliptical or round. Coarse chromatin granules are observed against a finely granular network of background. Chromocenters are also noticed. The nuclei are degenerating and comparatively hypochromatic. Compare the staining with that of the nuclei of scattered leukocytes.
2. A nucleolus is observed, though it is vague.
3. The cytoplasm reveals some vacuolization with inclusion of many leukocytes in the vacuoles.
4. Cells are in a clump overlapping one another and the borders are not distinct.
5. Polymorphonuclear leukocytes.

77. Endometrial adenoacanthoma — Squamous constituents (Endometrial aspiration smear) (The same case as of Plate 76)

a. Low magnification

At least two cells are considered definitely squamous elements.

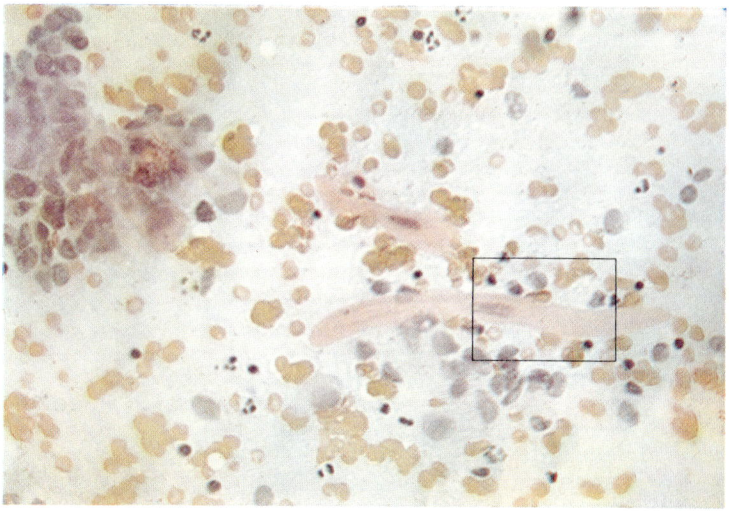

b. High magnification

1. A very long cell and an elongated triangular nucleus are observed. This is considered a squamous element.
2. Nucleus: Elongated and slightly hyperchromatic, but does not appear to be malignant.
3. Cytoplasm: Abundant, thick and homogeneous. The borders are sharp.

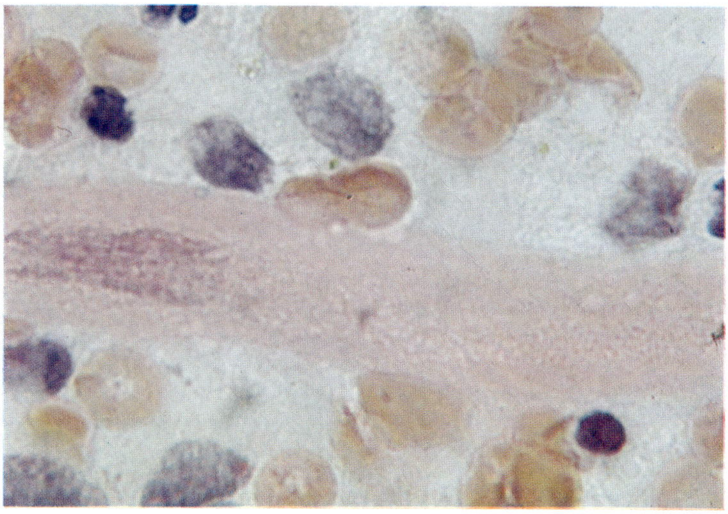

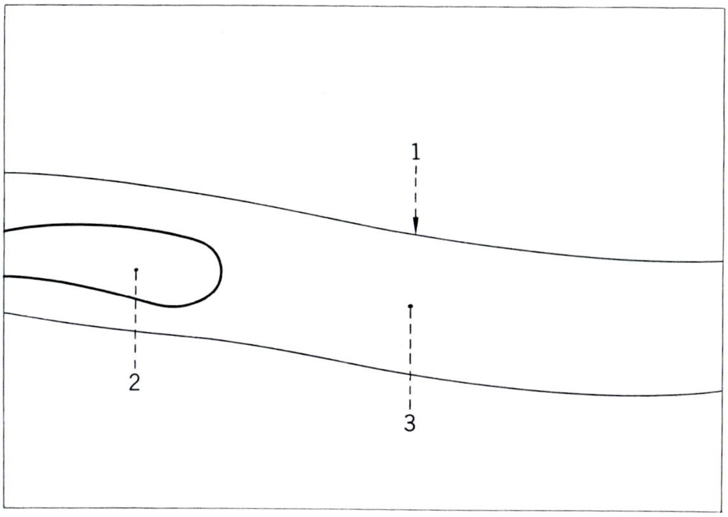

Other tumors of the female genital tract

Choriocarcinoma

1) Characteristics

1. Extremely large malignant cells with giant nuclei, often multinucleated.
2. Giant nucleoli are one of the most striking features of the tumor. These signs may be considered as important findings of the tumor, though it is still under doubt whether these cells are definitely of tumor origin.
3. Nucleus: The chromatin pattern is quite irregular and coarsely granular.
4. Cytoplasm: Irregular in shape, appearing thickened.
5. Since it is a hemorrhagic tumor, a large number of blood components are often present in the background.
6. Unless removed by endometrial aspiration or curettage, such cells rarely appear in vaginal smears.

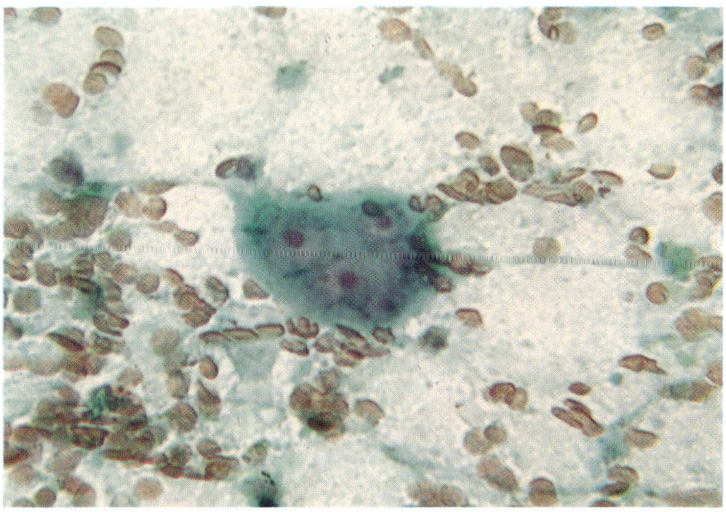

78. Choriocarcinoma (Endometrial aspiration smear)

a. Low magnification

A clump of cells with large nuclei is found.

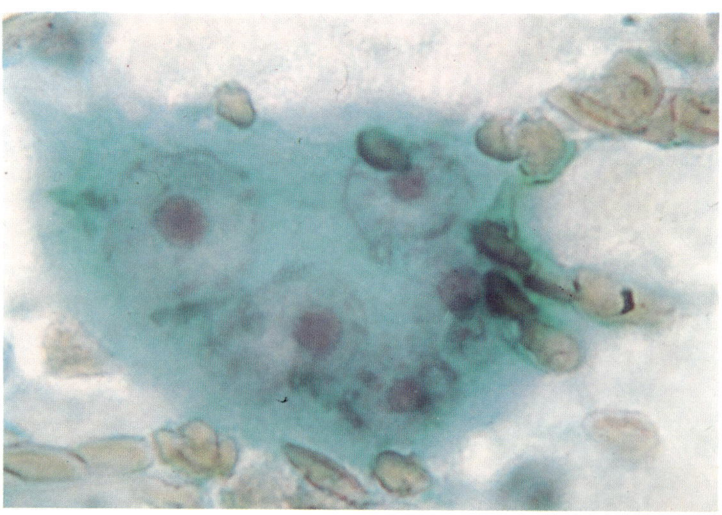

b. High magnification

1. Extremely large nuclei with centrally located giant nucleoli are seen. The giant nucleoli are stained clearly pinkish.
2. The nuclei are comparatively hypochromatic, and chromatin accumulates only around the borders.
3. No individual cellular border can be recognized, but boundaries of the cell cluster may be distinctly seen.

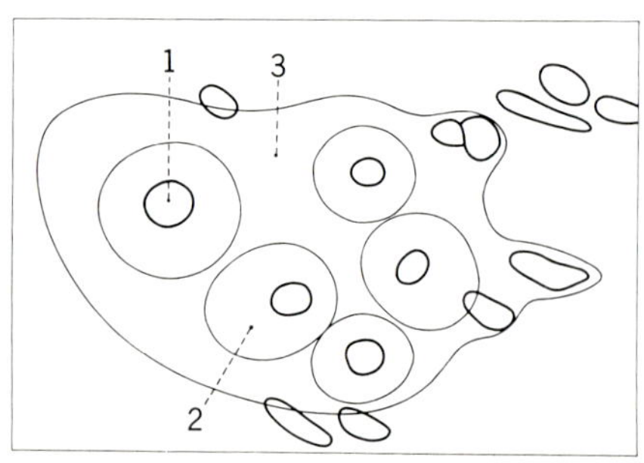

79. Choriocarcinoma (Endometrial aspiration smear) (The same case as of Plate 78)

a. Low magnification

Cells with giant and hyperchromatic nuclei are observed. Compare the nuclei with erythrocytes in the background.

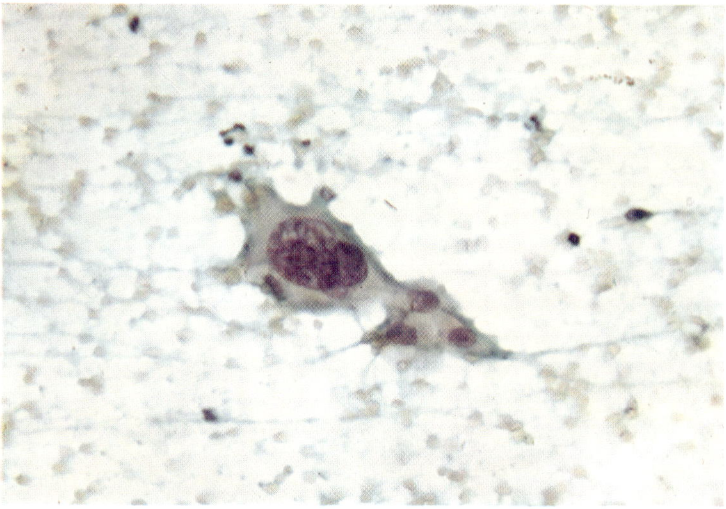

b. High magnification

1. The giant nuclei overlap one another. The chromatin pattern is irregular. Very coarsely granular or reticulated. Markedly hyperchromatic. No nucleolus is found in these cells.
2. The cytoplasm and its borders are fairly distinct (*cf.* Plate 112 which is from the same case).

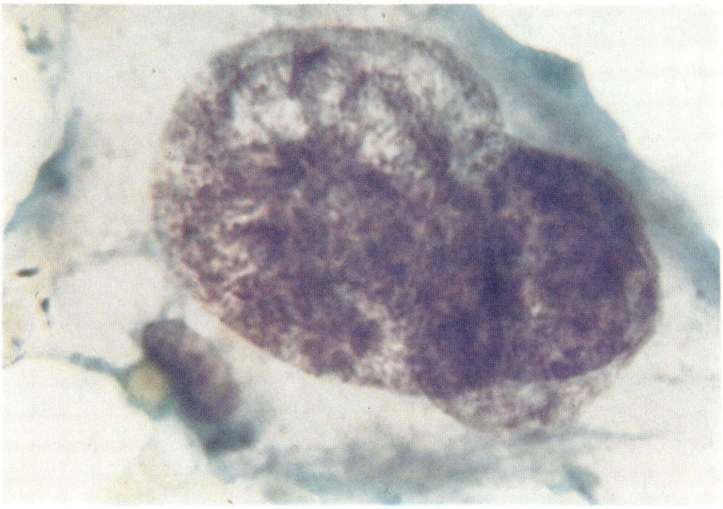

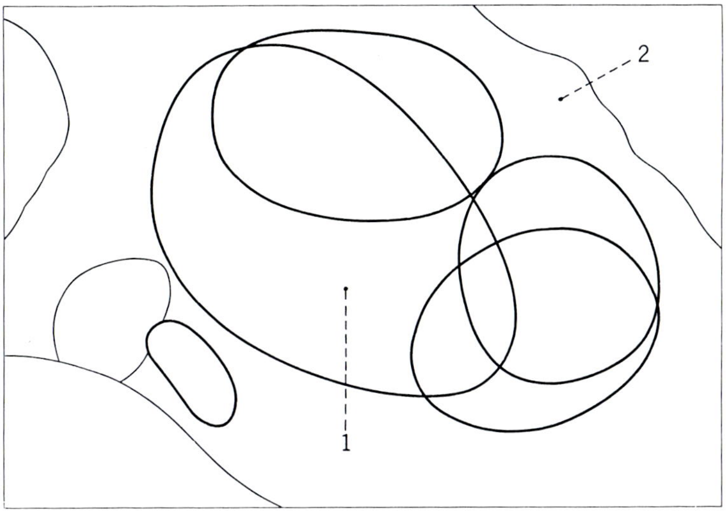

Destructive mole

The nucleus is not essentially different from that of a nucleus of choriocarcinoma, but is less hyperchromatic, and a nucleolus is not as noticeable. The nucleus is markedly enlarged and coarsely granular. Chromatin distribution may be more even than in choriocarcinoma. Chorionic villous pattern may be present. This is significant difference between choriocarcinoma and destructive mole.

80. Destructive mole (Endometrial aspiration smear)

1. Giant, round nuclei. Chromatin is coarsely granular but evenly distributed.
2. A rather indistinct but large nucleolus.
3. Cytoplasm: Abundant and the border is distinctly seen.
4. Numerous erythrocytes in the background.

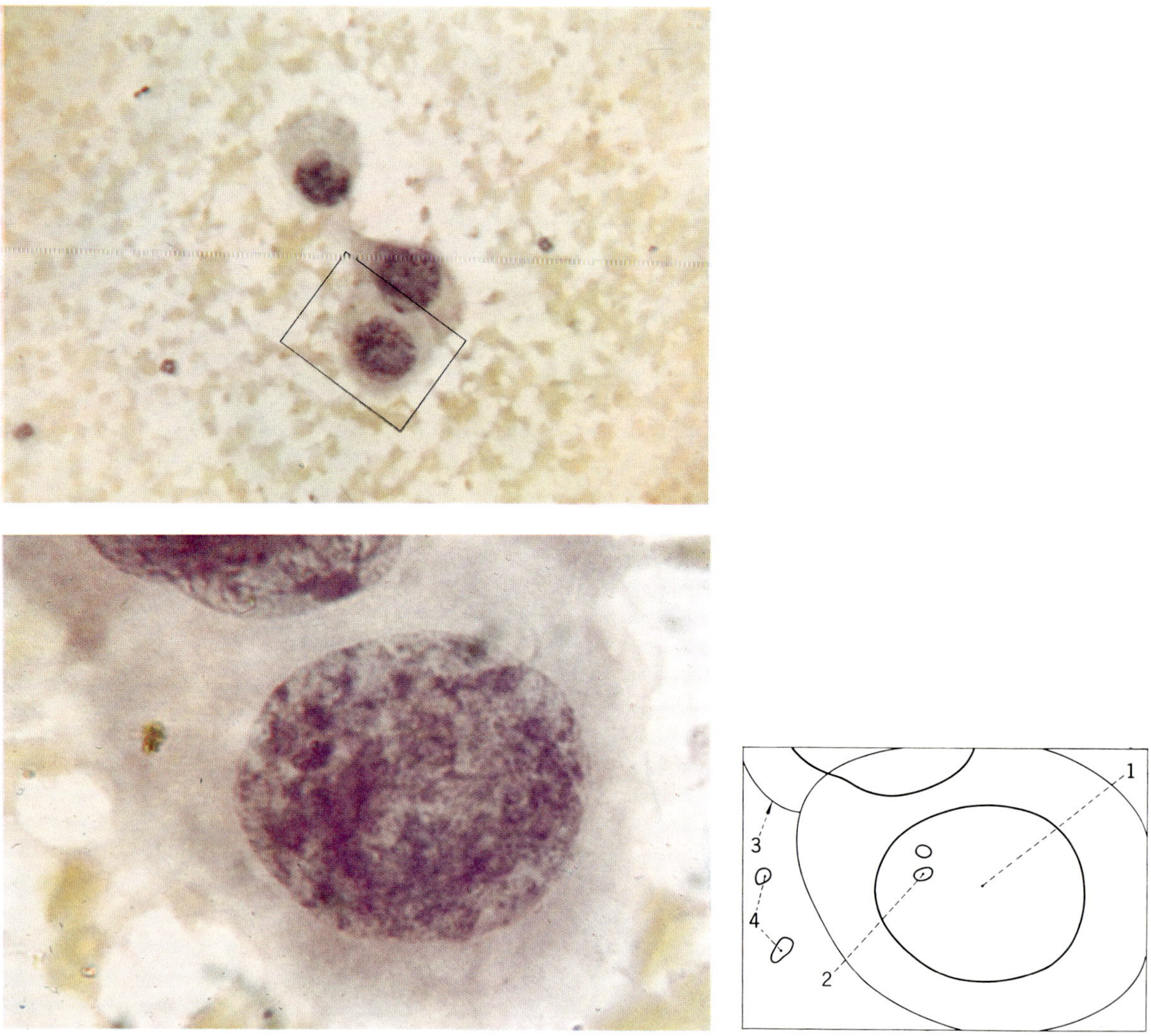

Hydatidiform mole

It is almost impossible to see the cells of hydatidiform mole in vaginal smears until abortion takes place. However, once abortion has begun, atypical trophoblasts may be accidentally found with blood in very early stage.

81. Hydatidiform mole (Smear of bleeding from the vagina immediately after abortion)

1. Nuclei : Round and comparatively small. Uniform in size.
2. Chromatin : Moderately ; not excessively hyperchromatic.
3. The nuclear rim appears denser.
4. A distinct chromocenter is observed.
5. A tendency to form a loose cellular cluster with overlapping nuclei is noticed.
6. Faintly remaining cytoplasm.

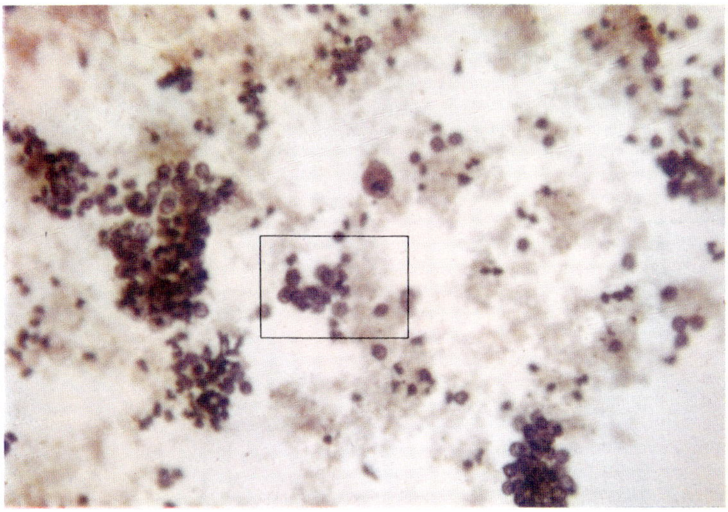

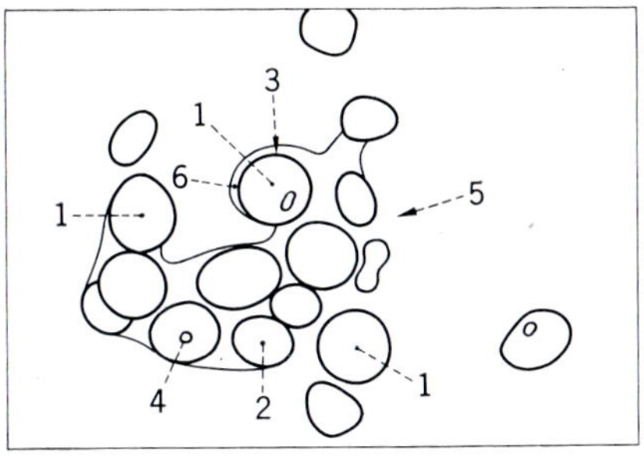

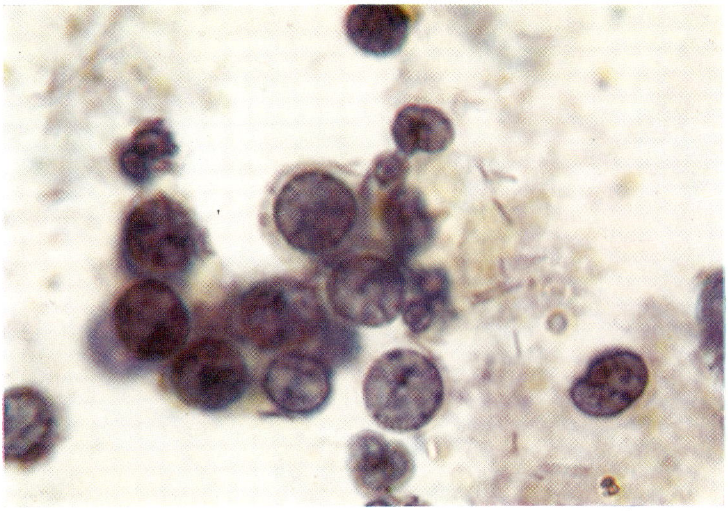

Malignant melanoma

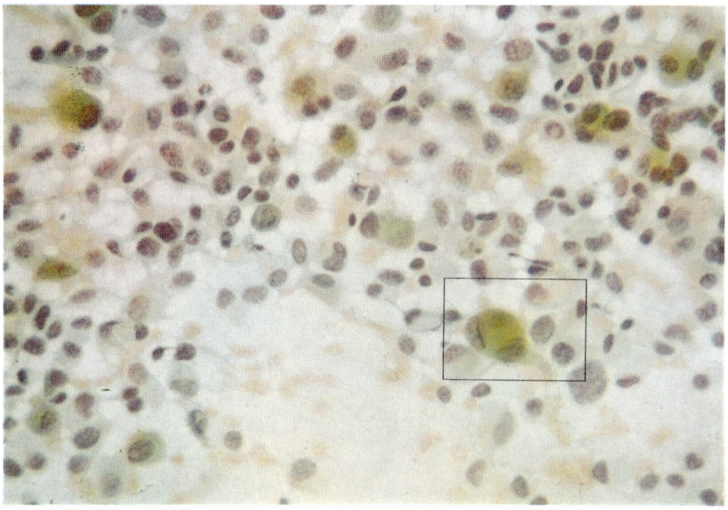

82. Malignant melanoma, primary site of origin — the vagina

This is a rare tumor in the vagina, but is easily diagnosed by the peculiar pigment in the cytoplasm.

a. Low magnification

Scattered large cells containing brownish pigment in the cytoplasm are found.

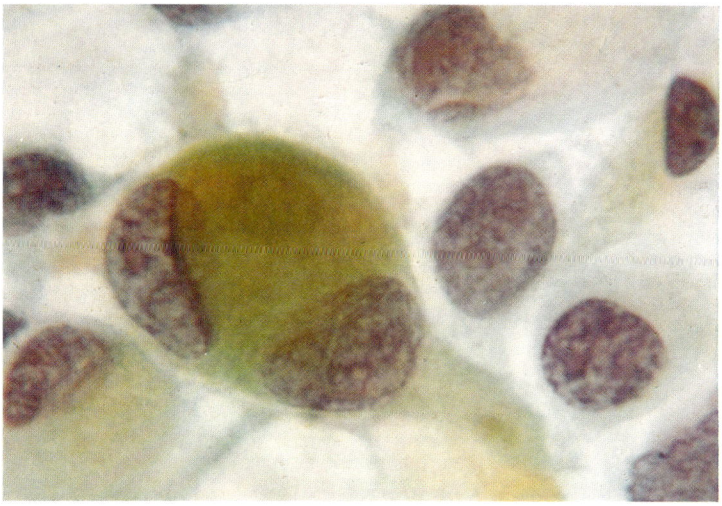

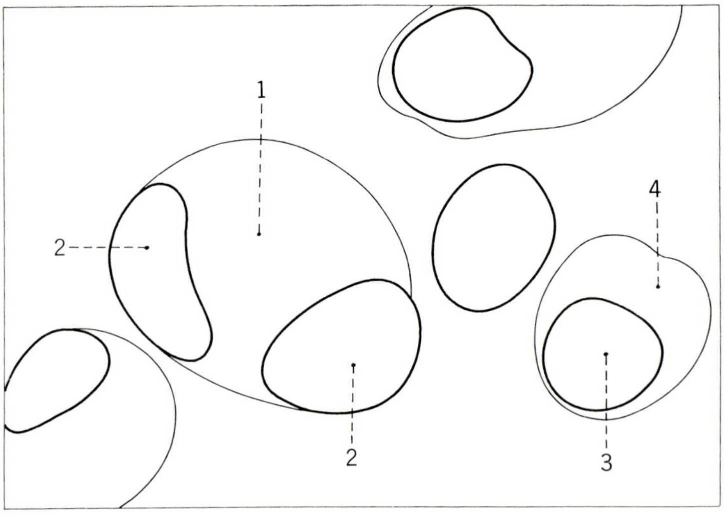

b. High magnification

1. A striking finding is the inclusion of brownish pigment in the cytoplasm.
2. Nuclei showing malignant features may be located eccentrically and appearing to be almost extruded from the cytoplasm.
3. This nucleus is also eccentrically situated and shows coarse granulation as well as a nucleolus.
4. The cytoplasm does not reveal the pigment in this cell.

Metastatic tumor

83. Metastatic tumor to the uterine cervix from carcinoma of the stomach

a. Low magnification

Scattered large malignant cells are present throughout the smear.

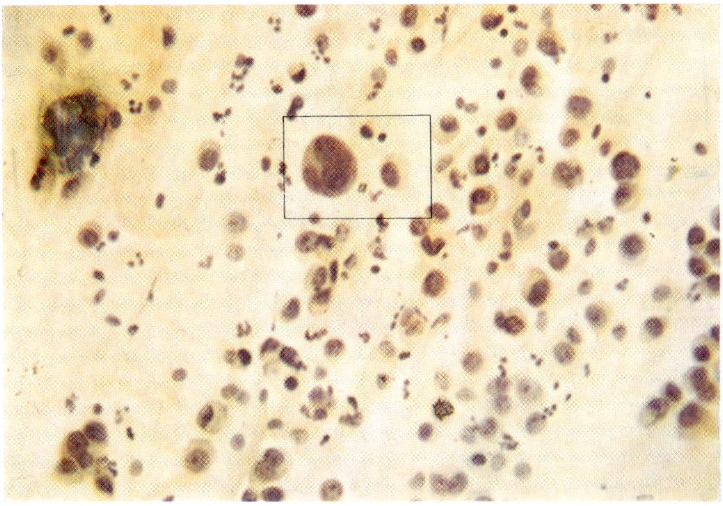

b. High magnification

1. This large cell consists of overlapping large nuclei located along the cellular boundary.
2. Nucleoli can be seen.
3. Chromatinic pattern is granular and dense.
4. Cytoplasm is thickened and its border looks clear.
5. Eccentrically located large nucleus showing hyperchromatism.
6. Cytoplasm appears thinner than that of large cell situated in the left.

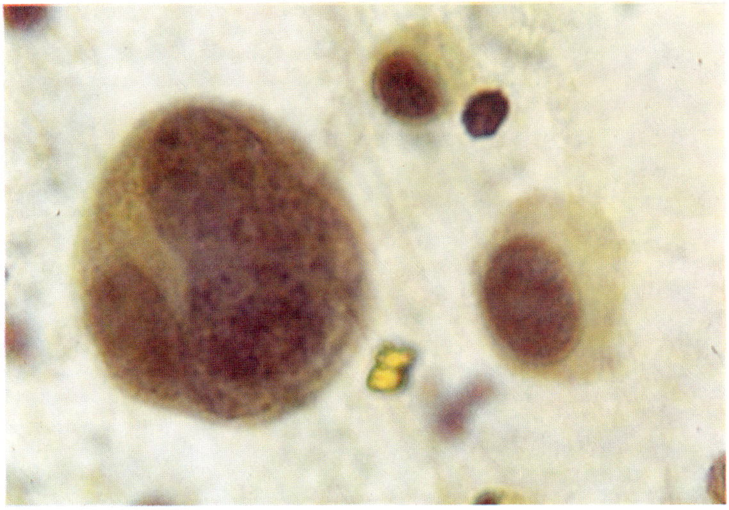

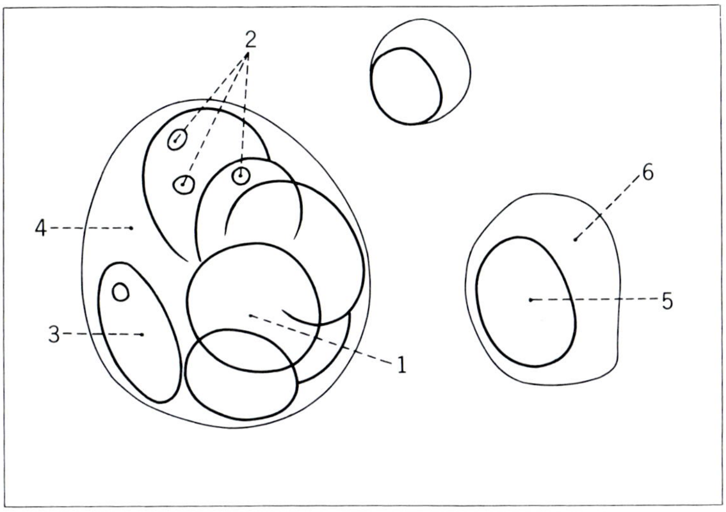

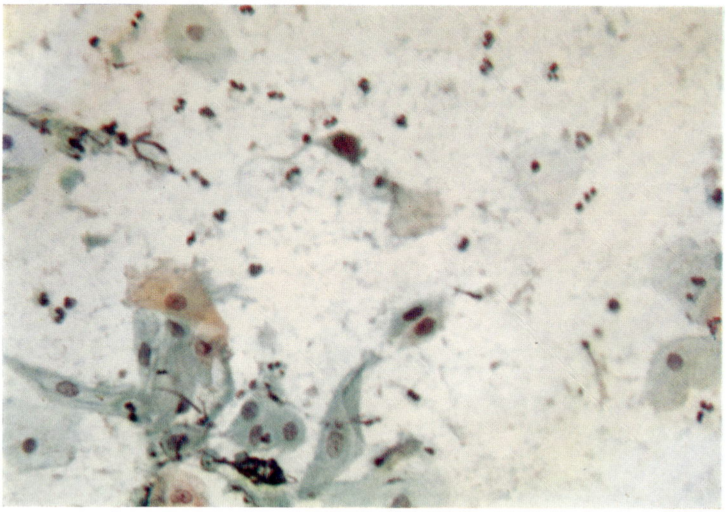

84. Metastatic tumor to the vagina from bronchogenic carcinoma

a. Low magnification

At least three malignant cells are observed. Benign squamous cells are also found in the same smear.

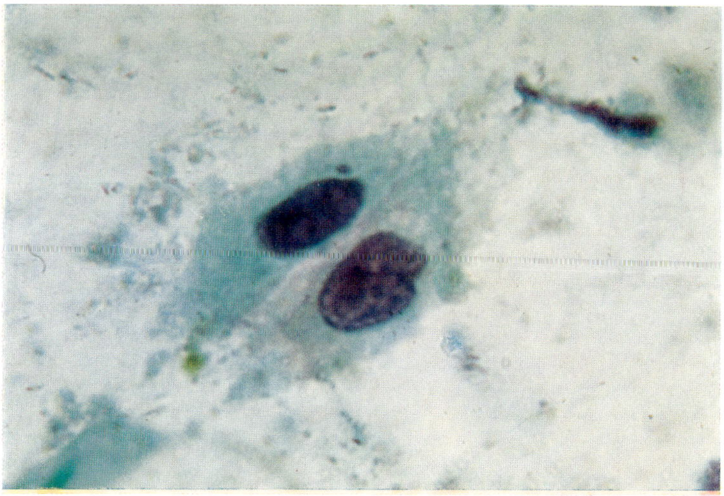

b. High magnification

1. Two large malignant cells are parallel and they each show a large nucleus in the center.
2. Nuclei are large, hyperchromatic and dense particularly in the borders.
3. A nucleolus can be seen in one of the nuclei.

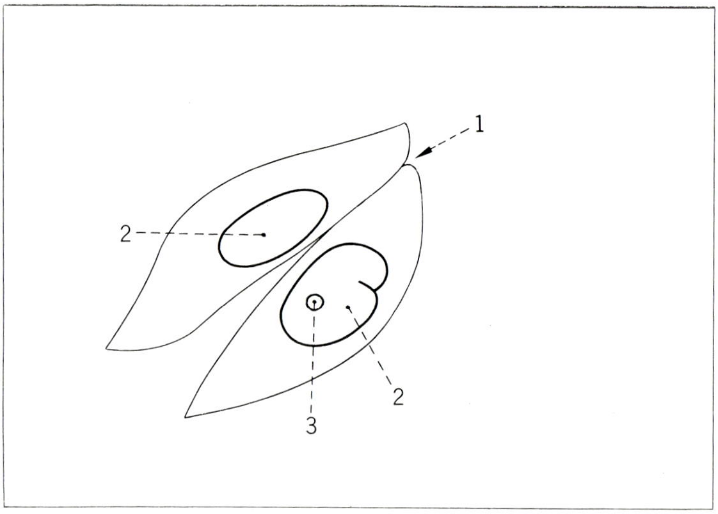

IX. Radiation Effects

Changes in benign squamous cells (Fig. 38)

1) General changes
   a. Enlarged: Sometimes twice to four times as large as a normal cell.
   b. Irregularly shaped: Elongated, tadpole-shaped or bizarrely amoeboid.
2) Nucleus
   a. Enlarged: The nuclear-cytoplasmic ratio is not greatly altered, since the cytoplasm is proportionally enlarged.
   b. Degenerated:
      1. Hyperchromatic or hypochromatic.
      2. The chromatin pattern is lost and chromatin is homogenized. Therefore, it appears indistinct. Sometimes, however, chromatin granules become more distinct.
      3. Irregularly shaped and contracted.
      4. A perinuclear halo may be found.
      5. Nuclear pyknosis, karyorrhexis and mitosis may be noticed.
      6. Vacuolated.
   c. Multinucleated: Sometimes more than three nuclei are observed.
   d. Nucleoli increase in number or size.
3) Cytoplasm
   1. Vacuolation is noticed. Vacuoles vary in size and often include inflammatory cell.
   2. Staining reaction is altered. Parabasal cells are occasionally stained brownish by Papanicolaou's stain. They sometimes appear indistinct.
   3. Fine threads, like pieces of hair, can be seen.
   4. Generally becomes hazy in appearance.
4) Background

Numerous polymorphonuclear leukocytes are observed within two weeks after the onset of irradiation, but they decrease when therapy is continued longer.

Changes in normal glandular cells

Essentially the same as the changes in squamous cells.

Changes in malignant cells

Essentially the same as the changes in benign cells. Usually, however, hyperchromatism of nuclei and abnormality in the nuclear-cytoplasmic ratio, which are the characteristics of malignant cells, persist to some extent. Malignant cells often disappear in the smear when irradiation is continued. Sufficient care is needed to distinguish malignant from benign cells since nuclear enlargement and hyperchromatism are also found in benign cells.

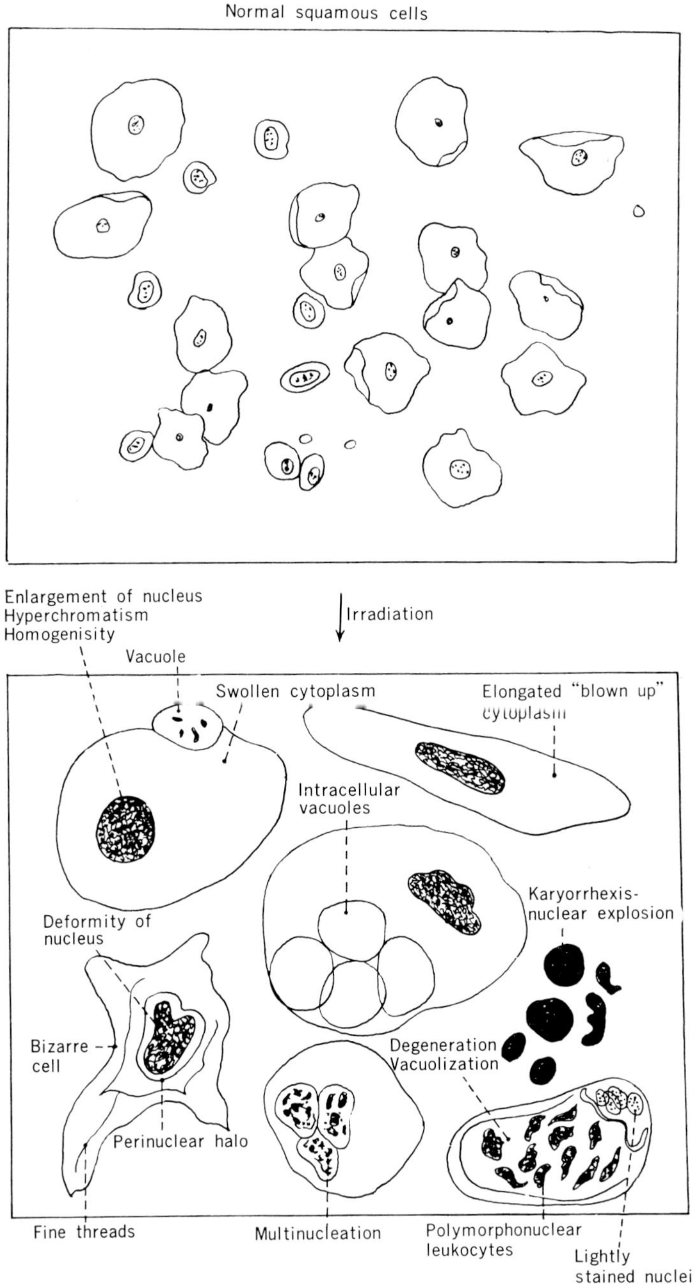
Normal squamous cells
Enlargement of nucleus
Hyperchromatism
Homogenisity
Irradiation
Vacuole
Swollen cytoplasm
Elongated "blown up" cytoplasm
Intracellular vacuoles
Karyorrhexis-nuclear explosion
Deformity of nucleus
Bizarre cell
Degeneration
Vacuolization
Perinuclear halo
Fine threads
Multinucleation
Polymorphonuclear leukocytes
Lightly stained nuclei

Fig. 38

85. Irradiated benign squamous cell—Intermediate cell (Vaginal smear)

a. Low magnification

A large, irregularly shaped cell is situated among numerous leukocytes and erythrocytes. Compare the size with those of leukocytes and erythrocytes.

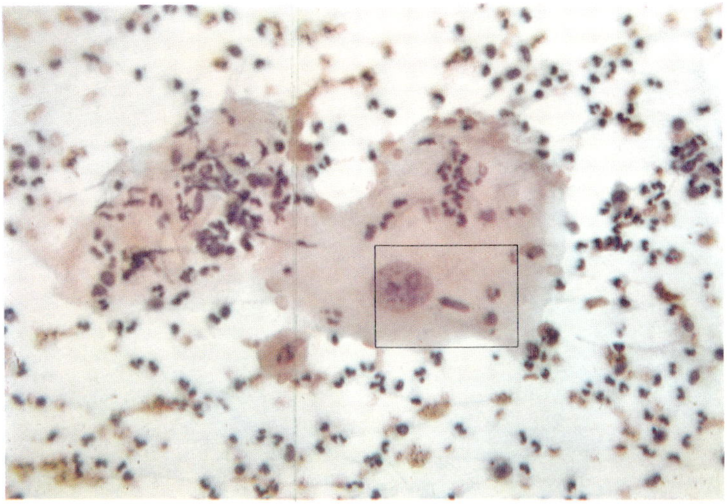

b. High magnification

1. Both the nucleus and the cytoplasm are markedly enlarged. The nuclear-cytoplasmic ratio is within normal limits.
2. The nucleus is hypochromatic and fine chromatin granules are definitely distributed.
3. Cytoplasm · Shows a fine thread-like delicate network.

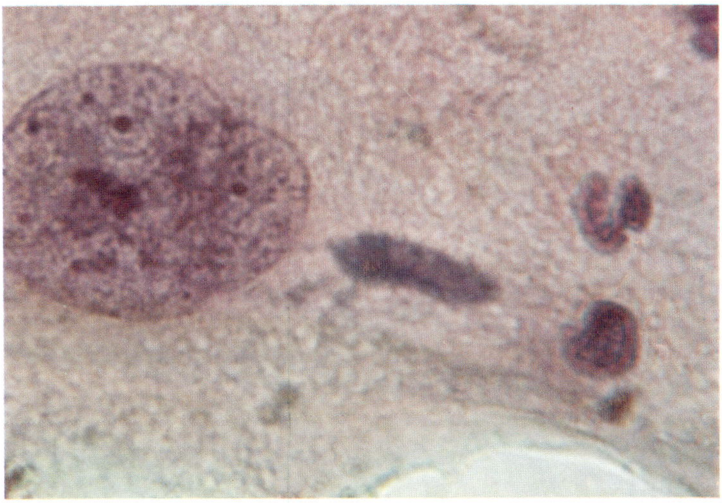

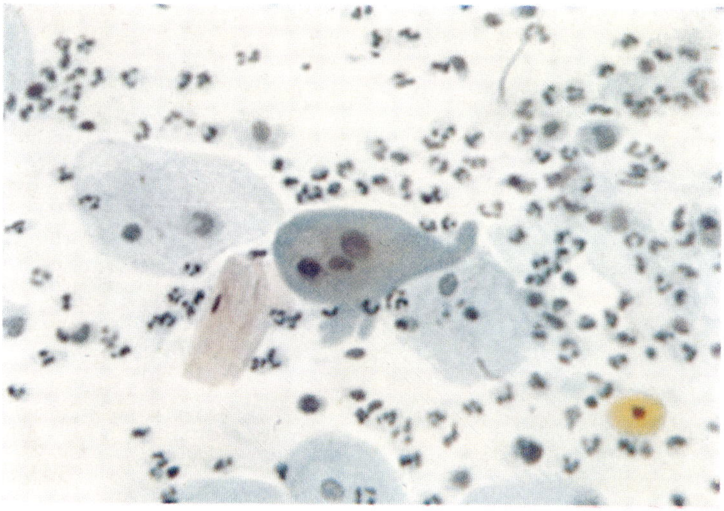

86. Irradiated benign squamous cell (Vaginal smear)

a. Low magnification

A large, bizarrely shaped cell is observed.

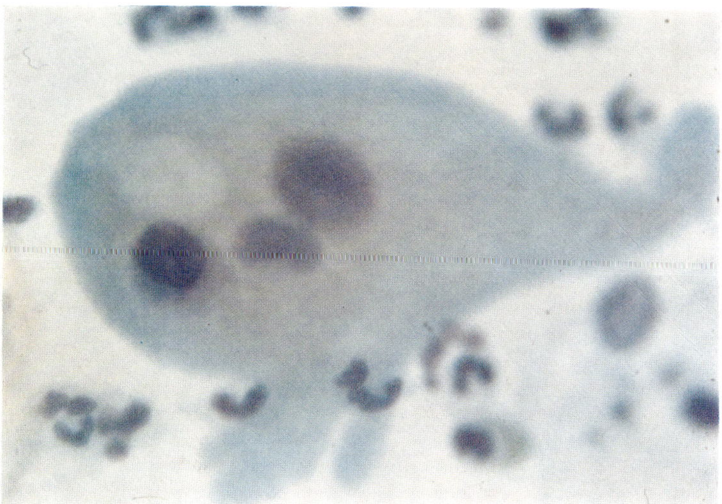

b. High magnification

1. The nucleus is hypochromatic and structureless.
2. A marked vacuolization.
3. The cytoplasm is bizarrely amoeboid in shape and appears denser. The border is sharply defined. In general hazily stained.
4. A different staining reaction is seen between the central part and peripheral of the cytoplasm.

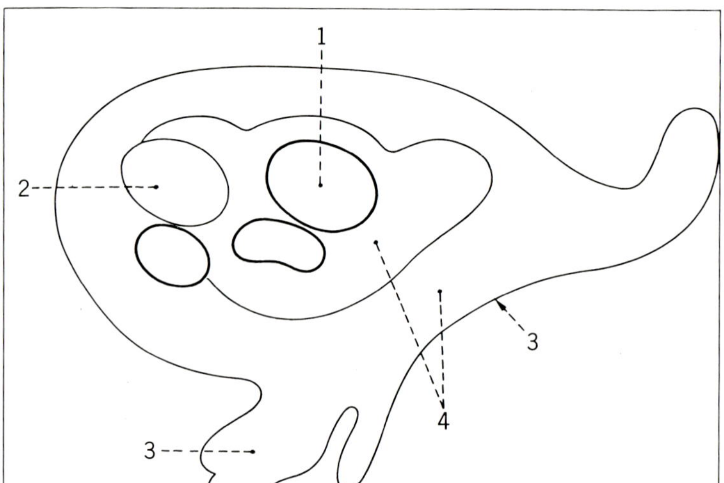

87. Irradiated malignant cells of epidermoid carcinoma(Vaginal smear)

a. Low magnification

Several large irradiated malignant cells are arranged in the center. Numbers of polymorphonuclear leukocytes are also seen.

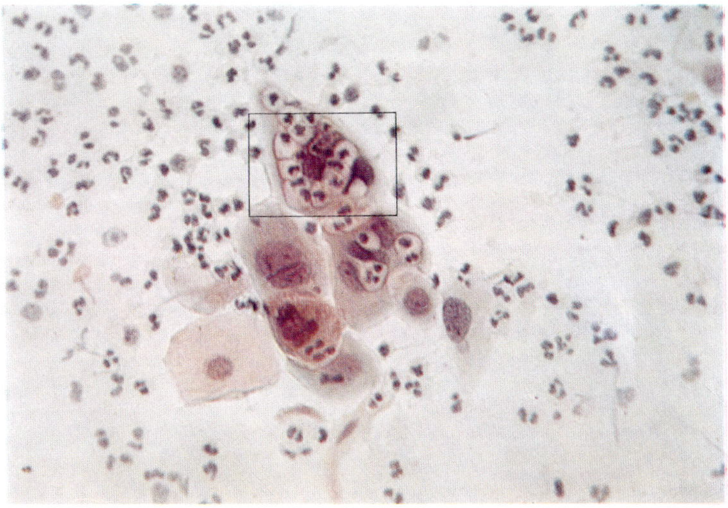

b. High magnification

1. Giant hyperchromatic nuclei are observed. The chromatin granules are coarse and very distinct. The nuclear membrane is more dense. The nuclear-cytoplasmic ratio is high. The shape is undoubtedly distorted by vacuolization.
2. Surrounding the nuclei several vacuoles are formed and leukocytes are included within the vacuoles.
3. Leukocytes are seen in the cytoplasmic vacuole.
4. The cytoplasm is fairly dense and appears structureless. The border is distinct.

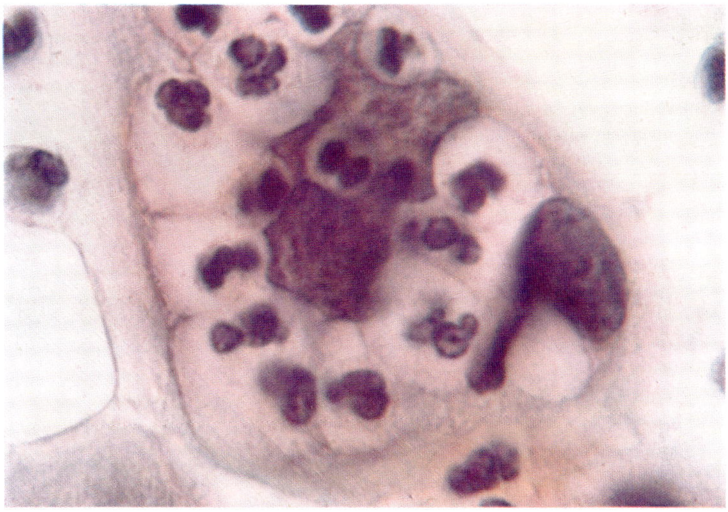

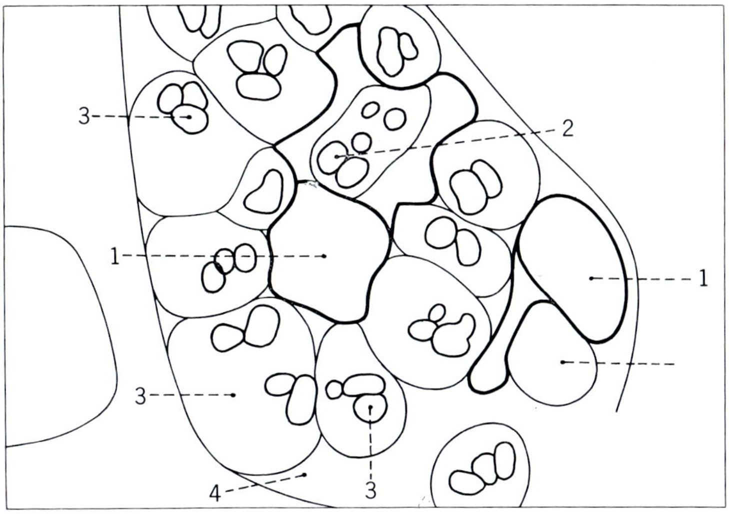

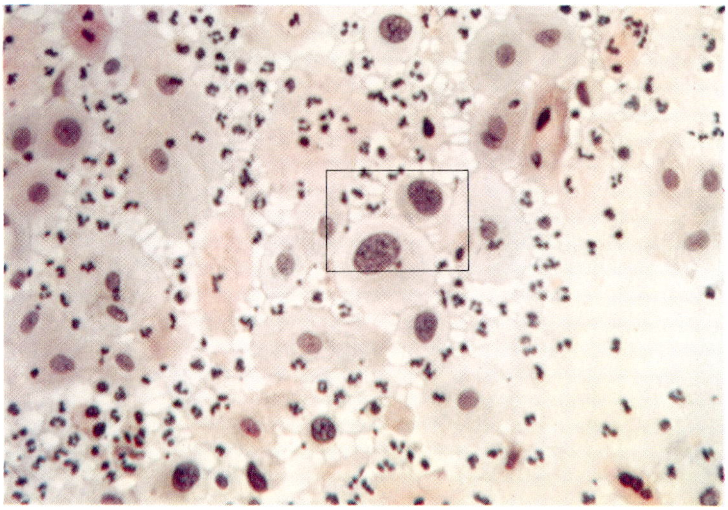

88. Irradiated malignant cells of epidermoid carcinoma — in comparison with irradiated benign cells (Vaginal smear)

a. Low magnification

Malignant cells and benign squamous cells are mixed. Both show marked radiation effects, but malignant cells can easily be distinguished by the giant nuclei. There are also numerous leukocytes.

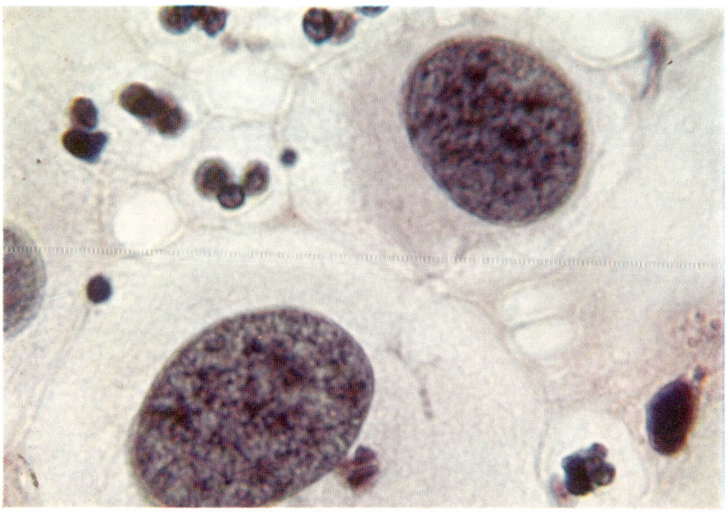

b. High magnification

1. The markedly hyperchromatic and coarsely granular nuclei of irradiated malignant cells.
2. A narrow perinuclear halo.
3. The cytoplasm is fairly abundant, structureless and appears misty.
4. Leukocytes in vacuole.

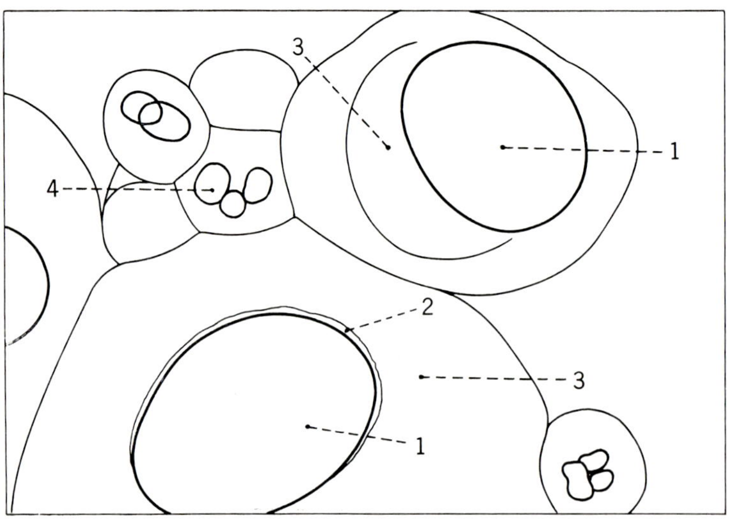

89. Irradiated malignant cells of epidermoid carcinoma (Vaginal smear)

a. Low magnification

There are markedly enlarged malignant cells. They are irregularly shaped and many vacuoles are observed.

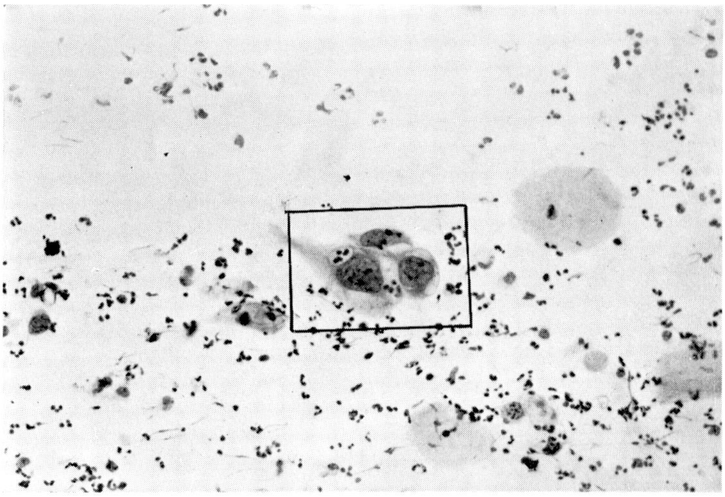

b. High magnification

1. There are markedly enlarged nuclei with coarse chromatin clumps. The nuclear membranes are thickened. Thus the characteristics of malignant cells persist.
2. Large cytoplasmic vacuoles.
3. Leukocytes in a vacuole.
4. This cell has a long tail.
5. Polymorphonuclear leukocytes.

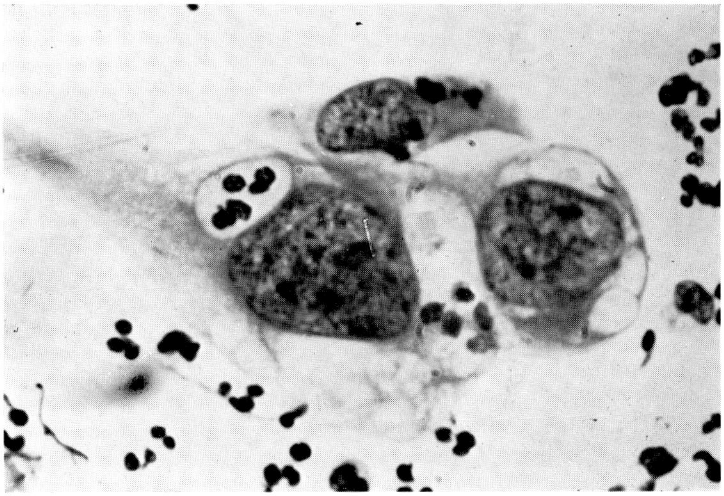

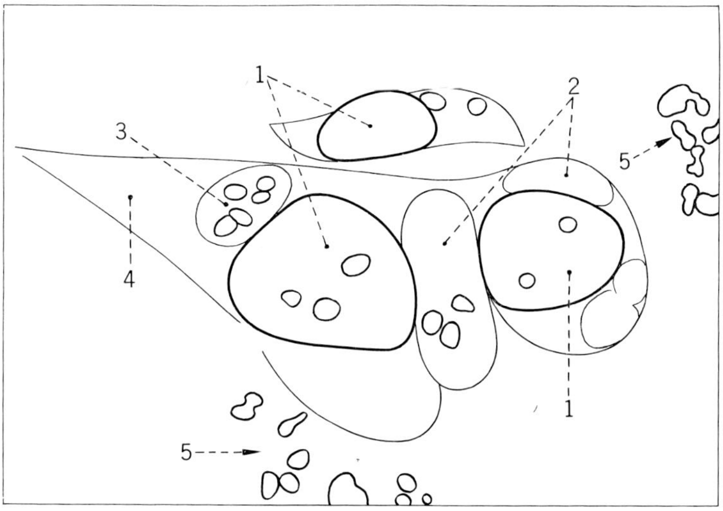

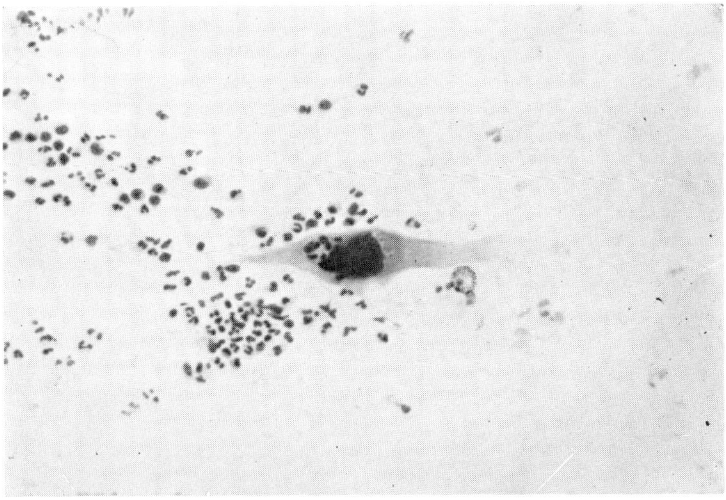

90. Irradiated malignant cell of epidermoid carcinoma(Vaginal smear)

a. Low magnification

A long, bizarre cell with a giant hyperchromatic nucleus is present.

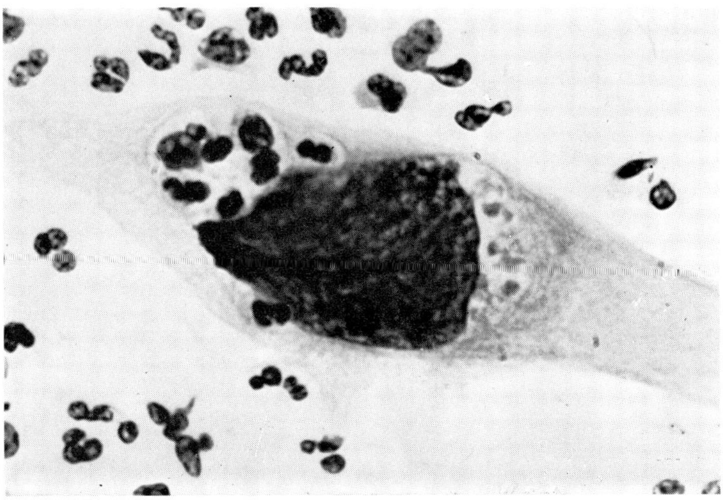

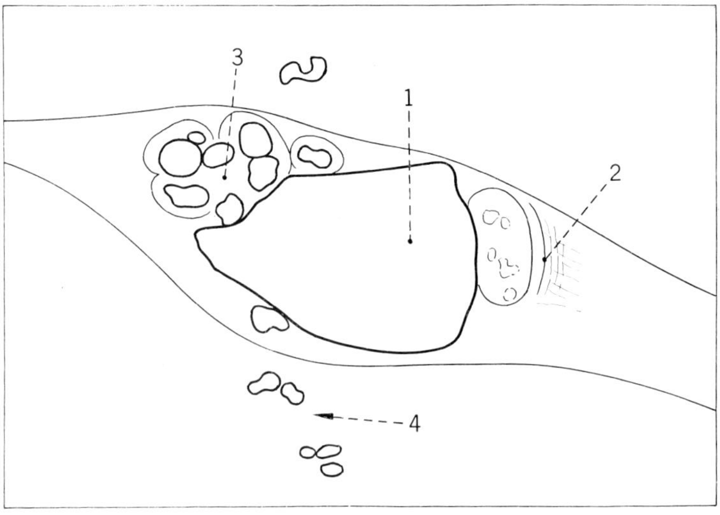

b. High magnification

1. The enlarged triangular nucleus. Remarkably hyperchromatic and chromatin stripes are noticed.
2. The cytoplasm is translucent and looks reticulated or finely fibrous.
3. Vacuoles contain leukocytes.
4. Polymorphonuclear leukocytes.

91. Irradiated malignant cells of adenocarcinoma (Vaginal smear)

a. Low magnification

Several enlarged cells are present, surrounded by numerous leukocytes. Their shapes are irregular and rather bizarre.

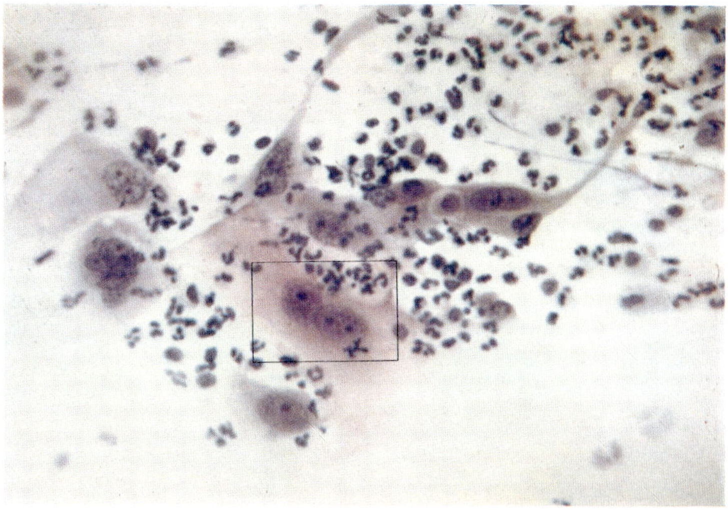

b. High magnification

1. Enlarged and elliptical nucleus showing evenly distributed, coarsely chromatinic granules.
2. Strikingly large nucleoli.
3. Cytoplasm appears thickened and becomes homogeneous, compared with the usual one of adenocarcinoma.
4. Leukocytes within vacuoles.

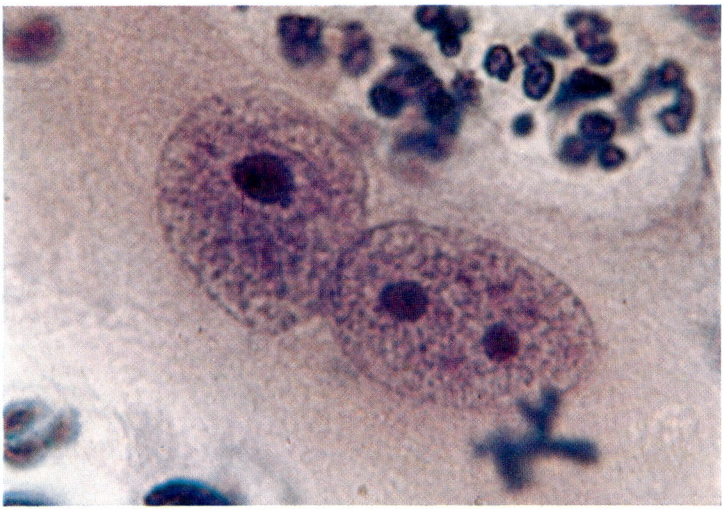

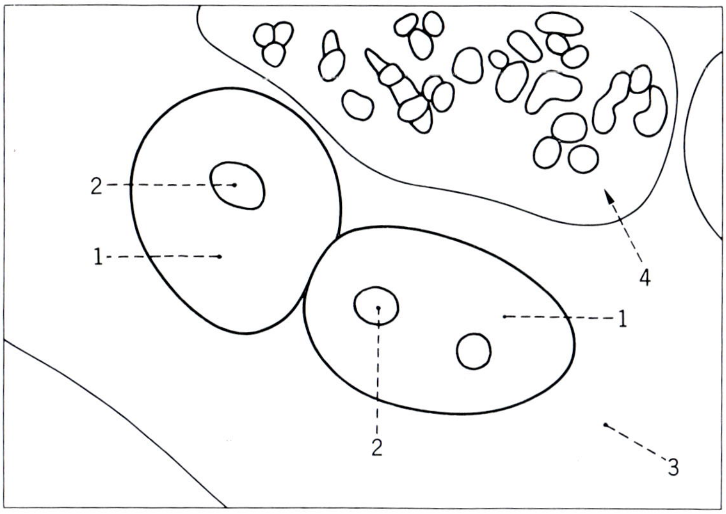

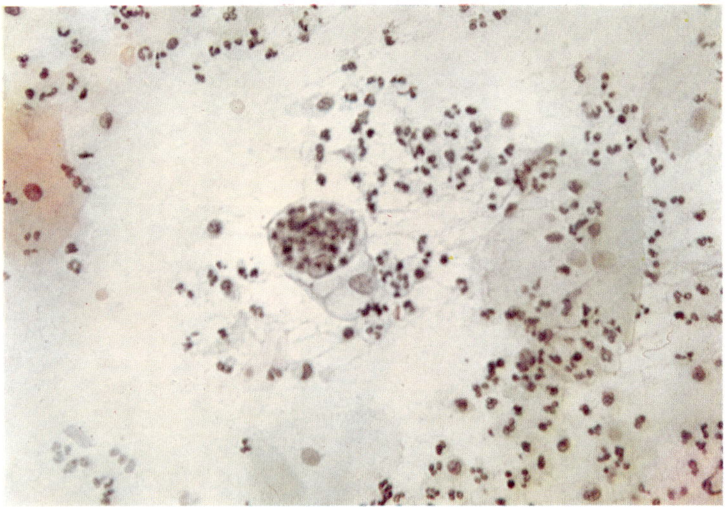

92. Irradiated malignant cells of adenocarcinoma (Vaginal smear)

a. Low magnification

Markedly vacuolated cells are observed among scattered irradiated benign squamous cells.

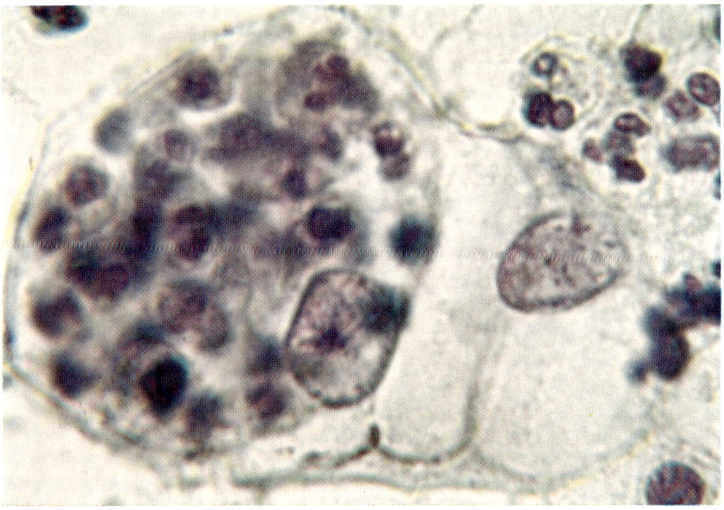

b. High magnification

1. The nuclei of the adenocarcinoma cells. The nuclear membranes are markedly thickened, and the inside is hypochromatic. They are comparatively small.
2. Indistinct nucleolus.
3. The cytoplasm is mostly occupied by vacuoles.
4. A large number of leukocytes are in a cytoplasmic vacuole. Such are the significant characteristics of an adenocarcinoma cell, and they may be intensified by irradiation.

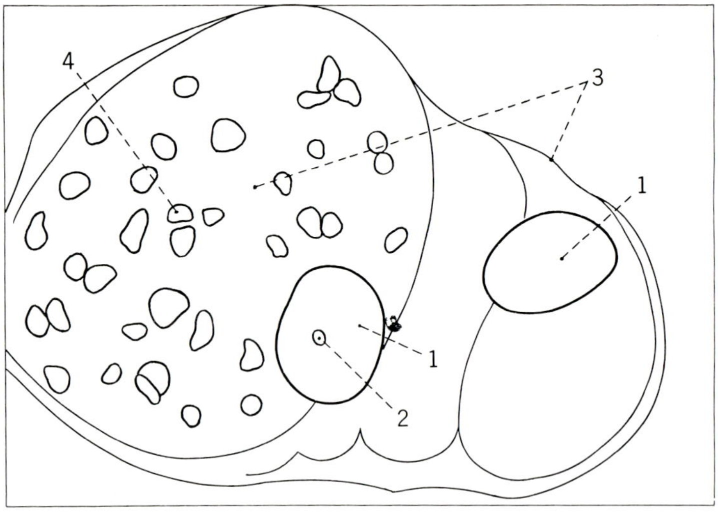

CYTOLOGIC EXAMINATION OF THE RESPIRATORY TRACT

I. Anatomy of the Respiratory Tract

The respiratory tract is divided into the upper and lower respiratory tract (Figs. 39 and 40).

The upper respiratory tract consists of the nasal cavity, oral cavity, rhinopharynx and larynx.

The lower respiratory tract is divided into the trachea, bronchi and lungs.

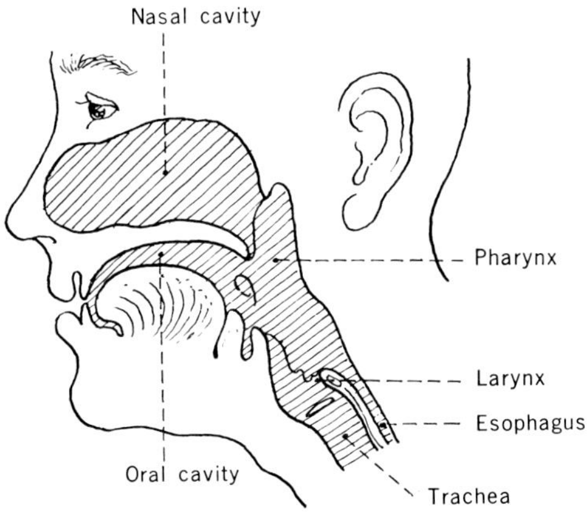

Fig. 39 Upper respiratory tract.

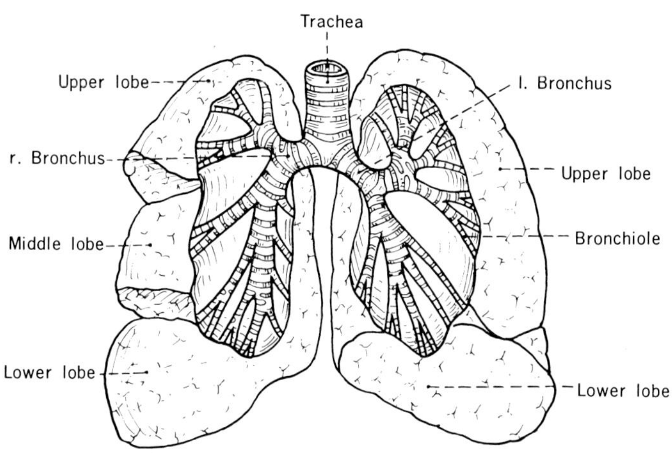

Fig. 40 Lower respiratory tract.

The important portions for the smear test of exfoliated cells are the lungs and bronchi. The trachea and oral cavity may also supply cells for smears. Therefore, it is necessary to know the characteristics of the epithelia of the source of cell supply.

1) Oral cavity

It is covered with stratified squamous epithelium. Because the oral cavity is moistened with saliva, its exfoliated cells are rarely dry. Their shapes are quite similar to those of the superficial squamous cells of the vagina and portio vaginalis. Sometimes comparatively small cells of the parabasal type, spindl cells and squamous pearls are observed.

2) Pharynx and larynx

The inner and lower parts of the pharynx are also covered with stratified squamous epithelium, and the upper part with ciliated respiratory epithelium. The larynx is covered with stratified ciliated epithelium mixed with goblet cells and stratified squamous epithelium.

3) Trachea and main bronchi

These parts are covered with ciliated respiratory epithelium. It is stratified and consists of three or more layers of columnar cells. The cells on the surface are definitely ciliated, but those in the lower layers are not distinctly ciliated. The ciliated epithelium contains some mucus-producing goblet cells. Basal cells (reserve cells) are observed at the base.

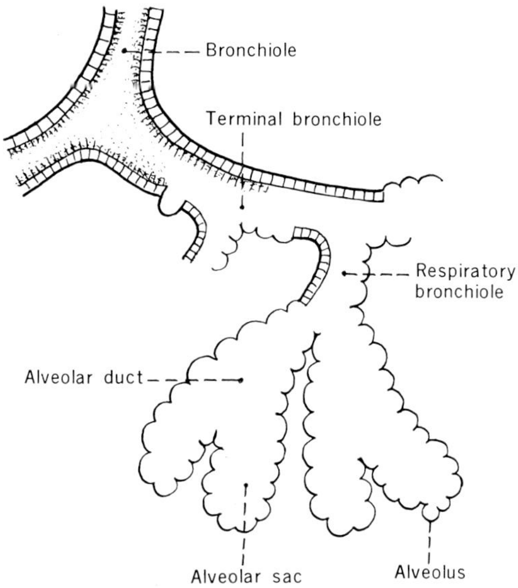

Fig. 41 From bronchus to alveolus.

4) Lungs (Fig. 41)

The trachea is divided into the right and left main bronchi, and each bronchus enters the lung at the hilus. The bronchi are then divided in the following order and finally terminate in the alveoli ;

Main bronchi → bronchi → bronchioles → terminal bronchioles → respiratory bronchioles → alveolar ducts → alveolar sacs → alveoli

a. Bronchi: Covered with pseudostratified ciliated columnar epithelium containing some goblet cells. Basal cells are observed at the base.

b. Bronchioles: 3 mm or less in diameter. Covered with single-layered ciliated columnar epithelium. No goblet cells are found.

c. Terminal and respiratory bronchioles: 0.5 mm or less in diameter. Covered with single-layered ciliated epithelium. The cells gradually become cuboidal, and then flat without cilia. No goblet cells are found.

d. Alveolar ducts, alveolar sacs and alveoli: Epithelial structures can be seen only with difficulty under the optical microscope. Only large or small, flat endothelioid cells are observed.

e. Alveolar macrophages: Attach to the basement membrane and protrude into the alveolus, and are markedly phagocytotic.

II. Benign Cells from the Respiratory Tract

Sputum

Theoretically, sputum can contain all types of cells from all parts of the respiratory tract, but in reality it contains only bronchial epithelial cells, alveolar macrophages, inflammatory cells, squamous cells of the saliva, and mucus.

1) Bronchial epithelial cells (Fig. 42)

Bronchial epithelial cells are, for the most part, ciliated columnar cells. They are, however, often degenerated and lack cilia. If a cell on the smear has cilia, it should be tentatively considered benign no matter how malignant the nucleus may look. These cells are occasionally stripped of cytoplasm.

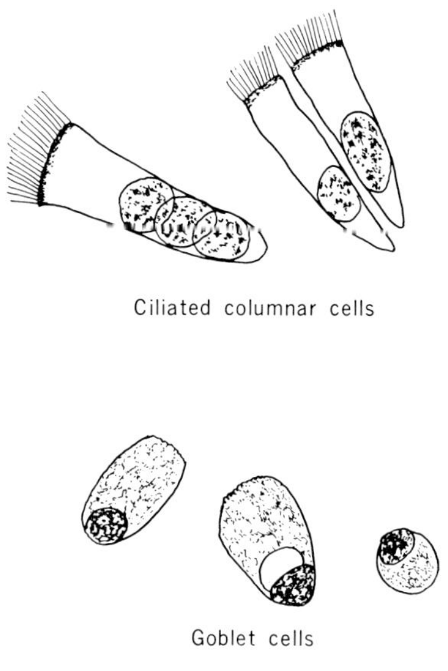

Fig. 42 Bronchial epithelial cells.

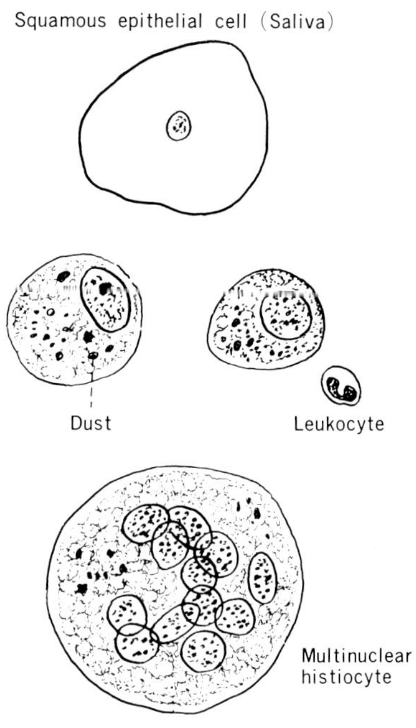

Fig. 43 Alveolar macrophages (dust cells).

a. *Nucleus:* Round or elliptical. The nuclear membrane is distinct, and a small nucleolus may be observed. Sometimes multinucleated.

b. *Cytoplasm:* Often characteristically elongated and triangular.

c. *Goblet cells:* A stocky cup-shaped cell with lacy cytoplasm containing mucus. Its nucleus is located near the cytoplasmic border.

2) Alveolar macrophages (Fig. 43)

a. *Nucleus:* Located near the cytoplasmic border. The nuclear membrane is distinct. Chromatin is comparatively evenly distributed. Sometimes multinucleated.

b. *Cytoplasm:* Abundant and spongy. Contains characteristic large or tiny brownish dust particles. Smaller than a normal squamous cells. Round or elliptical. Stained pale yellow, brown and pink.

c. An alveolar macrophage is also called a "dust cell" and is important in sputum smears as proof that the material is truly from the alveolus.

3) Inflammatory cells

They are mainly polymorphonuclear leukocytes, eosinophiles, lymphocytes and plasma cells.

Polymorphonuclear leukocytes are easily distinguished, but their nuclei are sometimes linear in shape as a result of karyorrhexis.

Lymphocytes are sometimes confused with oat cells, but the nuclei of the former are regular and uniform in size.

4) Squamous cells

They look like superficial squamous cells in vaginal smears. Cells of the parabasal type are also observed. They are mainly from the mucous membrane of the oral cavity and are present in saliva.

Saliva is often mixed in sputum, but sometimes such an extreme all the material is saliva without sputum. If the cells on the smear are mostly squamous cells and no dust cells are observed, the material should be considered to be saliva, not sputum, and has no diagnostic value.

5) Constituents other than somatic cells

These are mucus, fungi, Curschmann's spirals, vegetable cells and so on.

a. *Mucus:* Always observed as a light blue structureless background.

b. *Fungi:* Often candida albicans.

c. *Curschmann's spirals:* Observed not only in bronchial asthma, but sometimes in pulmonary epidermoid carcinoma (Fig. 44).

d. *Vegetable cells* (Fig. 44): Seen fairly often. These are puzzling to the unexperienced examiner, but they are easily identified by their extremely thick cytoplasmic membrane, which look multirefractive, their large size, polygonal shape and lack of normal cellular structures.

Bronchial secretion

It is obtained by bronchial aspiration or washing. The constituents are similar to those of sputum, but of course saliva is not mixed. The main constituents are ciliated bronchial epithelial cells.

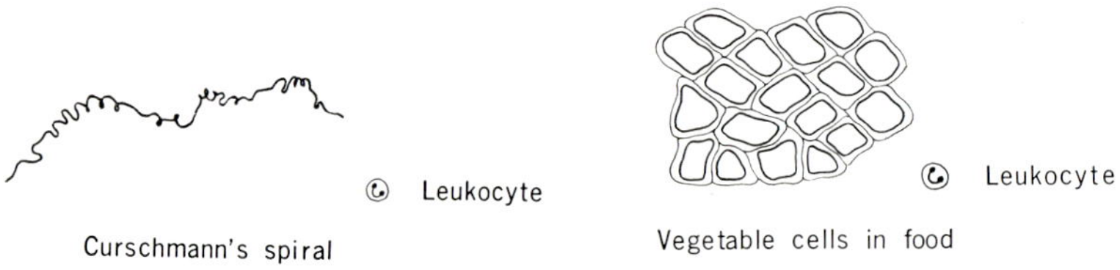

Fig. 44 Curschmann's spiral and vegetable cells.

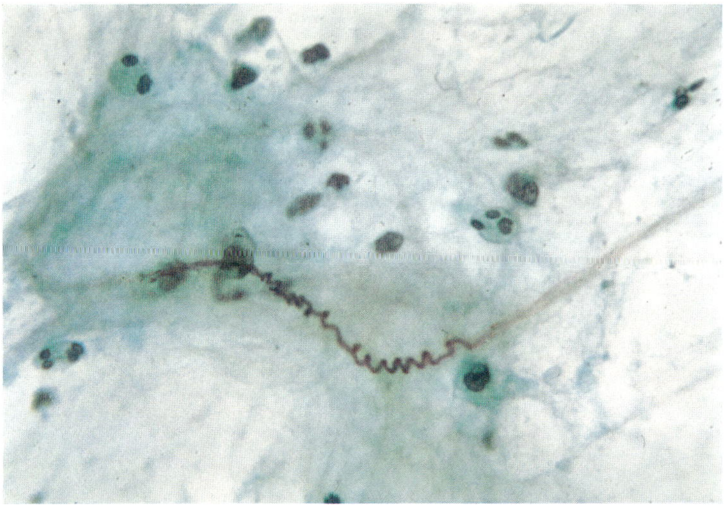

93. Curschmann's spiral
The characteristic coiled structure shows the dark central core of inspissated mucus and the lighter translucent periphery.

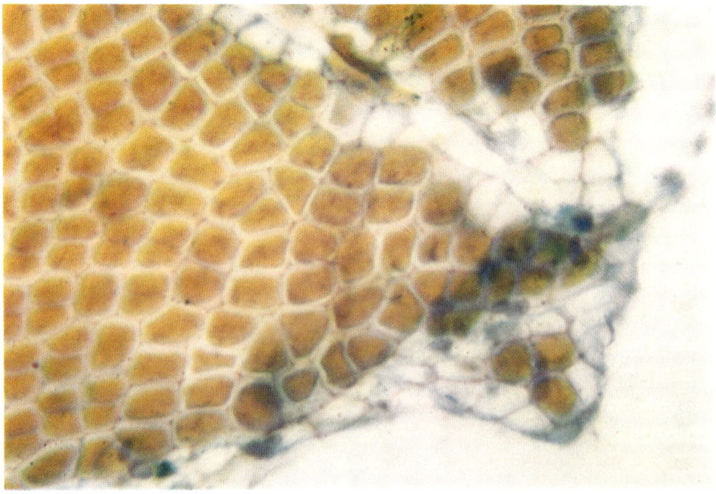

94. Vegetable cells
The striking features of the cells are the presence of their heavy cellulose membrane and the refractile granules within the cytoplasm.

95. Bronchial epithelial cells

a. Low magnification

A group of columnar epithelial cells is seen. Some of them look ciliated.

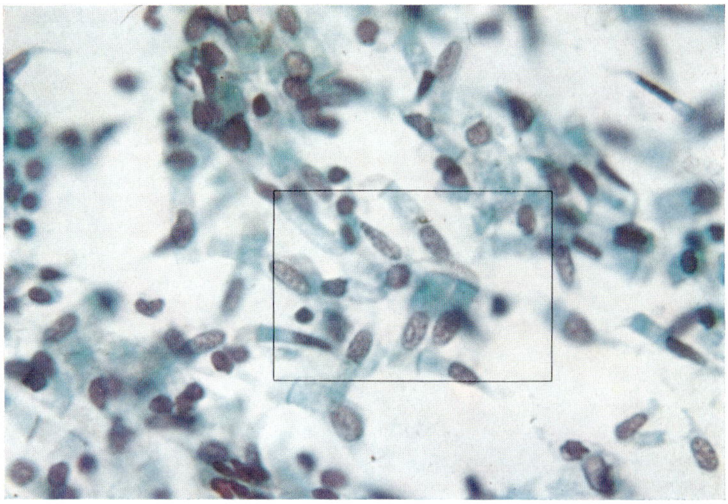

b. High magnification

1. Nucleus Elliptical. The nuclear membrane is thick and chromatin granules are quite easily seen.
2. Cytoplasm : Characteristically elongated. Faint cilia can be seen. One edge is sharp, as if it has been cut off. Note the lacy structure of the cytoplasm.
3. A goblet cell: The cytoplasm is foamy and the elliptical nucleus is pushed to the cytoplasmic border. The cytoplasm is characteristically pear-shaped.
4. A stripped nucleus. The cytoplasm is almost lost as a result of degeneration.

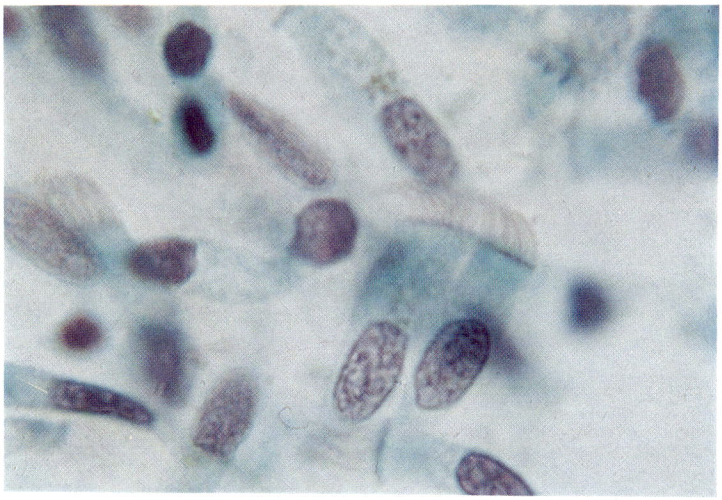

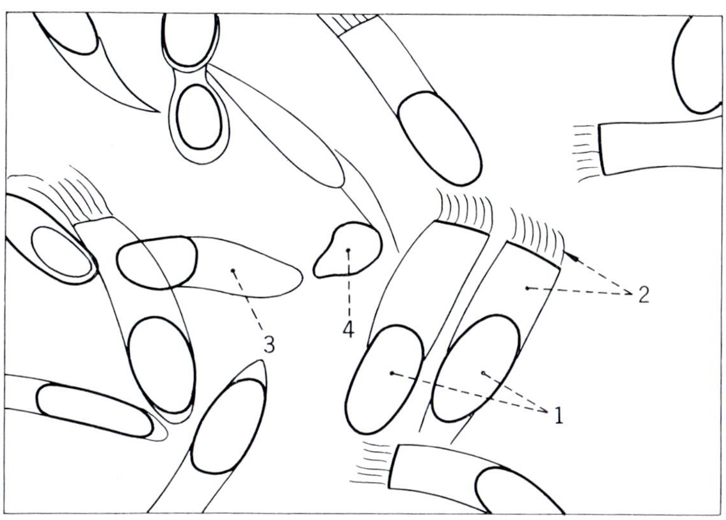

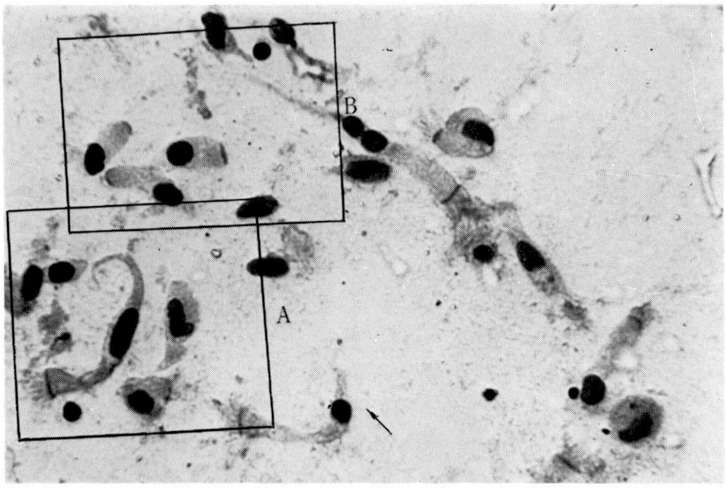

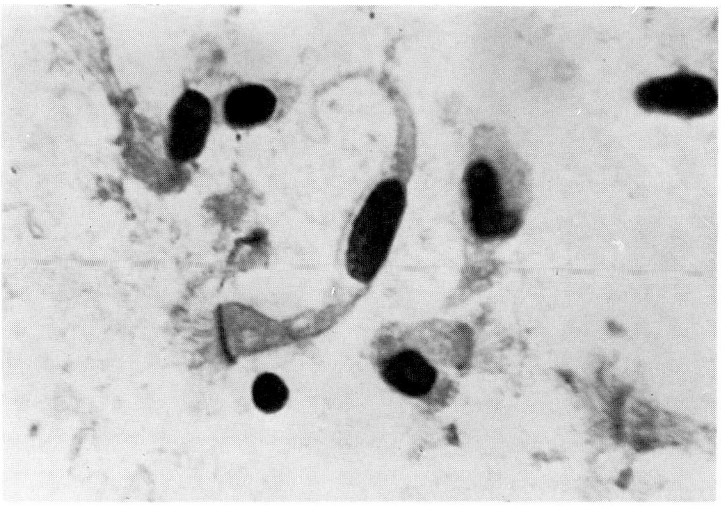

96. Ciliated columnar cells

a. Low magnification

More than ten columnar cells are seen. Some have distinct cilia. Smears contain a few dust cells.

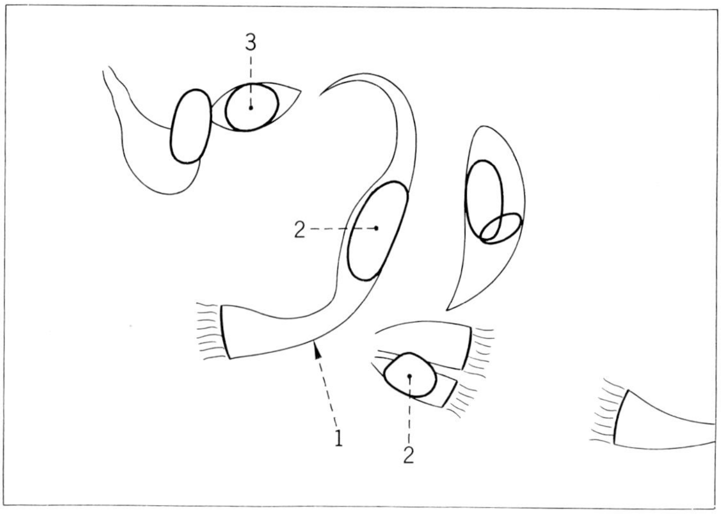

〔**A**〕

b. High magnification (A and B)

1. A bronchial ciliated columnar cell. Characteristically elongated and triangular. Note the widened end of the cytoplasm, the thickening of the cytoplasmic border in that region and the distinct cilia.
2. Nuclei : Located at the narrow end of the cytoplasm (arrow : low magnification) or in the middle. Long and fairly hyperchromatic.
3. A rather dense respiratory epithelial cell. Cilia are poorly seen. The nucleus is round or elliptical and comparatively hyperchromatic.
4. Goblet cells.

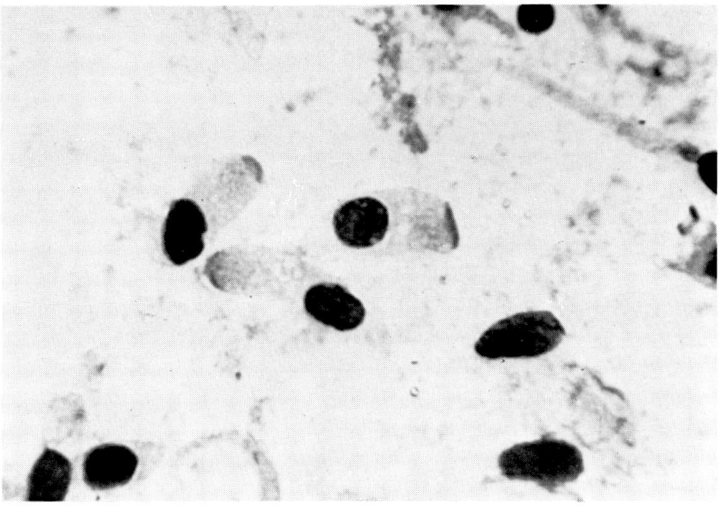

[B]

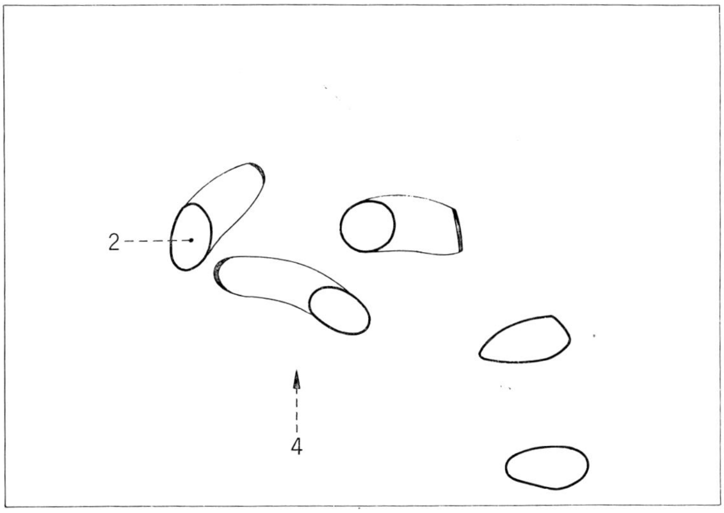

97. Alveolar macrophages — Dust cells

a. Low magnification

Scattered large, round, dark brown cells are seen in the field. The comparatively large nucleus is eccentrically located.

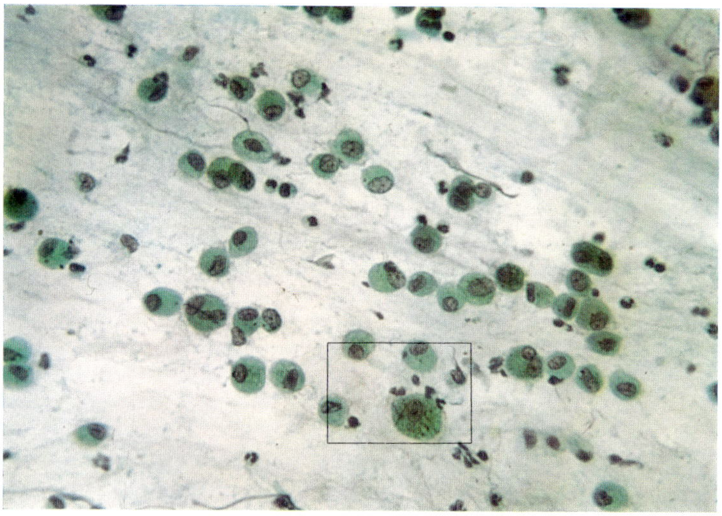

b. High magnification

1. Nucleus: Large and elliptical. Located near the cytoplasmic border. Stained pinkish-brown.
2. Cytoplasm: Characteristically vesiculated and lacy.
3. The border is not very distinct.
4. The most noticeable characteristic is large and small brownish particles of dust in the cytoplasm. The larger particles are easily identified, but the smaller ones give the appearance of a fine and brownish network.

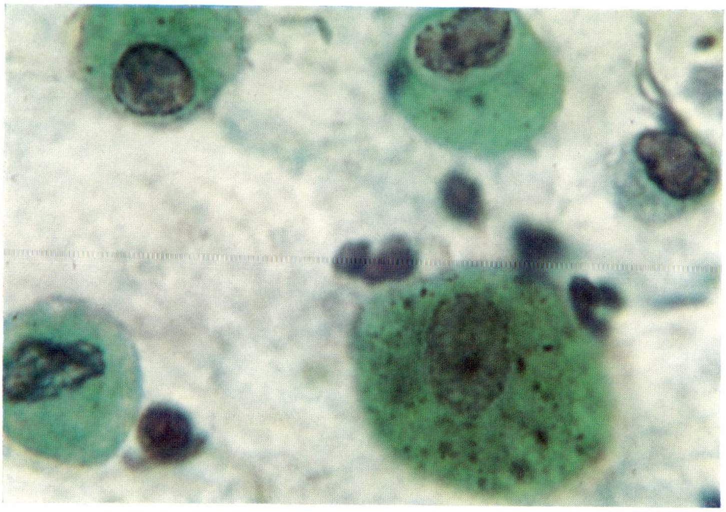

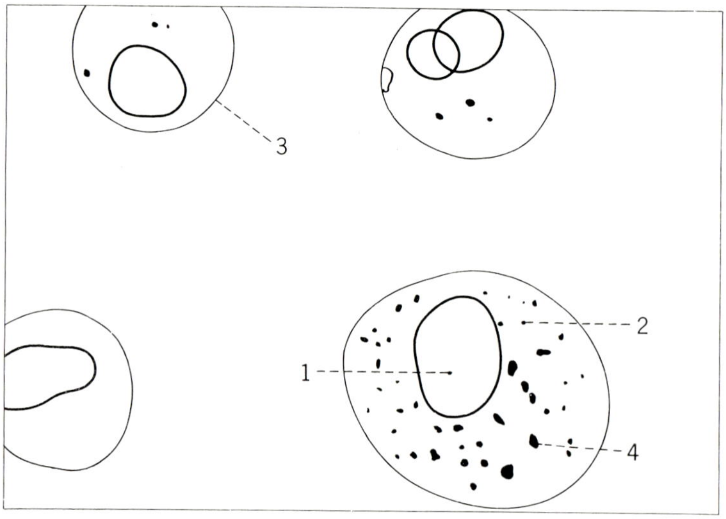

98. Squamous cells — Saliva
This is a specimen of saliva. At first sight it looks like a vaginal smear. The cells have nearly the same size and shape as vaginal superficial or intermediate cells. They are somewhat cornified and therefore some are eosinophilically stained. Sometimes vegetable cells are observed in the background. Such smear material comes from saliva, not from true sputum, and has no significance in the pulmonary cytodiagnosis.

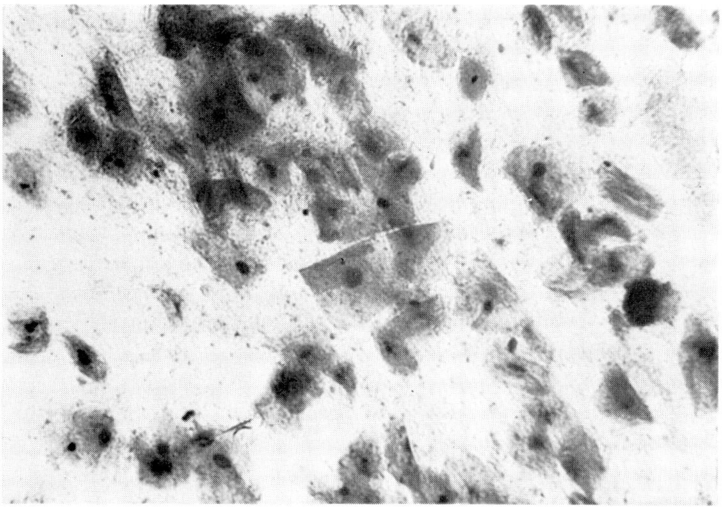

III. Benign Atypical Cells in the Sputum

Bronchial epithelial cells in inflammation

The nucleus is fairly hyperchromatic. Chromatin granules are coarse and sometimes a distinct nucleolus is observed. The nuclear membrane is quite thick. The cells tend to clump, and thus sometimes give a multinucleated impression. Even when the cell picture suggests malignancy, the cells can be distinguished from malignant ones by their cilia and their characteristic triangular shape.

Bronchial squamous metaplastic cells (Fig. 45)

1) Nucleus

Comparatively small and uniform in size. Rather hyperchromatic. Small chromatin masses are observed inside the nucleus and also around the nuclear membrane.

2) Cytoplasm

Scanty. Round or elliptical. Basophilically stained, but in advanced metaplasia sometimes eosinophilically stained and orange. The cells have a pavement like arrangement.

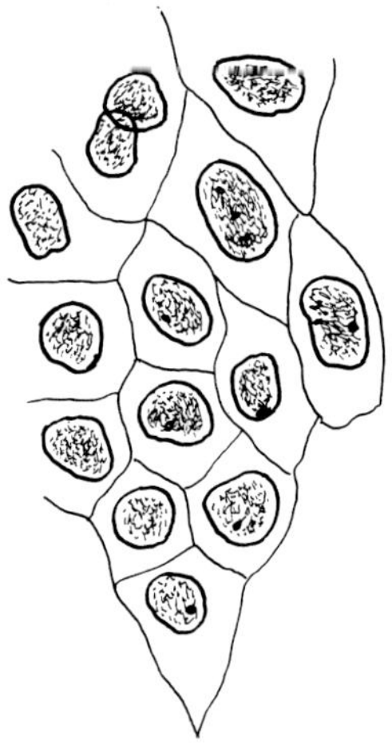

Fig. 45 Bronchial squamous metaplastic cells.

99. Bronchial squamous metaplasia

Four rather atypical metaplastic cells are observed.

1. Nuclei: Elliptical and uniform in size. Chromatin is evenly distributed. Chromocenters are observed.
2. Cytoplasm: Characteristically polygonal. Abundant. The cells may have a pavement-like arrangement. (Picture by courtesy of Dr. M. Takahashi)

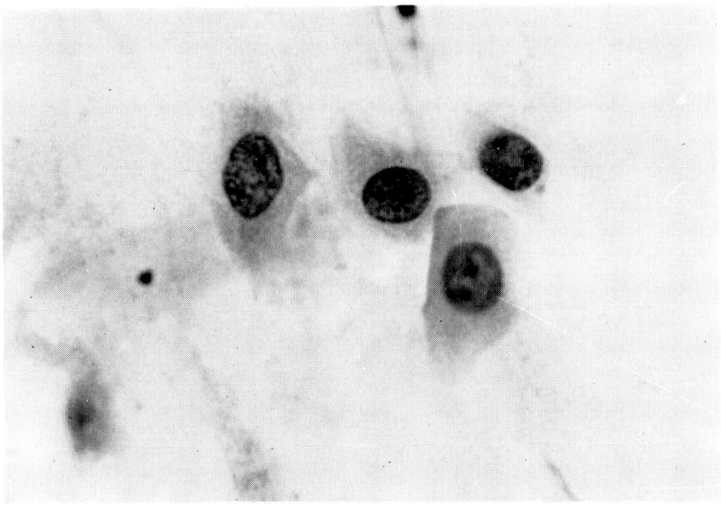

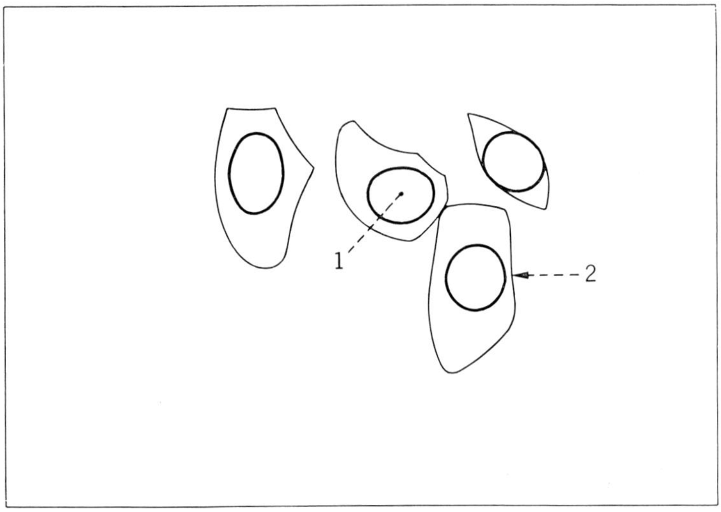

IV. Cytodiagnosis of Pulmonary Carcinoma

Classification of pulmonary carcinoma

- Primary carcinoma
  - Bronchogenic carcinoma
    - Squamous type
    - Large cell type
    - Small cell type (Oat cell carcinoma)
    - Adenomatous glandular cell type
  - Bronchial adenoma
  - Bronchiolar carcinoma or alveolar cell carcinoma
- Metastatic carcinoma

1. The commonest type is bronchogenic carcinoma, which originates from the bronchial surface epithelium.
2. Bronchial adenoma occurs mostly in the mucous glandular epithelium of the bronchial wall. It sometimes metastasizes and may be considered a malignant tumor It is, however, less malignant than bronchogenic carcinoma.
3. Bronchiolar carcinoma occurs in the pulmonary periphery. It sometimes forms an isolated tumor, but occasionally spreads throughout the whole lung. It is a type of adenocarcinoma.
4. Metastatic carcinoma is also often seen.

Characteristics of each type of pulmonary carcinoma

For malignant cells to exfoliate and appear in sputum or bronchial secretion, the bronchial epithelium must be replaced by tumor tissue or ulceration must be formed on the surface of the bronchial lumen. Bronchial adenoma, bronchiolar carcinoma and most metastatic carcinomas usually do not form ulcers in the bronchial lumen; therefore, these cells hardly appear in sputum.

1) Bronchogenic carcinoma, squamous type (Fig. 46)

Needless to say, they are similar to squamous cell carcinoma in other parts of the body (*cf*. pp. 86–87).

1. The nuclear-cytoplasmic ratio is high.
2. Nuclear chromatin increases, and therefore the nucleus is markedly hyperchromatic.
3. Coarse chromatin granules are unevenly distributed.
4. The nuclear membrane is thickened and irregular.
5. The nucleus may be multinucleated and varies in size. The outstanding feature of this type of cell is that the cytoplasm may be markedly keratinized and stained clear orange.

2) Bronchogenic carcinoma, large cell type (Fig. 46)

They have almost the same characteristics as the squamous type. The difference

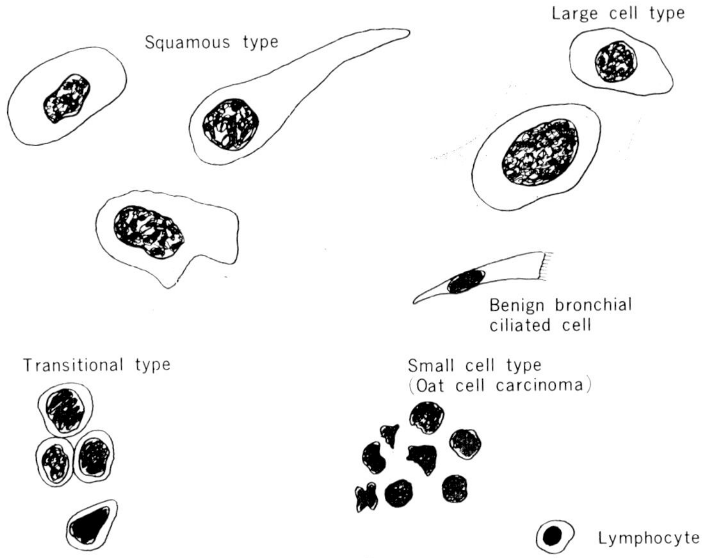

Fig. 46 Bronchogenic carcinomas.

is that the cytoplasm is not stained orange and that the cell size is larger than that of the squamous type.

3) Bronchogenic carcinoma, small cell type (Fig. 46)

Carcinoma of this type is called "oat cell carcinoma," and the cells are extremely small and are slightly larger than lymphocytes.

a. Nucleus :

1. Shape : Varies. Round, polygonal and others.
2. Size : Varies, though within limits.
3. Chromatin is extremely abundant. Concentrated and very hyperchromatic. The internal structure is poorly recognizable.
4. Sometimes necrotic.

b. Cytoplasm : Extremely scanty and hard to detect. Under low magnification, the cells look like numerous sesame seeds, overlapping one another. No characteristic cluster formation, however, can be seen.

4) Bronchogenic carcinoma, adenomatous glandular cell type and bronchial adenoma.

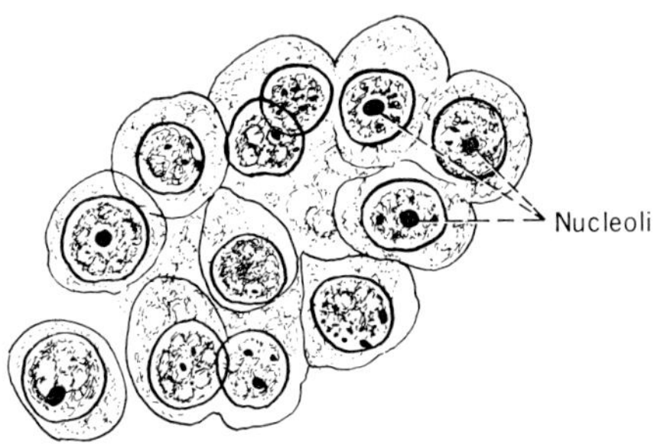

Fig. 47 Pulmonary adenocarcinoma.

The characteristics of these cells are similar to other adenocarcinoma cells. They are often columnar.

5) Bronchiolar carcinoma (Fig. 47).

Columnar cells appear in a cluster. They are comparatively uniform in shape and size, and chromatin is evenly distributed. Marked nucleoli are observed.

6) Metastatic carcinoma

Their appearance is somewhat similar to primary malignant cells. They can be diagnosed as malignant, but it is inadequate to point out the primary site of metastatic carcinoma only by smear without other diagnostic considerations.

100. Bronchogenic carcinoma, squamous type (Sputum)

a. Low magnification

A large, long, triangular malignant cell is observed. Numerous leukocytes and cellular fragments are seen in the background.

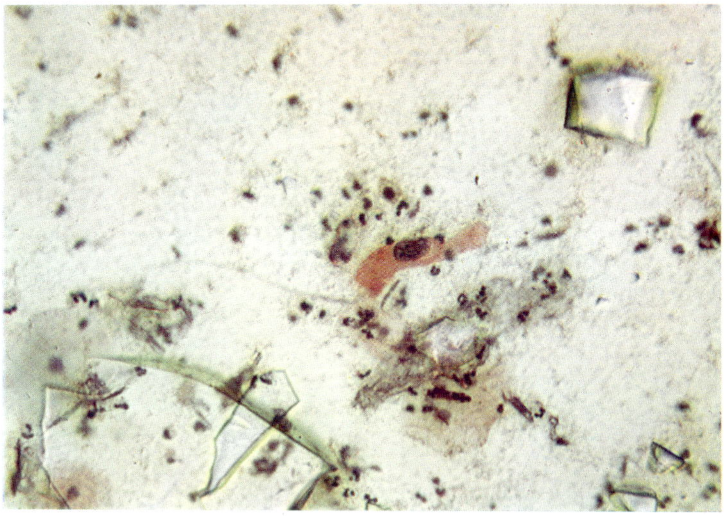

b. High magnification

1. Nucleus: Extremely large and rather irregularly shaped. Chromatin is increased and granular. The nuclear membrane is thickened Hyperchromatic.
2. Cytoplasm: Fairly abundant. Note the peculiar shape and the orange stain.
3. A cytoplasmic vacuole is observed near the nucleus.
4. This part of the cytoplasm appears thick, but the other end is thinner.
5. Leukocytes and necrotic cellular debris are observed.
6. Faint vacuoles are present.

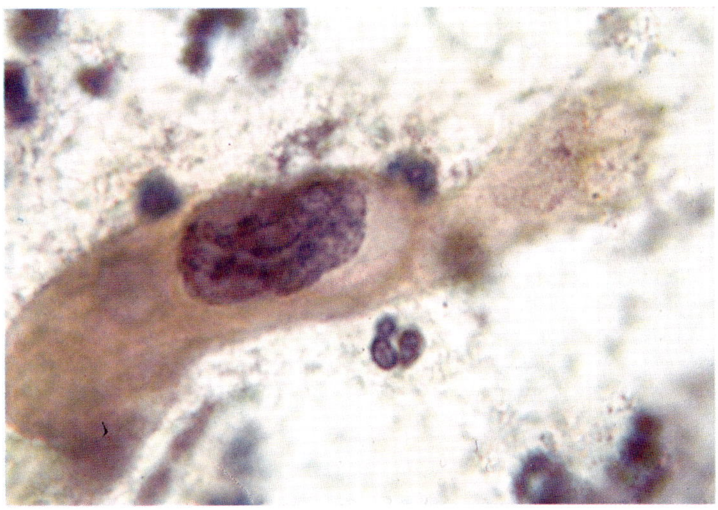

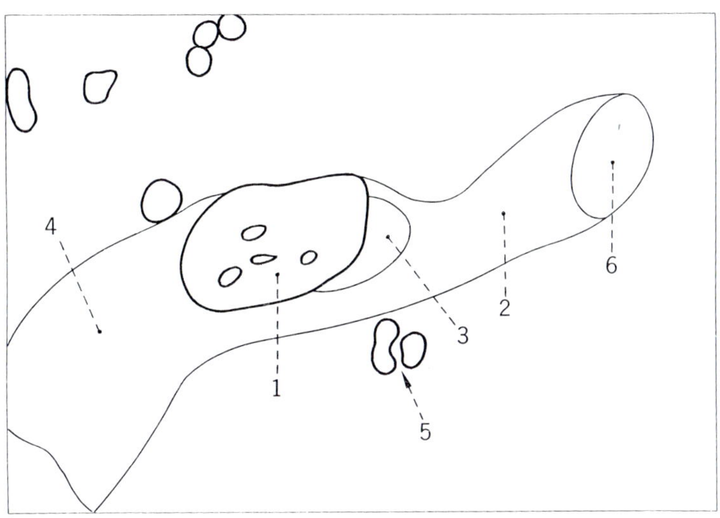

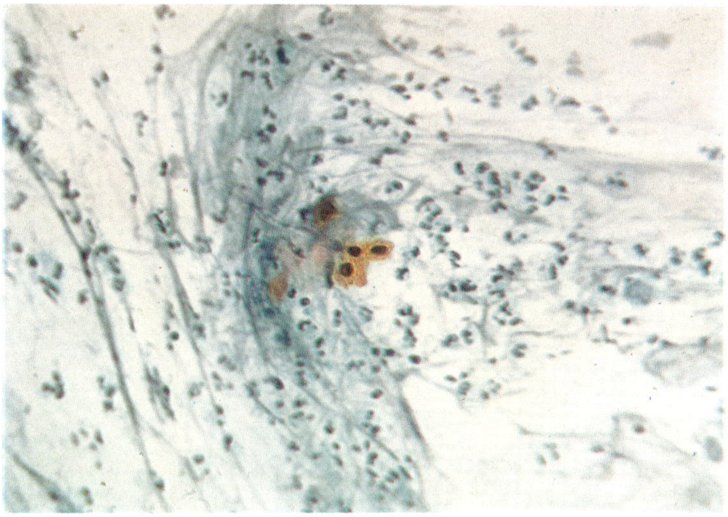

101. Bronchogenic carcinoma, squamous type (Sputum)

a. Low magnification

Large orange-colored cells are observed among numerous leukocytes, cellular debris and purulent material.

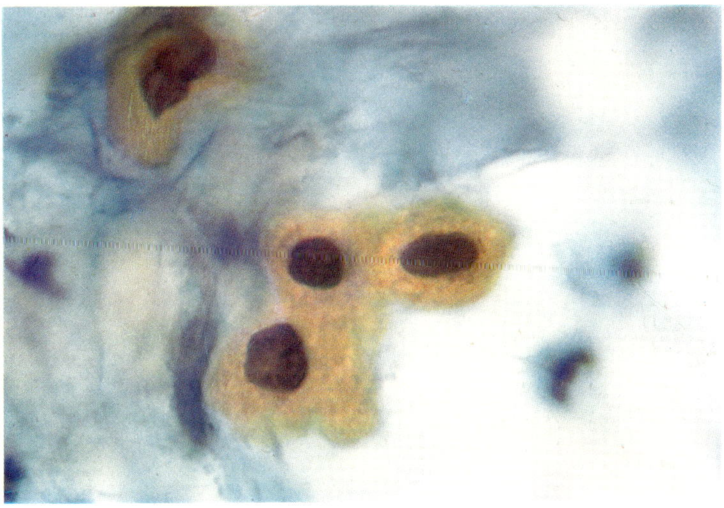

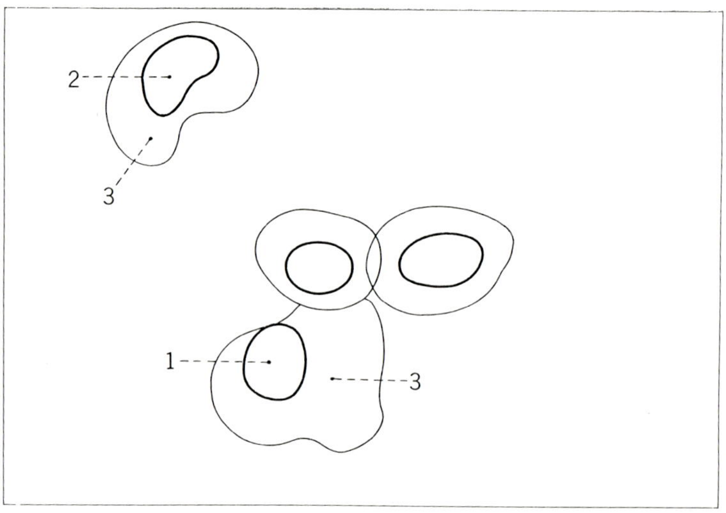

b. High magnification

1. Nucleus: Moderately large and hyperchromatic. Chromatin is comparatively evenly distributed.
2. A clump of chromatin is faintly seen.
3. Cytoplasm: Round, triangular or polygonal. Stained a peculiar yellowish orange. Rather thick and luminescent in appearance.

102. Bronchogenic carcinoma, squamous type (Bronchial washing)

a. Low magnification

An extremely large orange cell is observed on a comparatively clean background.

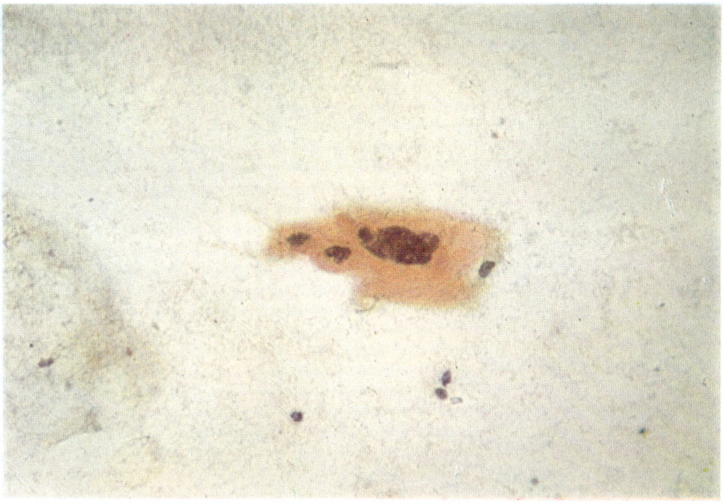

b. High magnification

1. A giant nucleus. Irregularly shaped. Chromatin is markedly increased and coarsely granular. Remarkably hyperchromatic.
2. Other large nuclei overlap the central giant nucleus. Their structures are similar.
3. A little smaller, but also has hyperchromatic nuclei. Chromatin granules are even more clearly seen.
4. A clear orange cytoplasm. Fairly abundant and thick. The border is comparatively distinct.

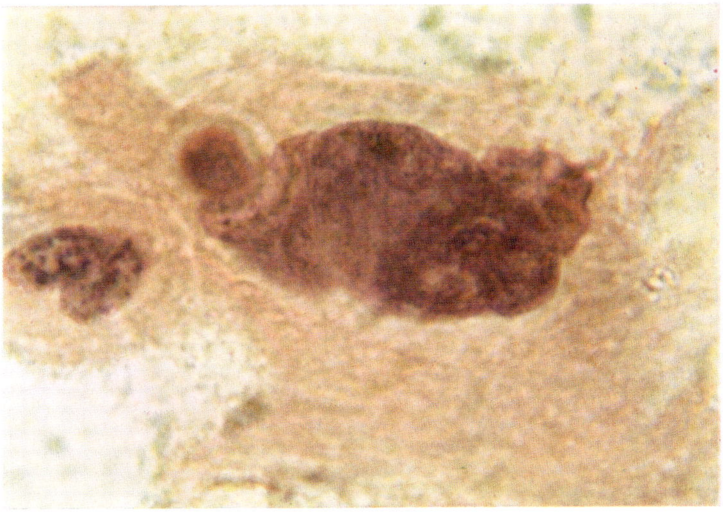

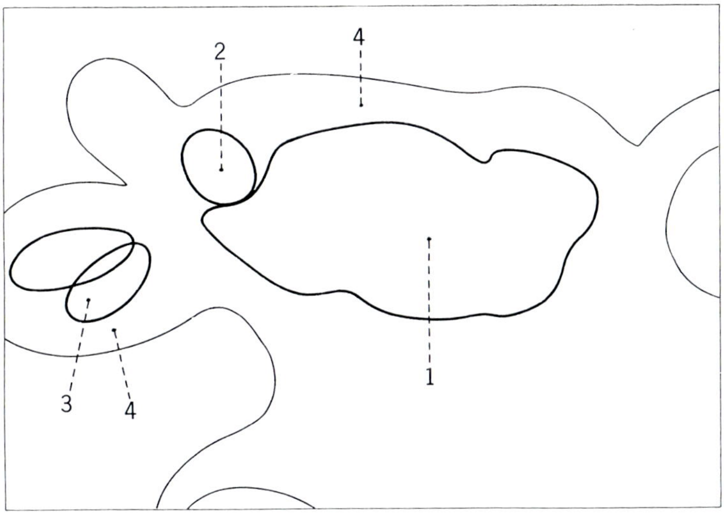

103. Bronchogenic carcinoma, large cell type (Bronchial washing)

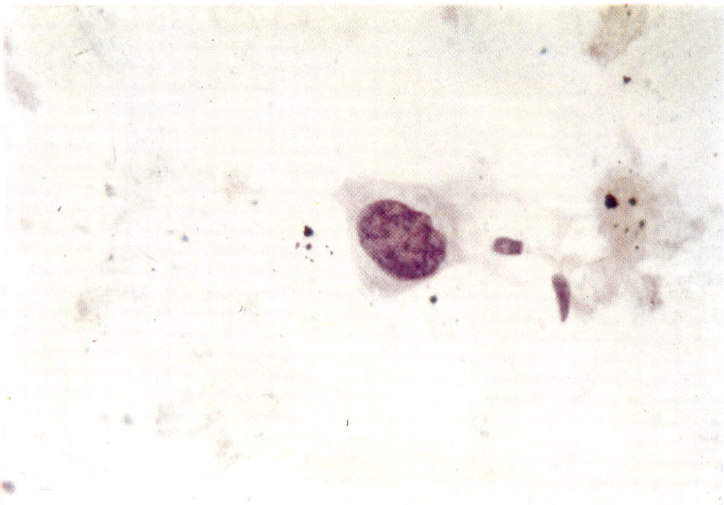

a. Low magnification

Extremely large cell is seen in the center. Note a clean background.

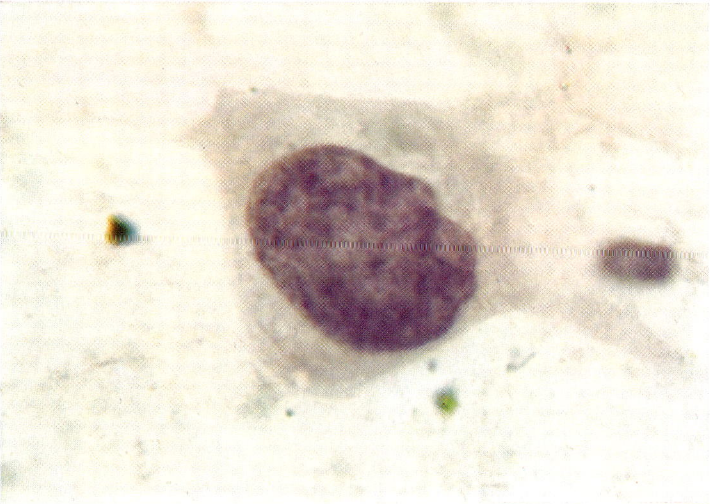

b. High magnification

1. There is a large, elliptically shaped nucleus that shows hyperchromatism and moderate irregularity of chromatin distribution.
2. Non-keratinized cytoplasm stained reddish blue and rather pale.

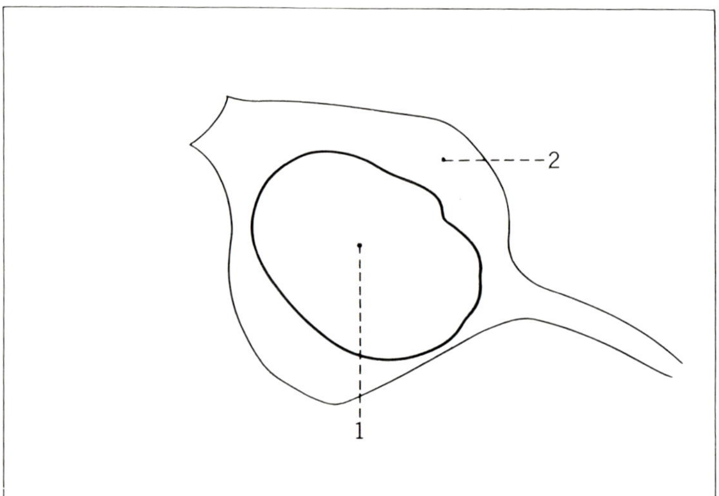

104. Bronchogenic carcinoma, large cell type (Bronchial washing)

a. Low magnification

There is a large malignant cell associated with scattreed benign bronchial cells.

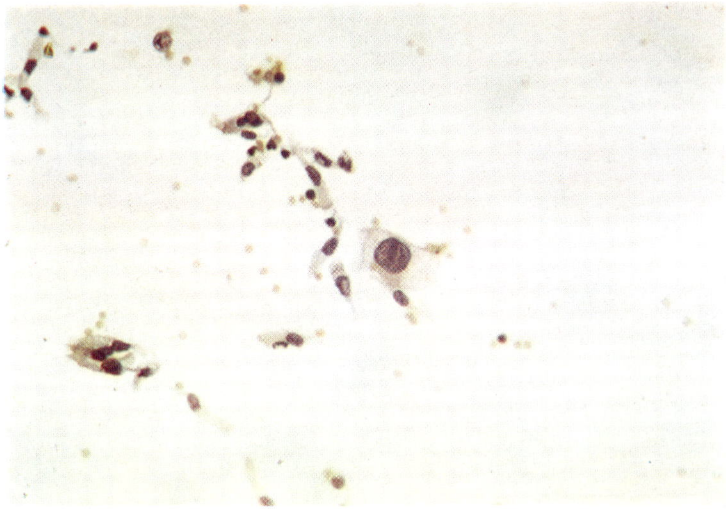

b. High magnification

1. The large round nucleus reveals moderate hyperchromatism and dense chromatinic network.
2. Surrounding the nucleus, abundant lacy cytoplasm is observed. Lightly basophilic.
3. Three benign bronchial cells are seen near the malignant cell.
4. Ciliated end of a bronchial cell.

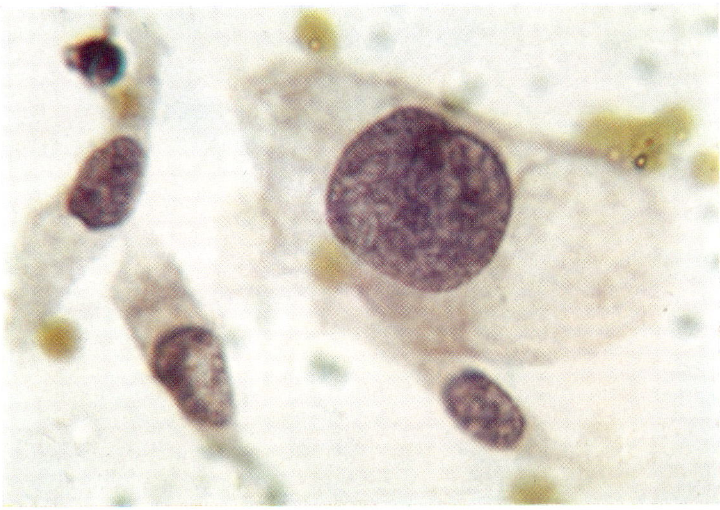

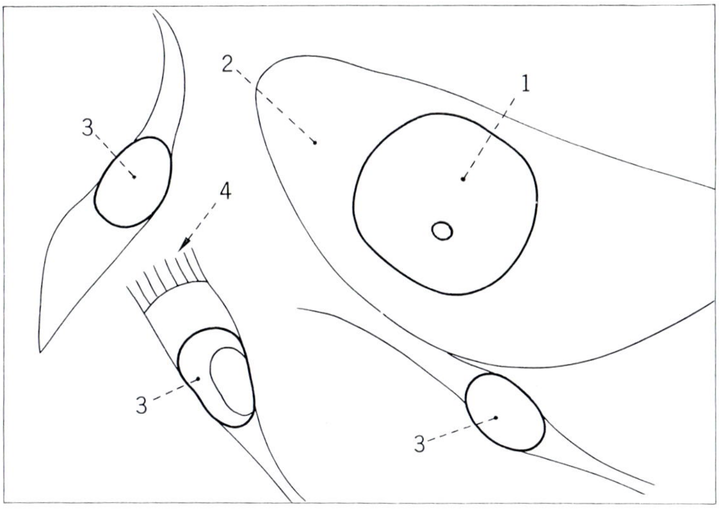

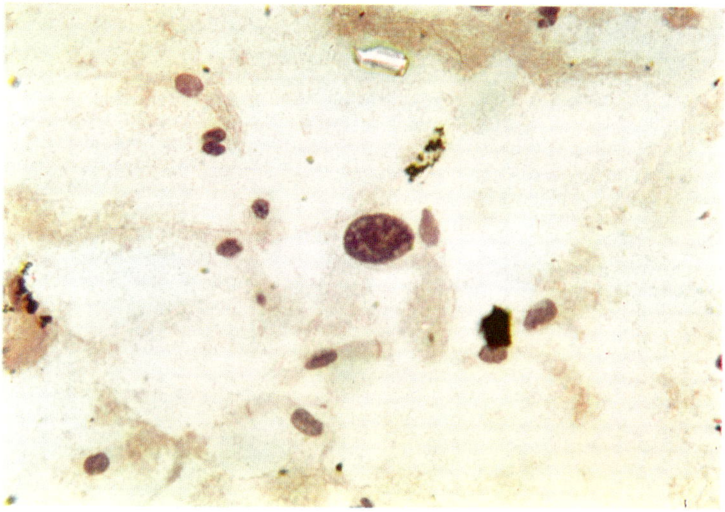

105. Bronchogenic carcinoma, large cell type (Bronchial washing)

a. Low magnification

A basophilic tadpole cell with an extremely large elliptical nucleus is observed among numerous respiratory columnar epithelial cells.

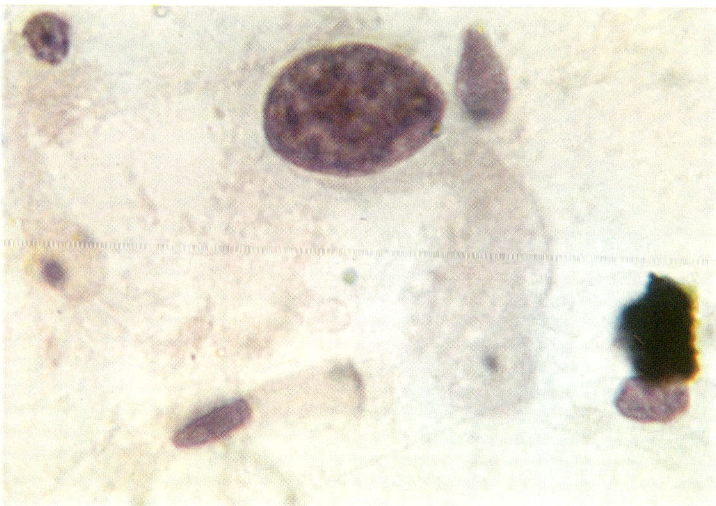

b. High magnification

1. The giant elliptical nucleus. Extremely hyperchromatic. The reticulated internal structure is not very clear.
2. Cytoplasm: Tadpole-shaped and rather vesicular. Stained basophilically light blue. It is in striking contrast to the clear orange cytoplasm of the squamous-type cell.
3. A typical benign, respiratory, ciliated, columnar epithelial cell. The cytoplasm is regularly shaped and the border is distinct.
4. Though the nucleus is hyperchromatic, a cell with such distinct cilia is not malignant.

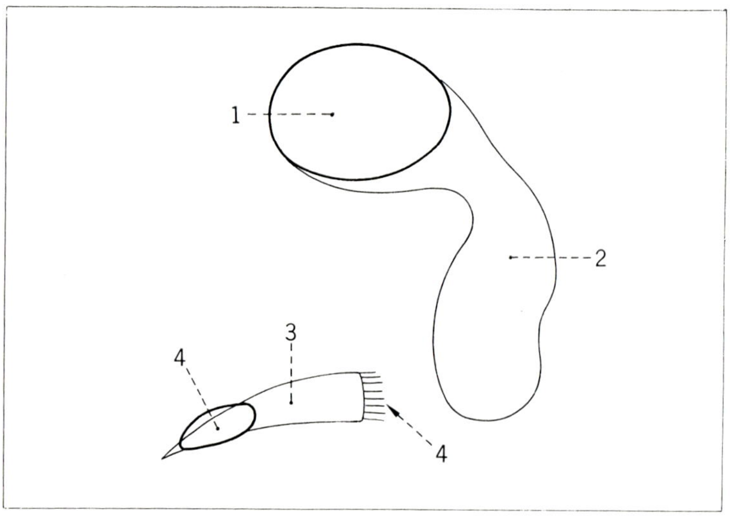

106. Bronchogenic carcinoma, transition to the small cell type (Sputum)

a. Low magnification

A group of nuclei of various sizes and shapes is observed against a background of leukocytes and cell debris.

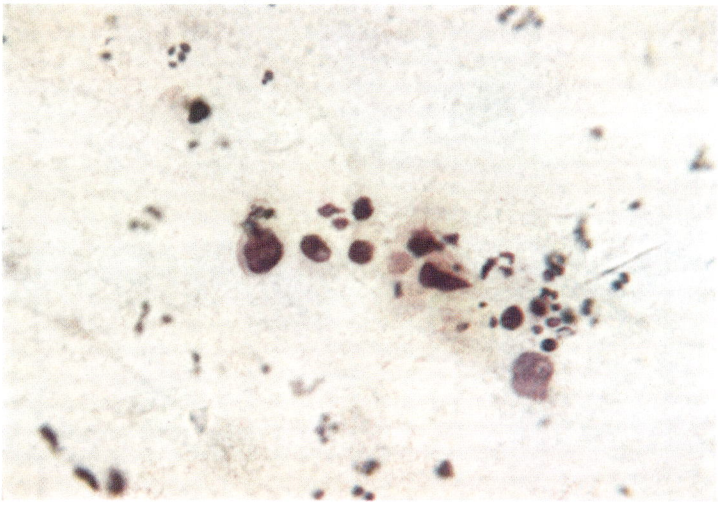

b. High magnification

1. Nuclei: Irregular in size or shape. Chromatin is extremely abundant and the internal structure is hardly seen. The chromatin looks like drops of India ink, and this is the main characteristic of these cells.
2. Cytoplasm: Very scanty or none. Therefore the nuclear-cytoplasmic ratio is very high. Thin and lacy. The border is indistinct.
3. Irregularly triangular and hyperchromatic nucleus.

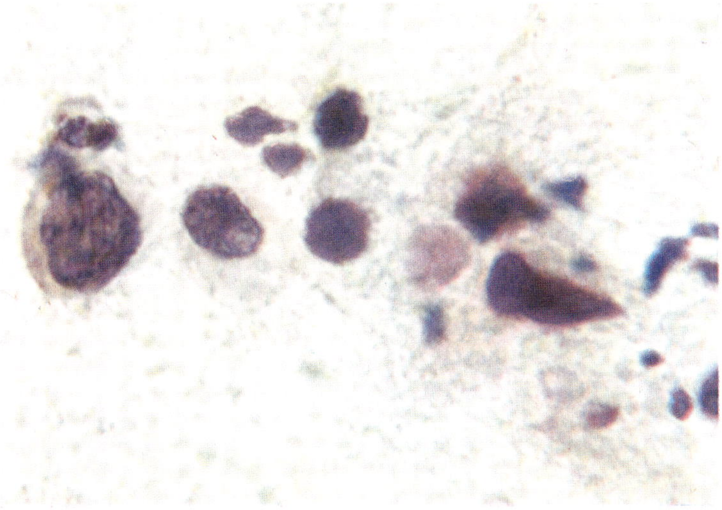

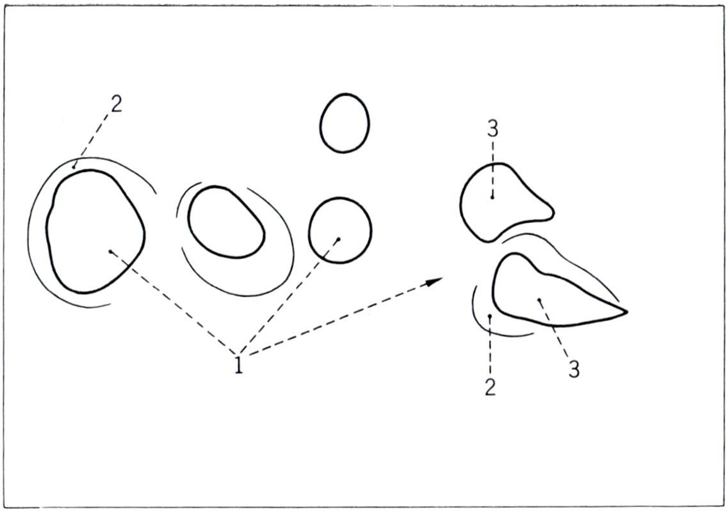

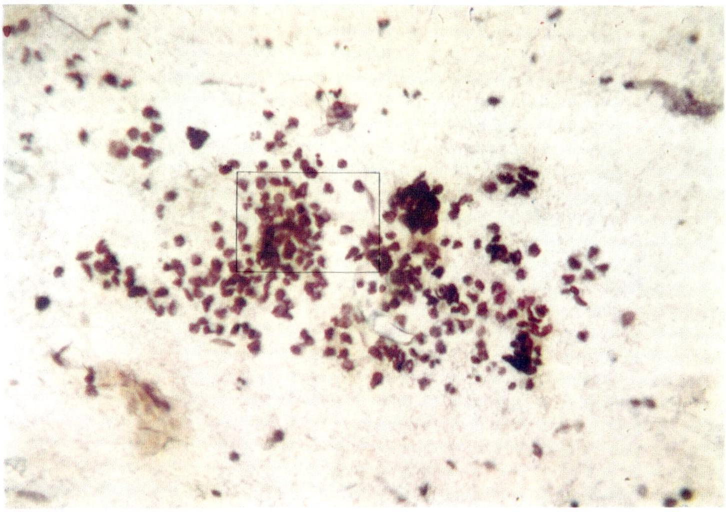

107. Bronchogenic carcinoma, small cell type — Oat cell carcinoma (Sputum)

a. Low magnification

This smear shows a peculiar pattern. Numerous small nuclei are scattered like sesame seeds or clumped in cluster against a necrotic background.

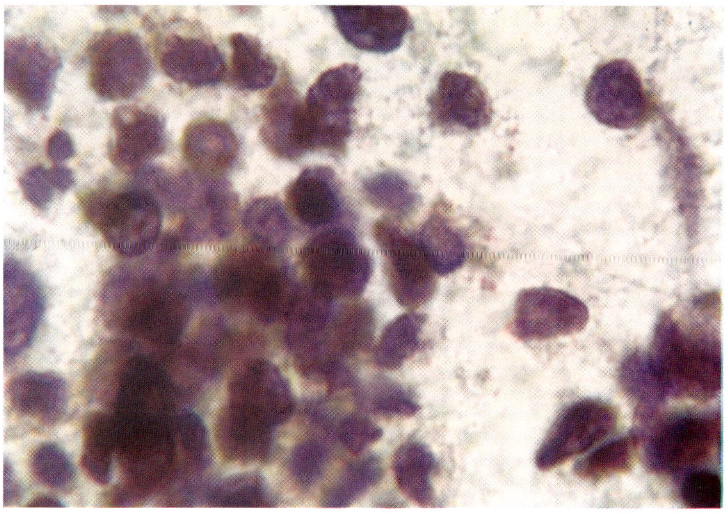

b. High magnification

There is no cytoplasm and the stripped nuclei overlap one another. Under low magnification, they look comparatively roundish, but high magnification reveals that they vary in size and shape. The microscope should be moved slightly up and down to observe all the cells, because they are overlapping.

1. Nuclei: Rather necrotic, and entirely structureless. Pyknotic and hyperchromatic due to marked increase in chromatin.
2. The borders are partly distinct but most are indistinct.

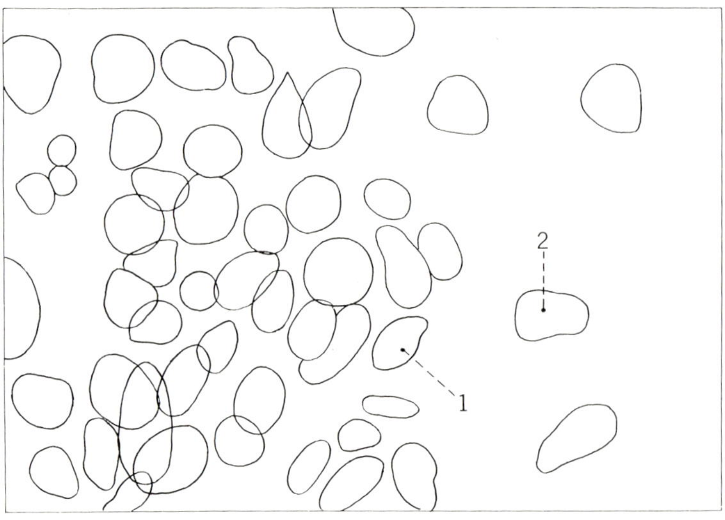

108. Terminal bronchiolar carcinoma (Sputum)

a. Low magnification

A cluster-shaped group of more than 20 large cells is observed in the middle.

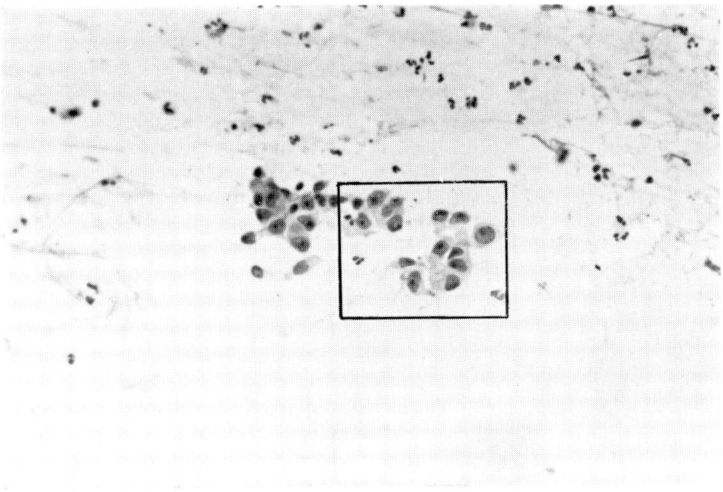

b. High magnification

1. The nuclei are round, elliptical or triangular. They are large but fairly uniform in size. The borders are comparatively smooth.
2. Coarse chromatin granules are clearly seen. Not very hyperchromatic.
3. Marked nucleoli.
4. Cytoplasm: Lacy and fairly abundant. The border is slightly indistinct. Sometimes rather column-shaped. Basophilic.
5. A tendency of glandular alveolar crowding is noticed.

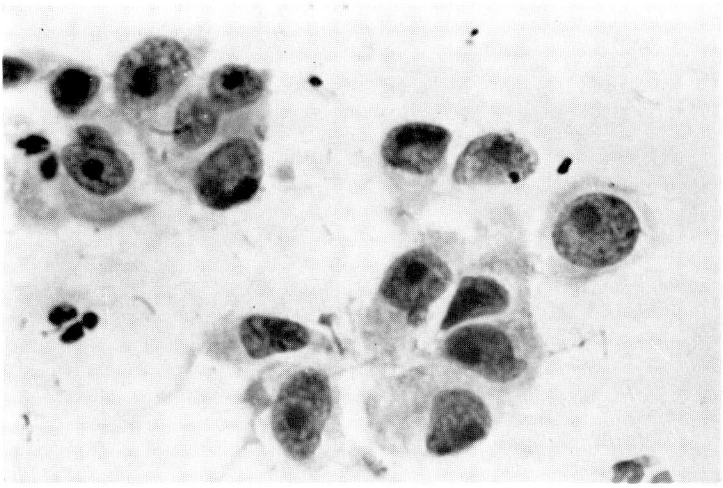

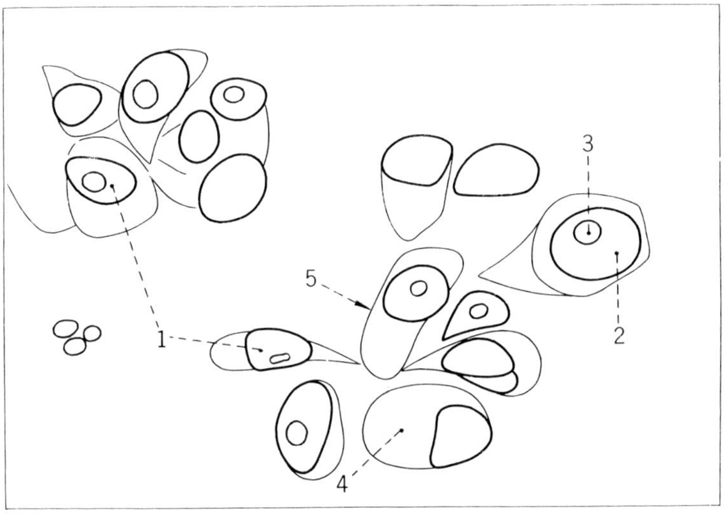

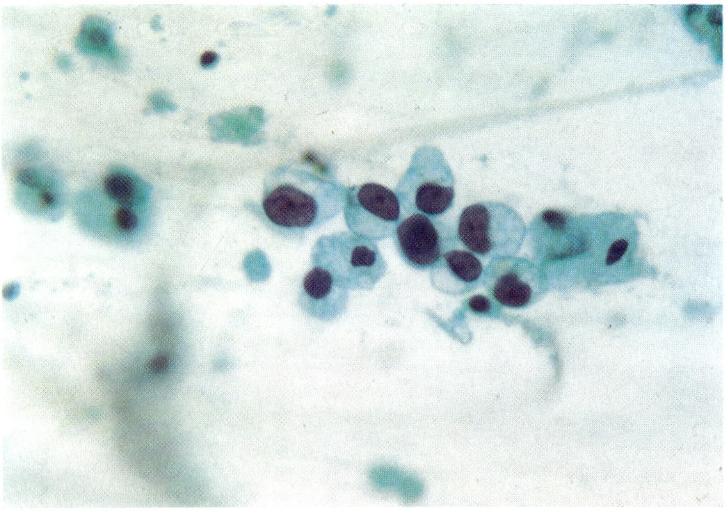

109. Terminal bronchiolar carcinoma (Sputum)

a. Low magnification

A cluster of cells with large nuclei and clear cytoplasm is present.

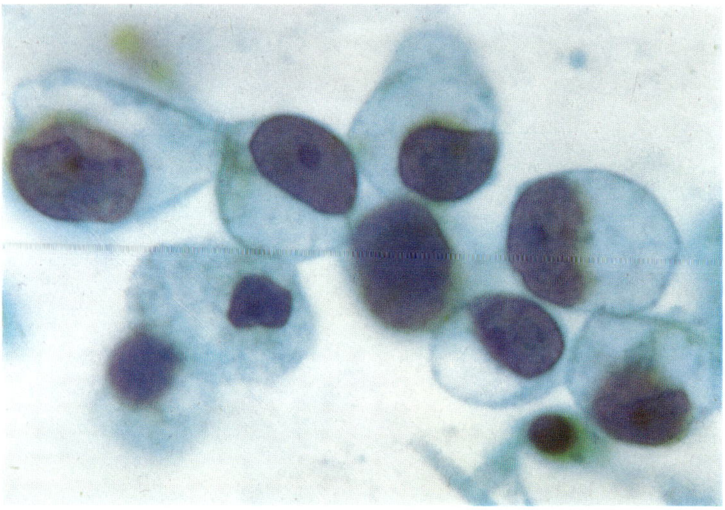

b. High magnification

1. Large elliptical nuclei reveal marked hyperchromatism and prominent nucleoli. The chromatin pattern looks even and finely granulated.
2. Cytoplasm Mostly round in shape, sometimes triangular, and lacy or finely foamy in appearance. Note the dense cytoplasmic border.

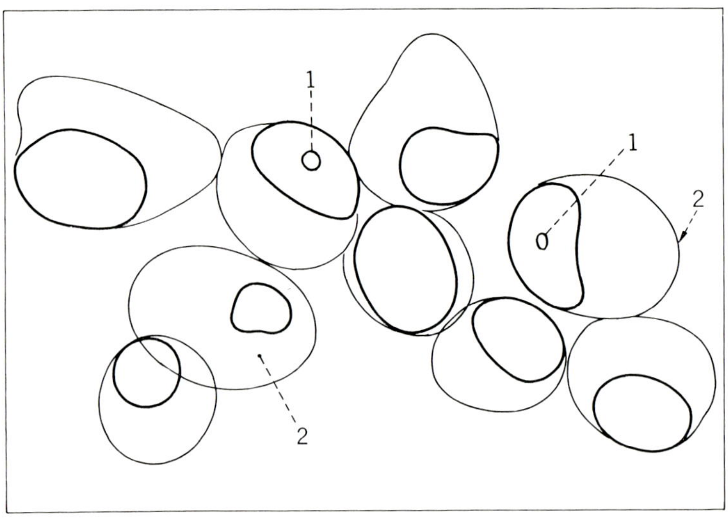

110. Metastatic malignant melanoma (Sputum) (Cell block specimen)

a. Low magnification

A group of round and brownish cells is present in the cell block specimen.

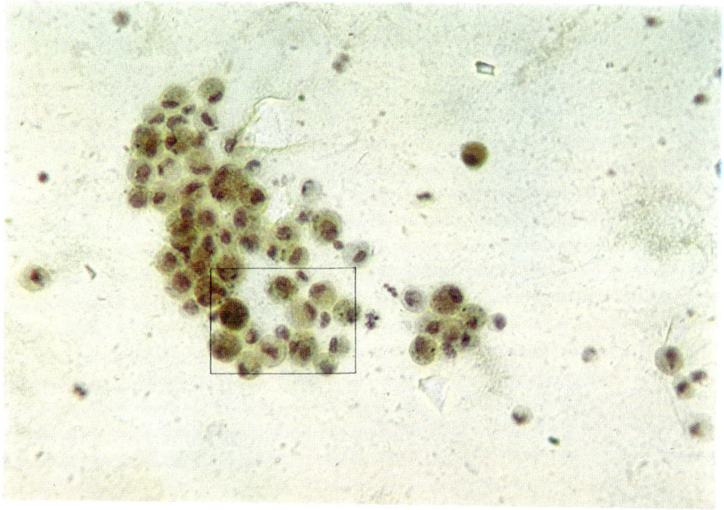

b. High magnification

1. Most of the round cells contain, more or less, brownish pigment in the cytoplasm.
2. Nucleus of the cell reveals a partial hyperchromatism.
3. Prominent nucleolus.
4. Irregularly shaped nuclei.
5. It is not certain whether all of these cells in the specimen originate from melanoma, or some are alveolar cells which only phagocytize melanin pigment.

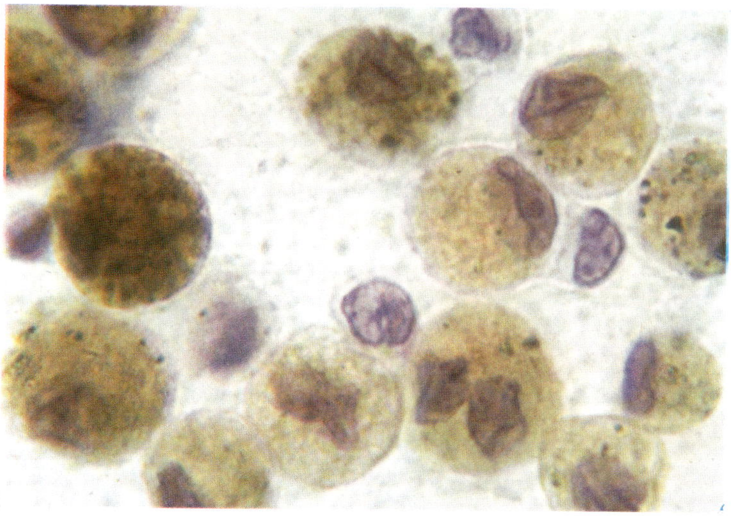

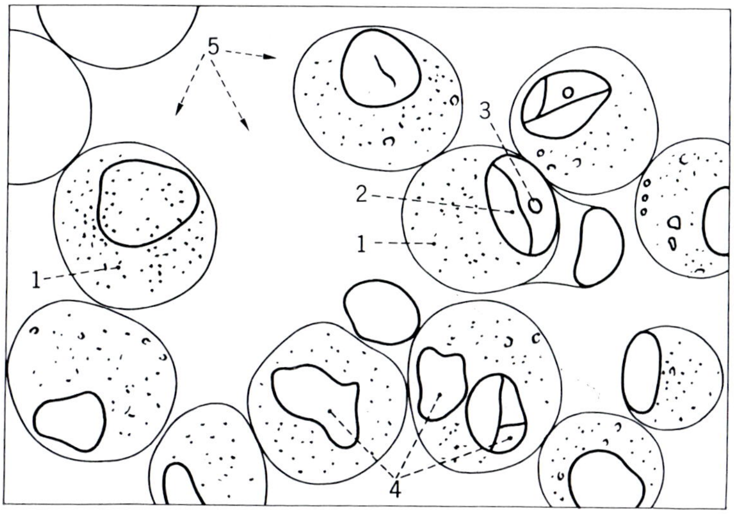

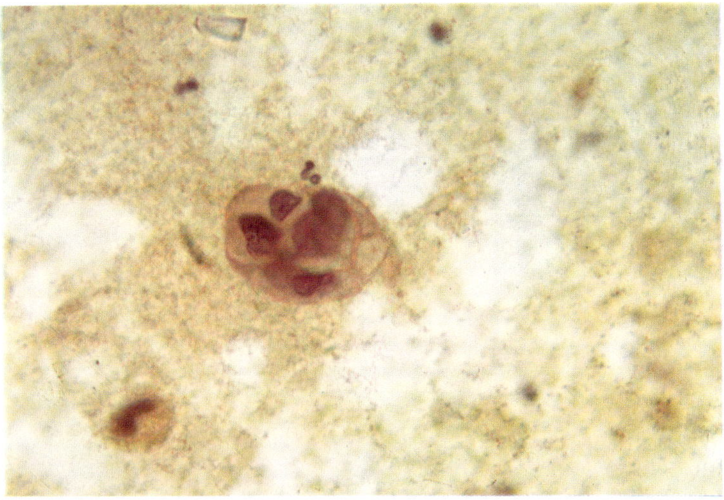

111. Metastatic carcinoma from carcinoma of the thyroid gland (Sputum)

a. Low magnification

A cell clustering is observed in the middle. This looks like a glandular structure. The nuclei are hyperchromatic.

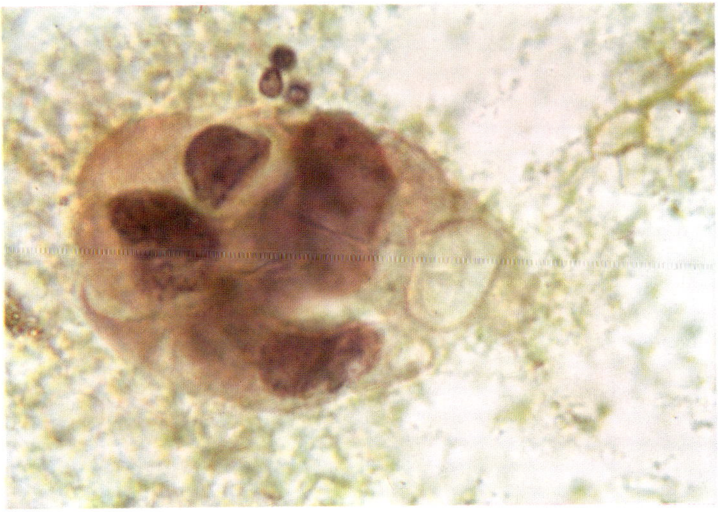

b. High magnification

1. A large, irregularly shaped hyperchromatic nucleus.
2. Another irregularly shaped nucleus. Chromatin is increased around the border.
3. Other nuclei are also irregular in shape, enlarged and hyperchromatic. Chromatin granules are visible in some nuclei.
4. Cytoplasm: Fairly distinct, but individual borderlines are only vaguely recognized.
5. Vacuole.

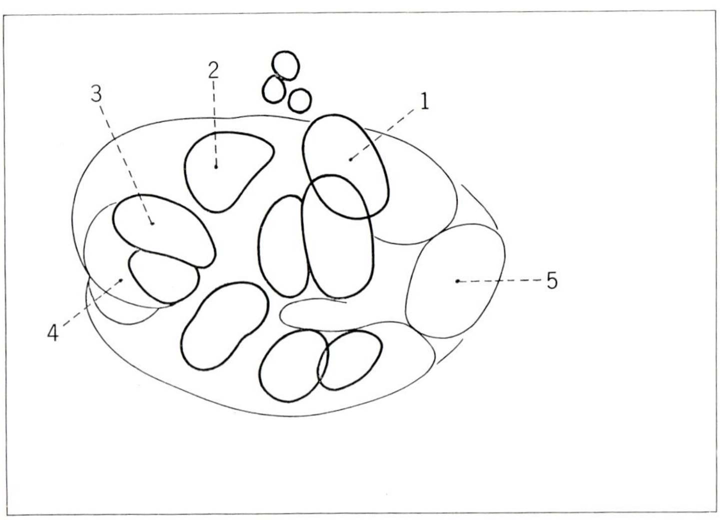

112. Metastatic malignant tumor from uterine choriocarcinoma (Sputum) (*cf.* pp. 112 and 113)

a. Low magnification

Two large cells with duplicate nuclei and a stripped nucleus are observed in the middle.

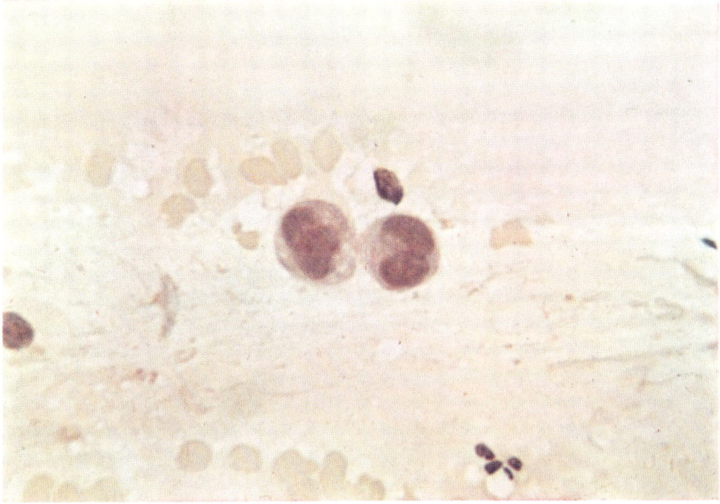

b. High magnification

1. Nuclei: Elliptical or kidney-shaped. Large and hyperchromatic. Chromatin is coarsely granular. There are no marked nucleoli and thickening of the nuclear membrane, which are noticed in the primary carcinoma.
2. Cytoplasm: Vesiculated and scanty. The border is comparatively distinct.
3. Erythrocytes.

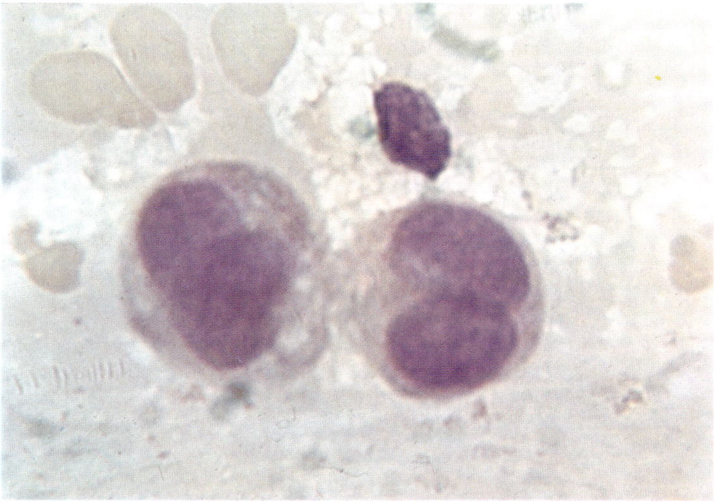

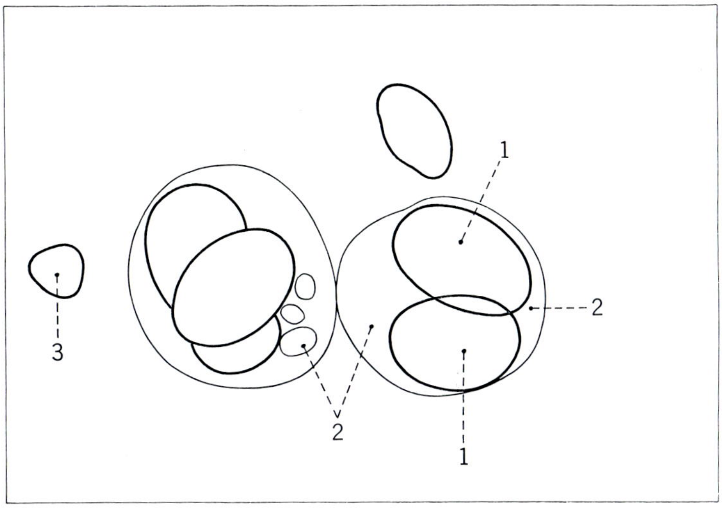

CYTOLOGIC EXAMINATION OF THE STOMACH

I. Anatomy of the Stomach

The anatomical location of the esophagus and the stomach is shown in Fig. 48.

Esophagus : The internal surface is covered with stratified squamous epithelium. Keratohyaline granules are observed, but cornification is not usually seen.

Stomach : Divided into the cardia, body and pylorus.

Superficial mucosal epithelium

The internal surface of the stomach is covered with single-layered, surface mucosal columnar epithelium and numerous small pits are observed in the mucosa. These are gastric pits into which gastric glands open. The mucosal epithelial cells of the internal surface of the stomach are the only gastric cells which exfoliate in the normal state.

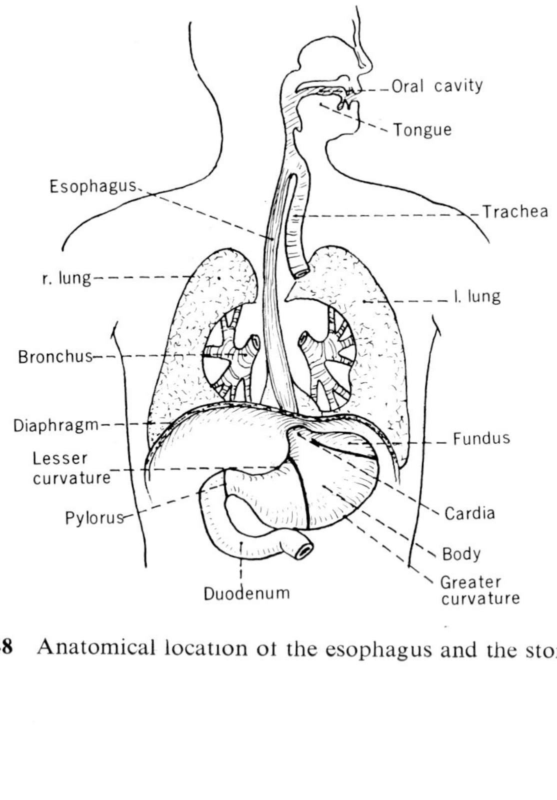

Fig. 48 Anatomical location of the esophagus and the stomach.

Gastric glands (Fig. 49)

Gastric glands are classified into the gastric glands of the cardia and corpus and those of the pylorus, and the intermediate region is called the transitional part.

Gastric glandular cells : A gastric gland proper is a single-tubular gland, and three kinds of cells are observed in the glandular tube : chief cells, mucous neck cells and parietal cells. These cells are rarely observed in smears.

Chief cells : Columnar. Light, basophilic stain. Pepsin-producing.

Mucous neck cells : The nucleus is flattened and eccentric. Mucus-producing.

Parietal cells : Irregularly polygonal. Eosinophilically stained. Acid-producing.

Pyloric glandular cells : Similar to mucous neck cells. Lightly stained. Mucus-producing. Chief and parietal cells are not observed at the pylorus. Pyloric glandular cells do not appear on the smear as they do not exfoliate.

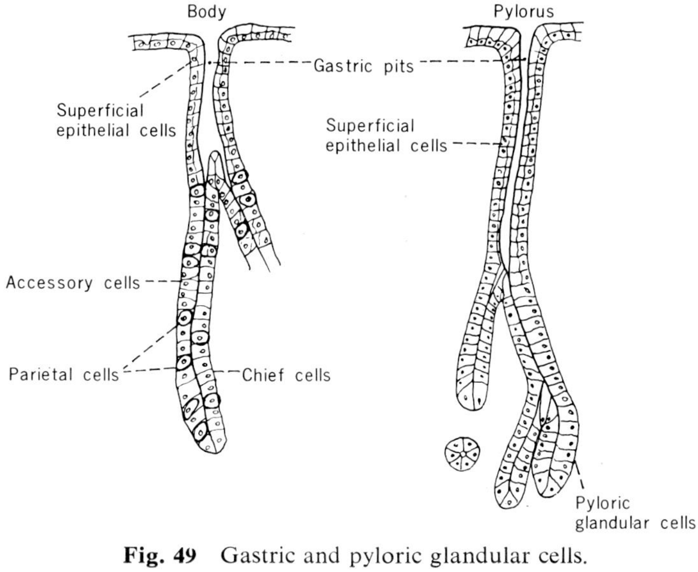

Fig. 49 Gastric and pyloric glandular cells.

II. Benign Cells in Gastric Smears

Superficial epithelial cells (Fig. 50)

a. Nucleus: Chromatin is minutely granulated and evenly distributed. Round or elliptical and uniform in size. Eccentric. A nucleolus is not clearly seen.

b. Cytoplasm: Columnar and vesiculated. Faintly basophilic or eosinophilic stain. Sometimes vacuolated.

These cells often exfoliate in a plate-form. In such cases, each cell is round or polygonal and the borderlines of individual cells are distinct. Thus they look like a honeycomb.

The cytoplasm is often degenerated and the nucleus is sometimes stripped. When the nucleus degenerates, the chromatin pattern is lost and the nucleus looks hazy.

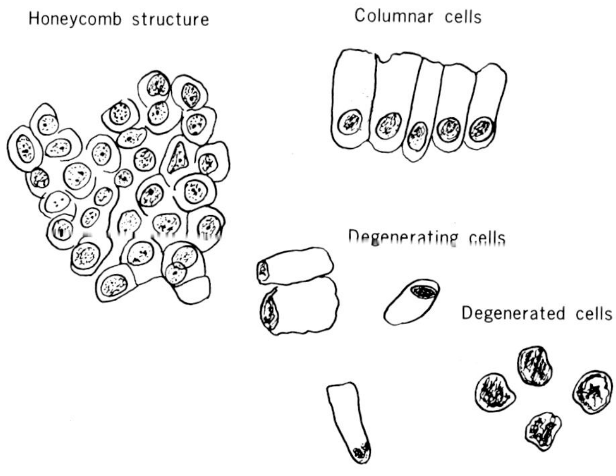

Fig. 50 Superficial epithelial cells.

113. Superficial epithelial cells

a. Low magnification

Plate-like clusters of cells arranged in pavement form are observed.

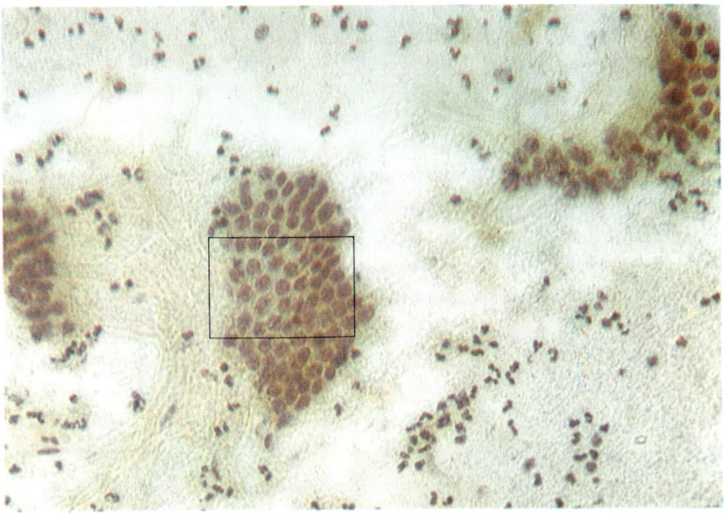

b. High magnification

1. Nuclei: Round or elliptical. Uniform and homogeneous. Small chromatin granules are observed.
2. Cytoplasm: Abundant and lacy. The borderlines of individual cells are fairly well recognized in this case.
3. A honeycomb structure is slightly visible.

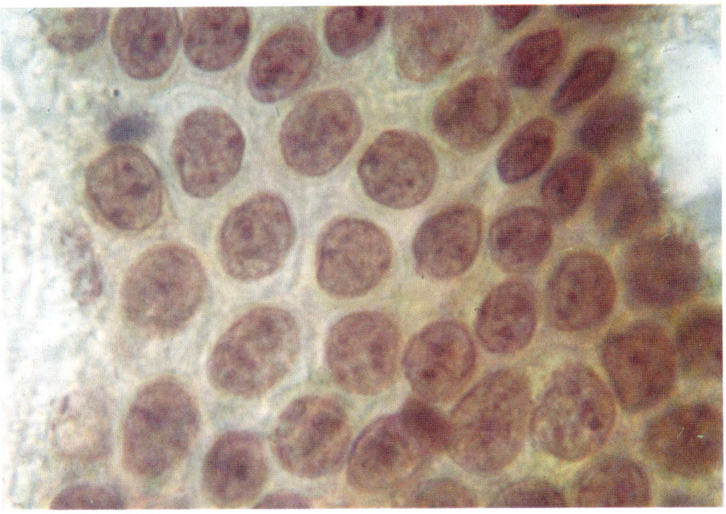

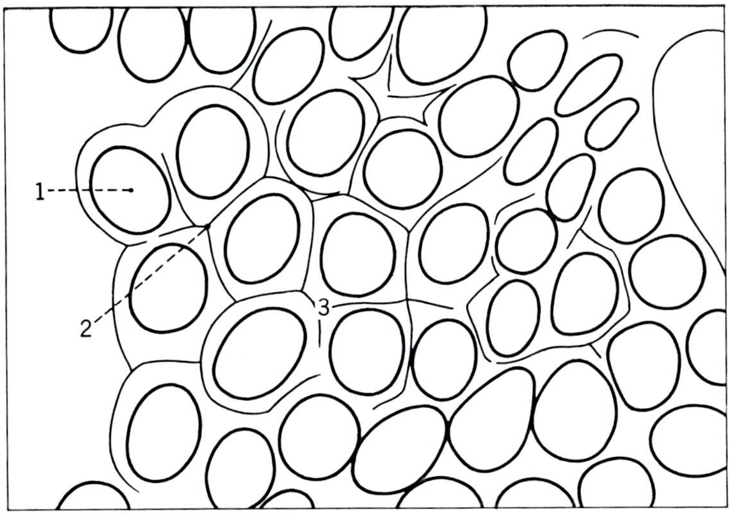

Other cells in gastric smears

a. Squamous cells: From the oral cavity and the esophagus. Mainly superficial cells (*cf.* p. 139)

b. Cells from the respiratory tract: Nasal mucosal epithelial cells and bronchial epithelial cells. They can be distinguished from gastric mucosal epithelial cells by the longer shape and cilia which are often observed (*cf.* pp. 135–137).

c. Alveolar macrophages: Distinguished easily by the peculiar dust in the cytoplasm and the kidney-shaped eccentric nuclei (*cf.* p. 138).

These cells exfoliated and swallowed into the stomach may appear in gastric smears made by the aspiration method.

d. Blood components: Polymorphonuclear leukocytes, lymphocytes, plasma cells and erythrocytes. These are often seen because of hemorrhage during aspiration.

e. Vegetable cells (*cf.* p. 134), muscle fibers, bacteria and fungi are sometimes observed.

III. Cytodiagnosis of Gastric Carcinoma

Classification of gastric carcinoma

About sixty percent of gastric carcinomas begin in the pylorus, twenty percent from the area along the lesser curvature, and the rest from the fundus along the greater curvature. They are classified into the following three types.

Ulcerative type (60%): Mucus-producing adenocarcinoma which forms a glandular alveolar structure. Signet-ring cells are observed.

Polypoid type (25%): Adenocarcinoma which mainly forms a papillary alveolar structure. It secretes little mucus.

The above two types of adenocarcinoma are subdivided into several stages according to the degree of differentiation.

Diffuse type (10–15%): Also called "scirrhous carcinoma." Composed mostly of connective tissue——stroma. Sometimes malignant cells are hard to find.

Cellular characteristics of gastric carcinoma (Fig. 51)

1) Different from the benign type, gastric malignant cells in groups often overlap one another.

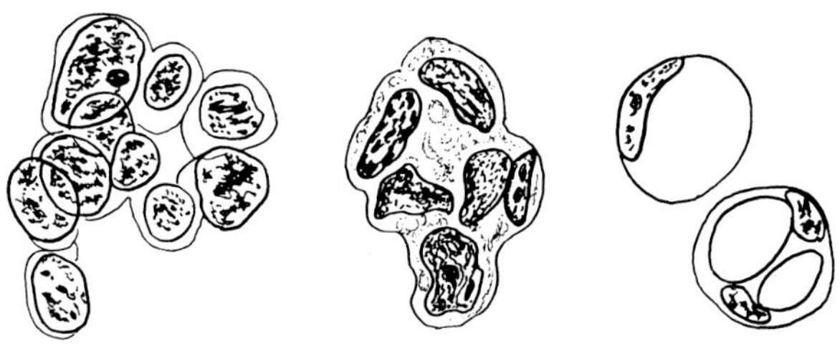

Fig. 51 Gastric carcinoma cells.

2) Nucleus
   1. Chromatin is unevenly distributed.
   2. Hyperchromatic due to an increase in chromatin.
   3. The nuclear border is thick and irregular.
   4. The nuclear-cytoplasmic ratio is high.
   5. No uniformity in size.
   6. Nucleoli are enlarged and fairly prominent.
   7. In degeneration, the nucleus is partly hyperchromatic, but the other part is homogeneous and hypochromatic, and irregularly shaped.
3) Cytoplasm
   1. Basophilic.
   2. Vacuolated.

3. The border is indistinct.
4. A signet-ring cell formation is observed. The nucleus is pushed to the cytoplasmic border by a large cytoplasmic vacuole, and the nucleus becomes crescent-shaped.

4) Changes in the background

The background looks very dirty, and contains bacteria, inflammatory cells (mainly polymorphonuclear leukocytes), an increased number of histiocytes, blood and fragments of necrotic cells.

114. Gastric adenocarcinoma — well preserved

a. Low magnification

A group of abnormal cells is observed, mixed with mucus.

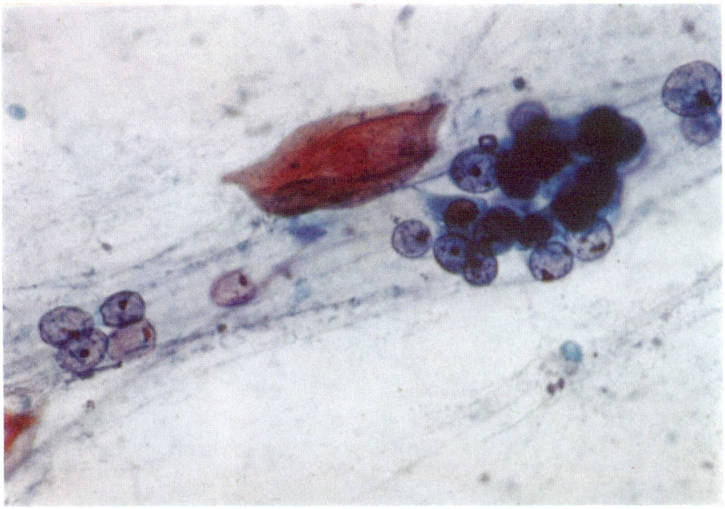

b. High magnification

1. The nuclear borders are very clear. Distinctly ring-shaped. The inside is rather hypochromatic.
2. The nuclei are not uniform in size.
3. Mostly round in shape.
4. Prominent nucleoli.
5. The cytoplasm is scanty and the border is indistinct.

(Pictures by courtesy of Dr. Y. Sakai)

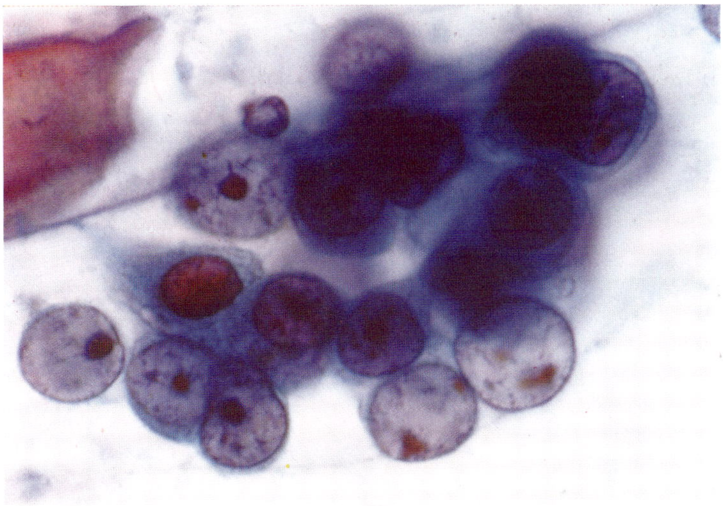

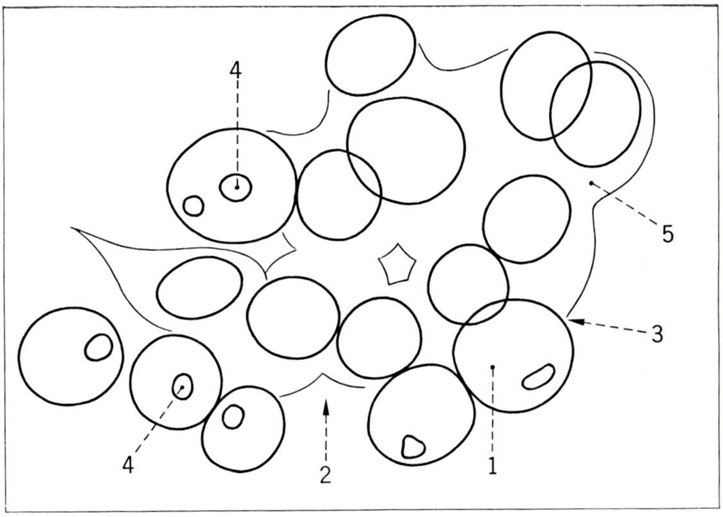

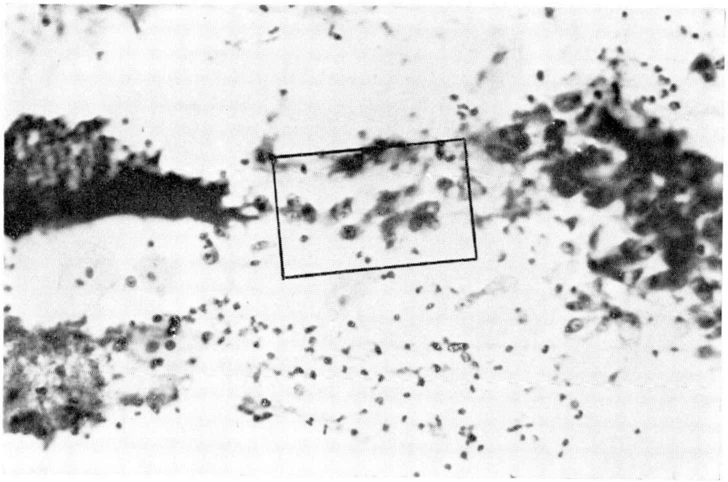

115. Gastric adenocarcinoma — well preserved

a. Low magnification

The cells are well preserved and have a fairly abundant and distinct cytoplasm.

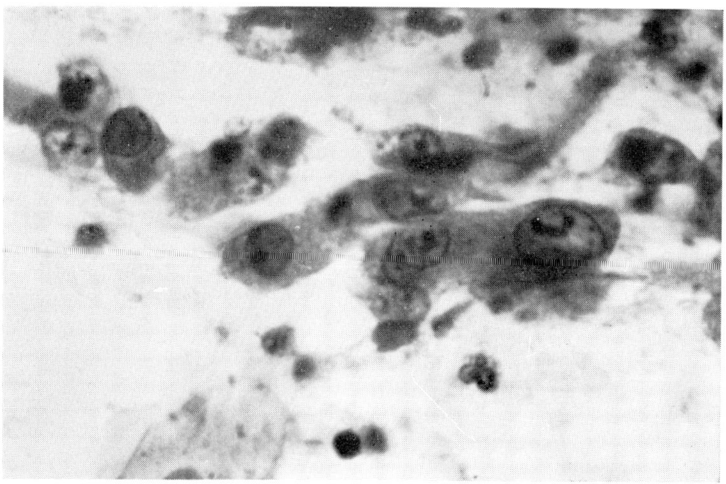

b. High magnification

1. The cytoplasm is clearly column-shaped and the border is comparatively definite.
2. The nuclei vary in size. The borders are markedly sharp.
3. Marked nucleoli.
4. A normal squamous cell.
5. Polymorphonuclear leukocytes.

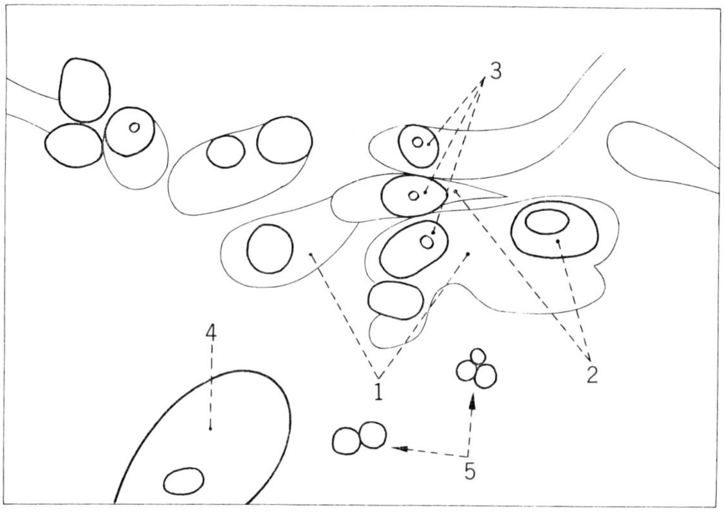

116. Gastric adenocarcinoma — mucus-producing

a. Low magnification

A tight clump of vacuolated cells is observed.

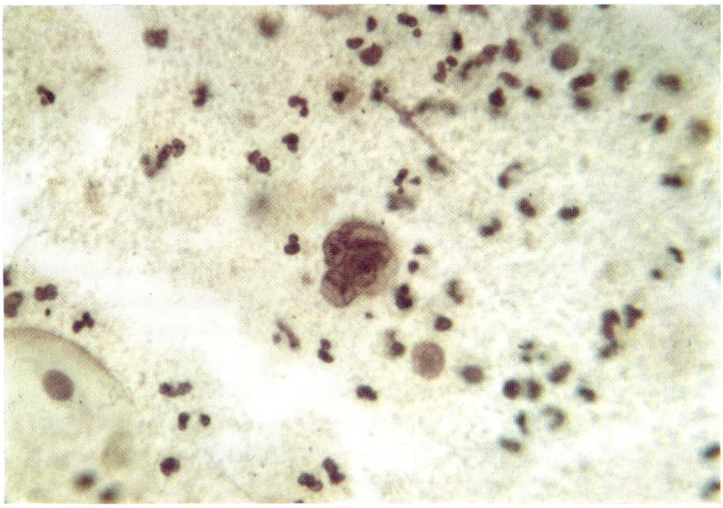

b. High magnification

1. The nucleus is considerably hyperchromatic, particularly along the nuclear rim.
2. Mild vacuolation.
3. Inside of the nucleus shows rather hypochromatism.
4. Cytoplasmic border is moderately distinguishable, though no individual cellular boundary can be seen.
5. Prominent nucleoli.
6. Leukocytes.

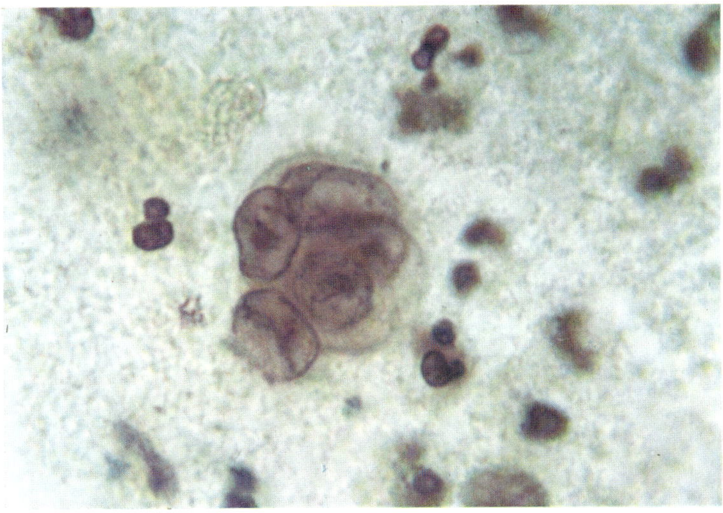

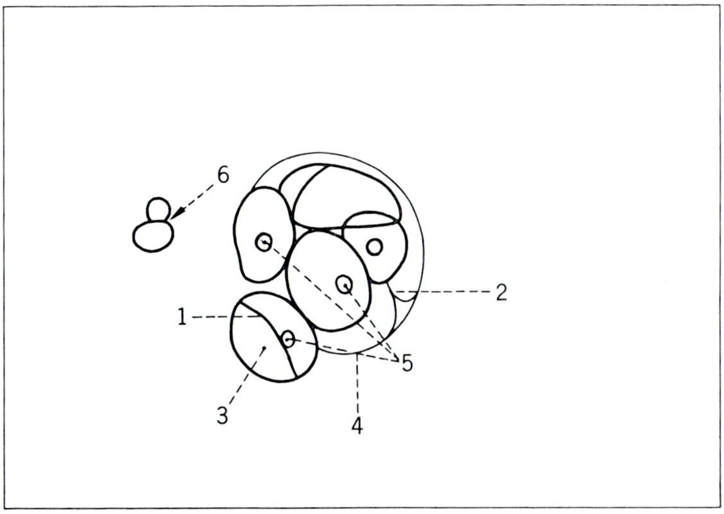

117. Gastric adenocarcinoma — Signet-ring cell

a. Low magnification

At least three distinct signet-ring cells are observed among isolated abnormal cells. The background is clean.

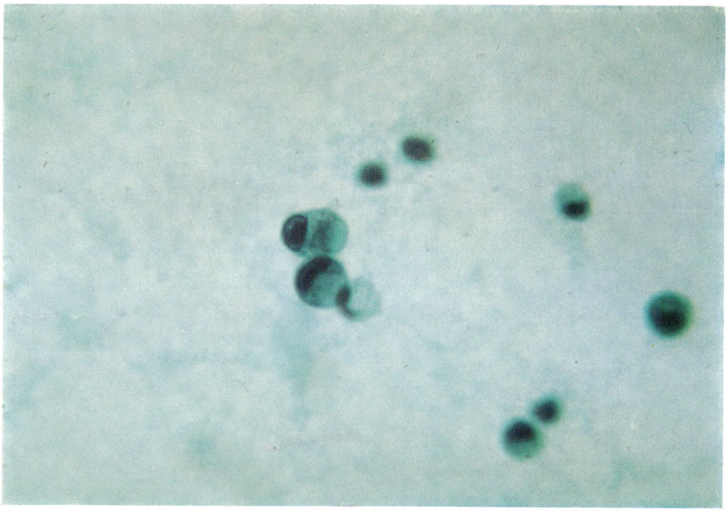

b. High magnification

1. They look vaguely vacuolated.
2. Nucleus: Very hyperchromatic. The border is thickened. Crescent-shaped. Pushed to the cytoplasmic border. No definite chromatinic pattern can be seen.
3. Cytoplasm: The border is comparatively prominent.

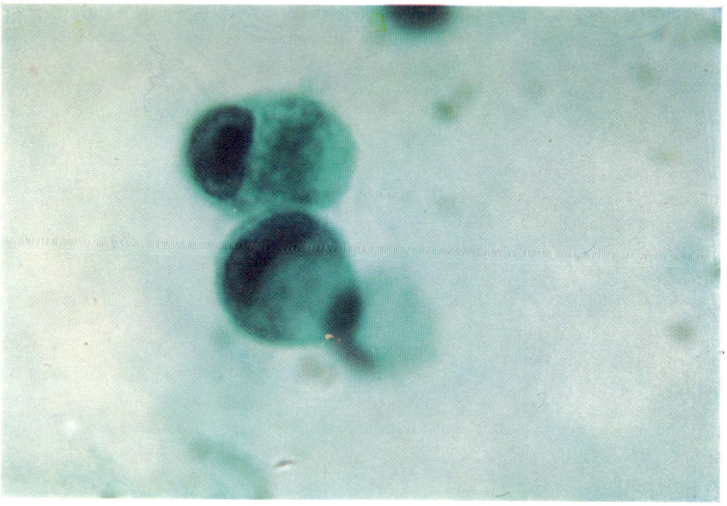

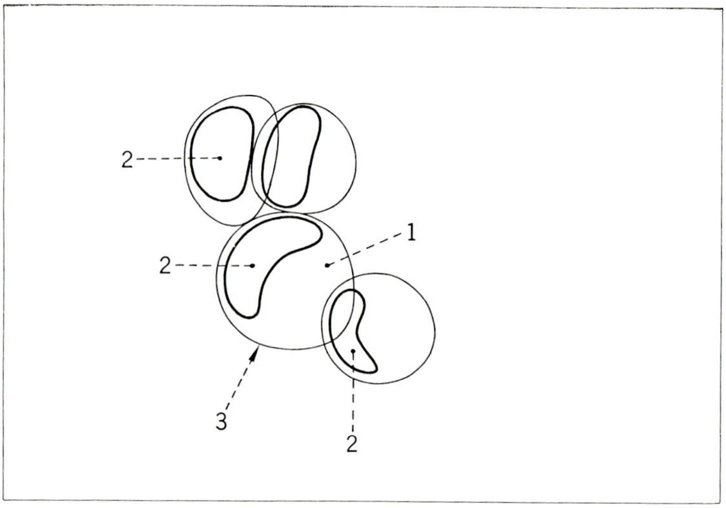

118. Gastric adenocarcinoma

a. Low magnification

More than five malignant nuclei can be seen. A squamous cell is present close to the group.

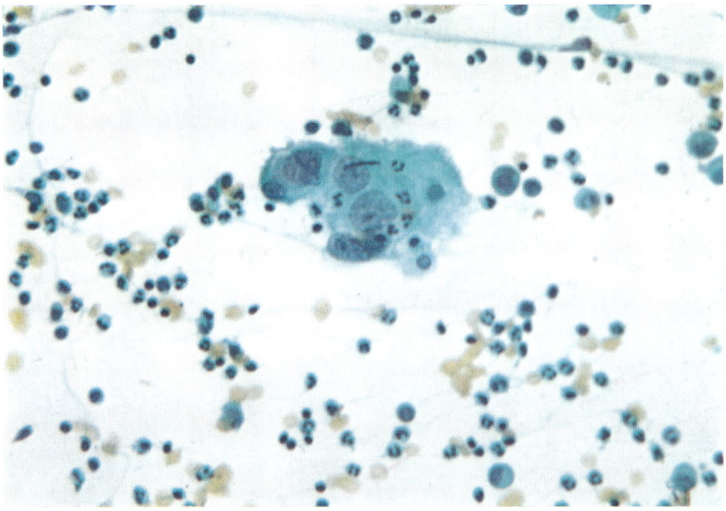

b. High magnification

1. Prominent nucleoli.
2. Chromatin granules are coarse and evenly distributed. No chromatin accumulation can be found in the nuclear borders.
3. Cytoplasm is abundant, but the boundaries are indistinct.
4. Leukocytes.

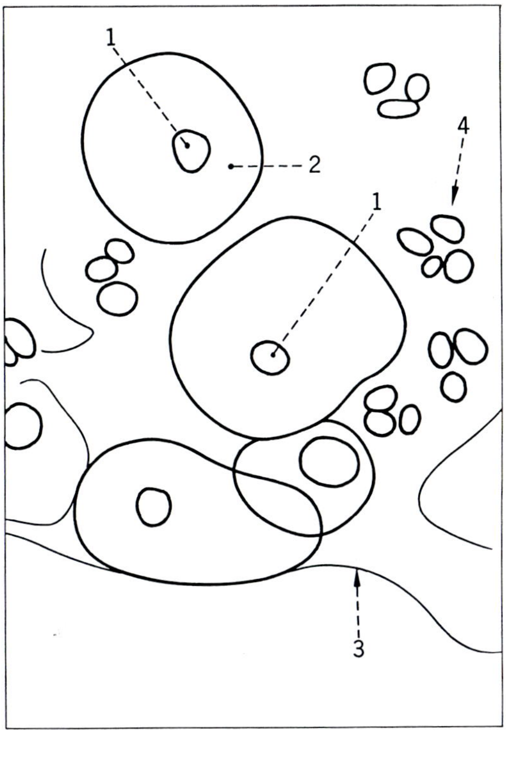

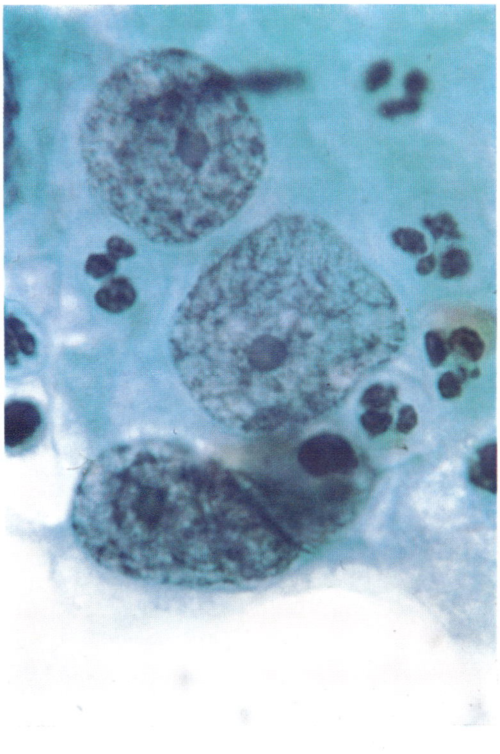

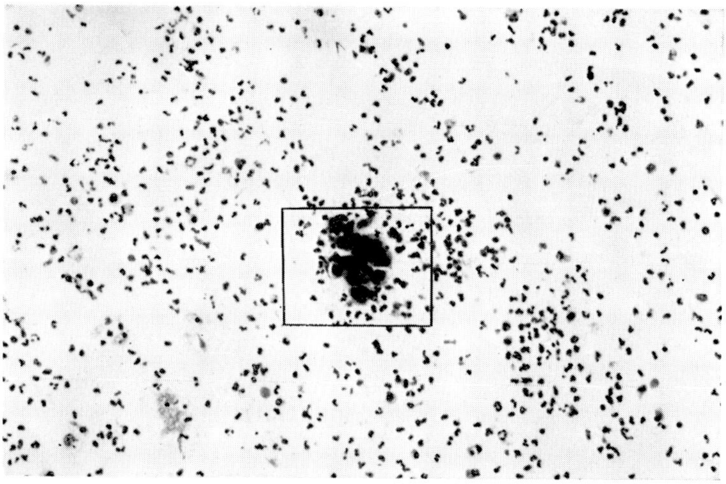

119. Gastric adenocarcinoma — degenerate

a. Low magnification

A messy cellular group is observed among numerous inflammatory cells and fragments of necrotic cells.

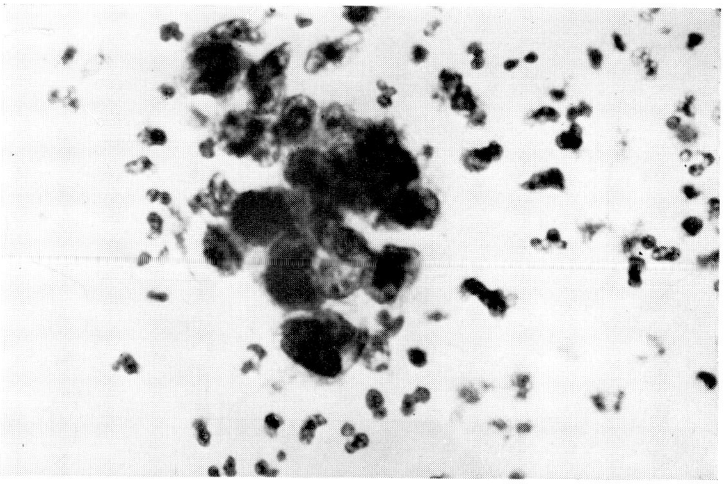

b. High magnification

1. The nuclei are markedly irregular in shape due to degeneration and no detailed internal structure is observable.
2. Chromatin granules are not observed and the nucleus is homogenously hyperchromatic.
3. The cytoplasm is barely visible and the nuclei are almost stripped.
4. Polymorphonuclear leukocytes.

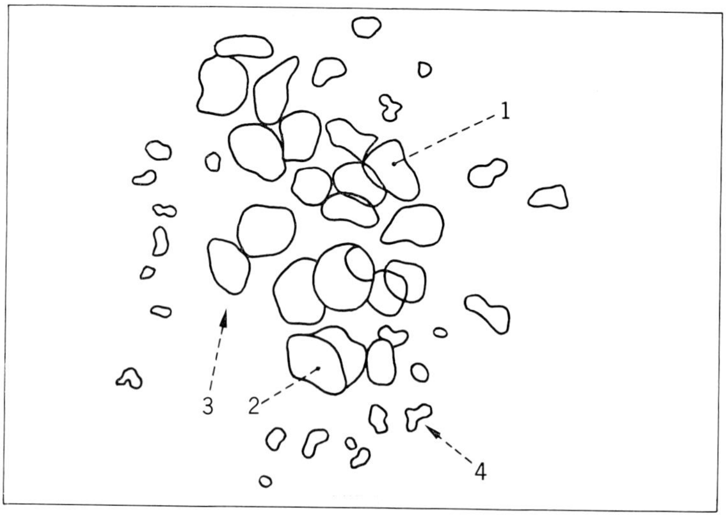

120. Gastric adenocarcinoma

a. Low magnification

A group of nuclei surrounded by little cytoplasm.

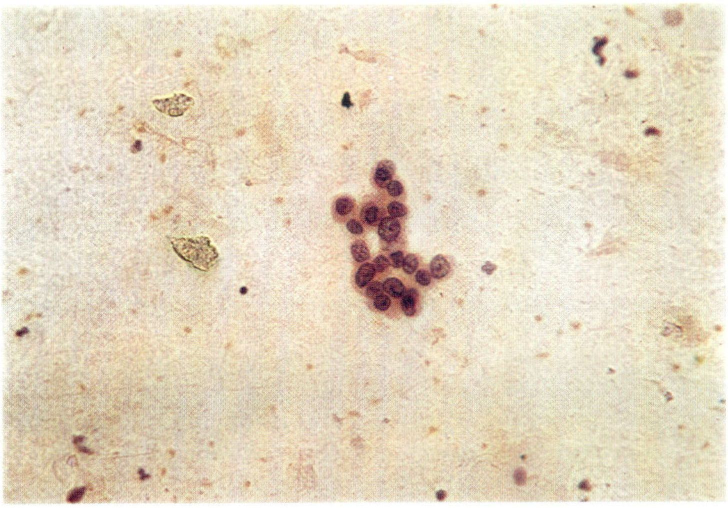

b. High magnification

1. Nuclei: Round or elliptical and markedly hyperchromatic. Chromatin is abundant and forms a network.
2. Comparatively coarse chromatin granules. Chromatin accumulates around the border.
3. The cytoplasm is very scanty and the borderlines of individual cells are indistinct. Outline of the cell cluster is visible.

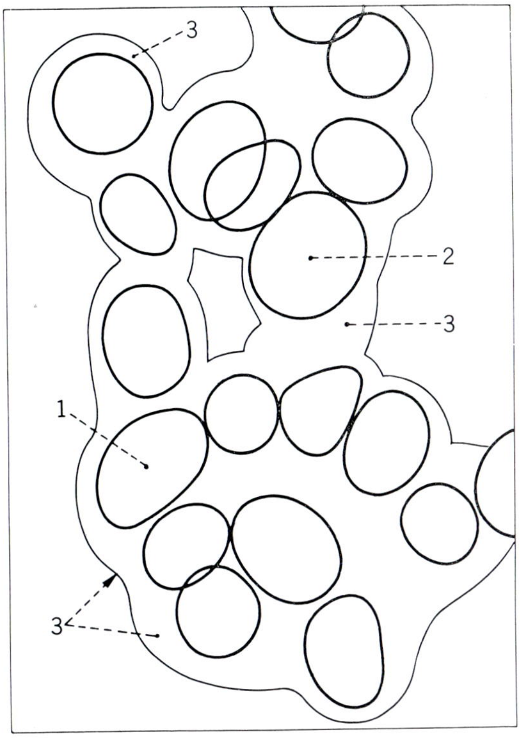

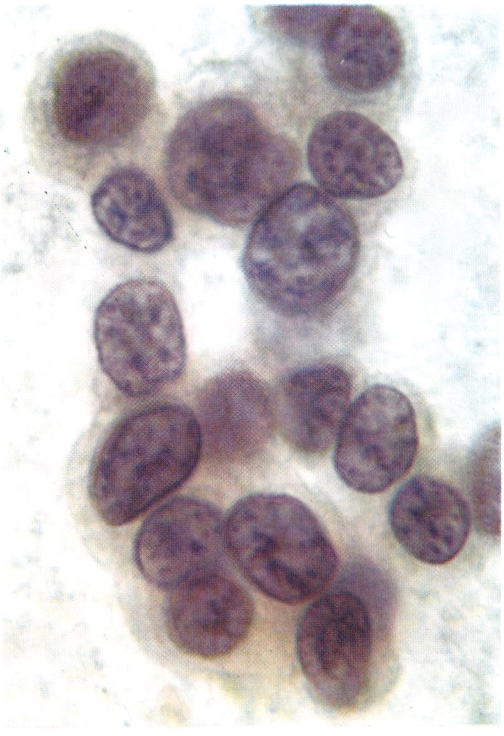

CYTOLOGIC EXAMINATION OF THE URINARY TRACT

I. Anatomy of the Urinary Tract

The urinary tract is extraperitoneally situated and divided into four parts: two kidneys and ureters, the urinary bladder and the urethra. Therefore, in principle, the urinary sediment can contain cells exfoliated from the kidneys, renal pelves, ureters, urinary bladder and urethra.

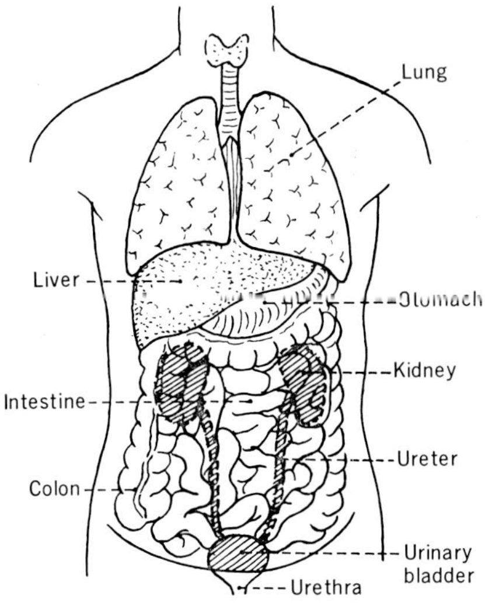

Fig. 52 Anatomical location of the urinary tract (extraperitoneal).

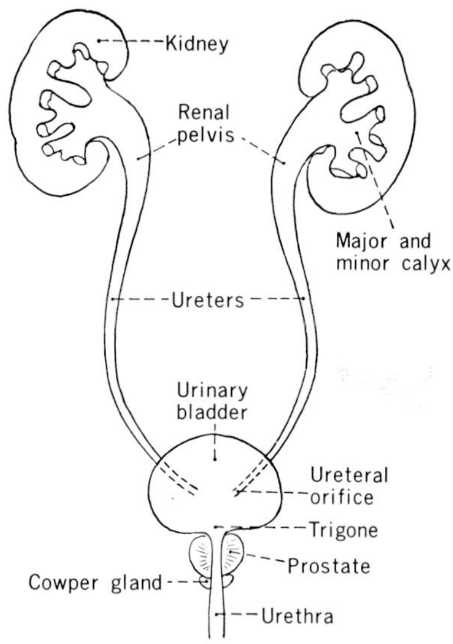

Fig. 53 Anatomy of the urinary tract.

In the male: Cells may come from the prostate, deferent duct, seminal vesicles and Cowper's gland, but in reality they are almost impossible to identify.

In the female: Often cells from the female genital tract are mixed with cells exfoliated from urinary tract in voided urine.

Renal calyx, renal pelvis, ureter, urinary bladder and a part of urethra: The internal surfaces are covered with transitional epithelium which is thought to represent a transition between stratified squamous and columnar epithelium. This epithelium can change its thickness according to the amount of urine in the bladder. For instance, when the bladder is empty and contracted the epithelium consists of 6–8 layers, but only 1–3 layers when the bladder is filled with urine. The deeper layers consist of cubic or columnar cells, but several of the upper layers consist of irregularly polygonal cells. The cells of the uppermost layer are larger than the cells underneath. Numerous lightly stained mucous cells are present at the trigone of the bladder.

Urethra :

In the male : Prostatic region — transitional epithelium. Membranous region — stratified columnar epithelium. External meatus of urethra — stratified squamous epithelium.

In the female : Mainly stratified squamous epithelium and sometimes pseudo-stratified columnar epithelium.

II. Characteristics of the Urinary Smear

1. The number of cells varies in urinary smears. Even in the same person, sometimes there are very few cells, while at times they are numerous.
2. Cells exfoliated into urine are somewhat deformed because urine is hypertonic.
3. Generally, while voided urine contains only a very few cells, catheterized urine contains numerous exfoliated cells, probably because cells are exfoliated from the mucosal surface by passage of the catheter.
4. Cells and their nuclei in catheterized urine vary in shape. They are easily distinguished from malignant cells by the characteristics of their nuclei.
5. Vaginal cells are often mixed in voided urine of females.

III. Benign Cells in Urine

Transitional cells (Fig. 54)

1) Nucleus
   1. Centrally located.
   2. Varies in size and shape.
   3. Variously stained (from hypochromatic to hyperchromatic).
   4. Distinct nuclear membrane.
   5. Minute or medium chromatin clumps.
   6. Chromatin granules are more abundant around the border.
   7. Often double or multinucleated, particularly in catheterized urine.

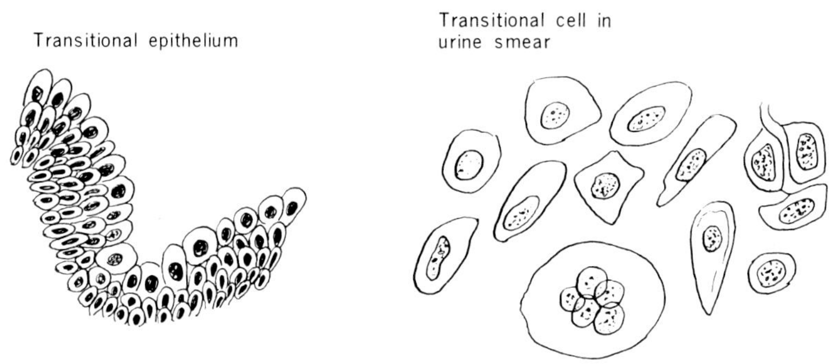

Fig. 54 Transitional cells.

2) Cytoplasm
   1. Abundant and not uniform in size.
   2. Mostly round or elliptical but sometimes markedly deformed.
   3. Orange or pink stain.
   4. The border is distinct.
   5. Minutely granulated.
   6. Vacuolated.

Squamous cells

Superficial squamous cells are often present, especially in urine containing vaginal cells (*cf.* p. 11) or among the cells exfoliated from the lower portion of the urethra.

Histiocytes

Their differentiation from transitional cells is not always easy. The typical histiocyte shows the same characteristics as those from other parts of the body (*cf.* p. 36).

Multinucleated giant cells

Occasionally observed in catheterized urine. Sometimes numerous nuclei gather at the center of a giant cell, but this is not atypical in urinary smears.

Cells with acidophilic cytoplasmic inclusions

Fairly common. These cells are often degenerate and the nuclei are pyknotic.

Small columnar cells

The nucleus is usually eccentric. They come mostly from the urethra.

Prostatic cells

These cells are rarely observed in ordinary urine, except following prostatic massage.

The smears obtained by prostatic massage are classified into two types.

1) The scanty cell type

This is characteristic of smears obtained from youths. The cells are small, eosinophilic or basophilic and low columnar or cuboidal. The nuclei are round, stained moderately and occupy 2/3 of the cell. The chromatin pattern is indistinct and looks homogeneous.

2) The abundant cell type

The nuclei are larger than that of the sparse cell type. The nuclear membrane may be heavy and distinct. The cell most commonly observed is called a "foam cell," which has a small hyperchromatic nucleus and a round, eosinophilic, abundant and finely reticulated cytoplasm.

Leukocytes, spermatozoa, crystals and corpora amylacea may be also encountered in the prostatic secretions. The corpus amylaceum is round or elliptical and eosinophilic, and has a homogeneous or concentrically laminated structure.

Seminal vesicle cells

Large histiocytes are often observed. They are markedly phagocytic and include spermatozoa and yellow-pigment granules in the cytoplasm.

The epithelial cells are rather small, deeply stained and homogeneous. The nucleus is located near the cytoplasmic border, homogeneous and comparatively hyperchromatic and occupies at least 1/2 of the cell. Bizarre cells are common. They are rather large. The nucleus is evenly hyperchromatic and deformed according to the shape of the cell. Yellow-pigment granules are occasionally present in the cytoplasm. These cells are common in males over fifty.

Blood components, bacteria, fungi, spermatozoa, crystals and trichomonas vaginalis may be occasionally encountered.

121. Transitional cells

a. Low magnification

A group of several polygonal cells is seen at the center. Variety of nuclear shapes is noticed even under low magnification.

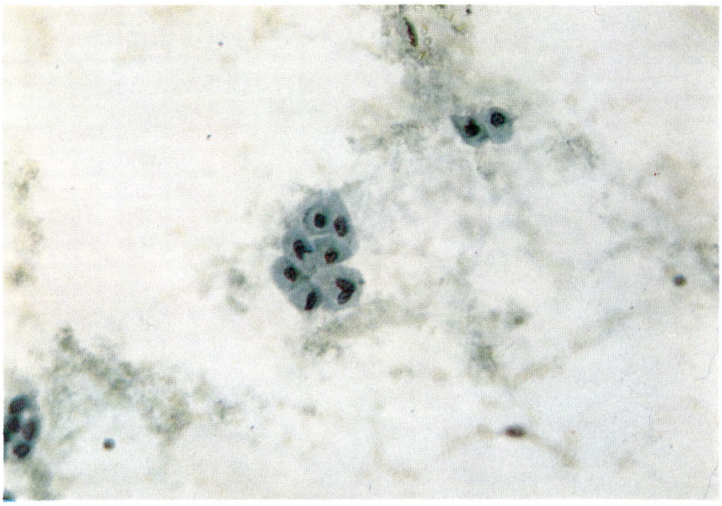

b. High magnification

1. Nuclei: Elliptical, polygonal and variously shaped.
2. Chromatin: Occasionally hyperchromatic.
3. A coarsely granulated nucleus.
4. Cytoplasm: Round, elliptical or polygonal. The border appears thickened.
5. Lacy or vesiculated.

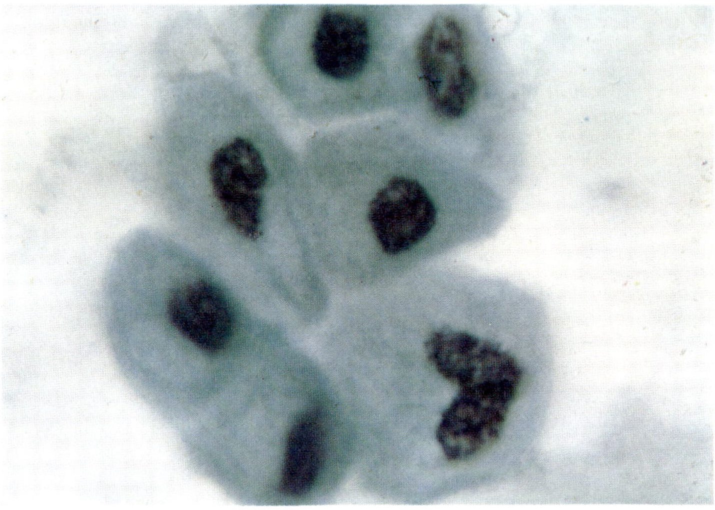

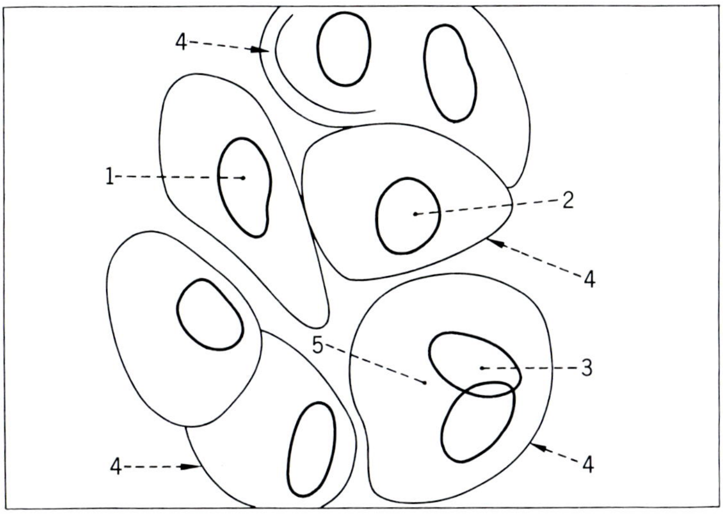

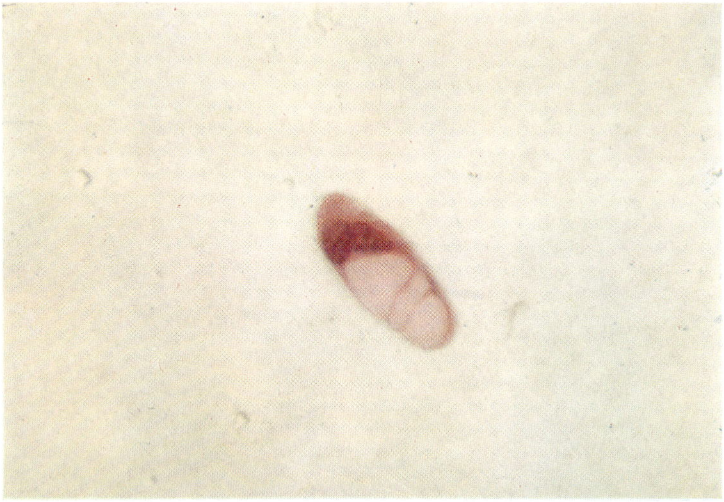

122. Transitional cell — Radiation effect

a. Low magnification

A single vacuolated atypical cell is observed. Cytoplasmic vacuolation is marked. The nuclei are fairly hyperchromatic.

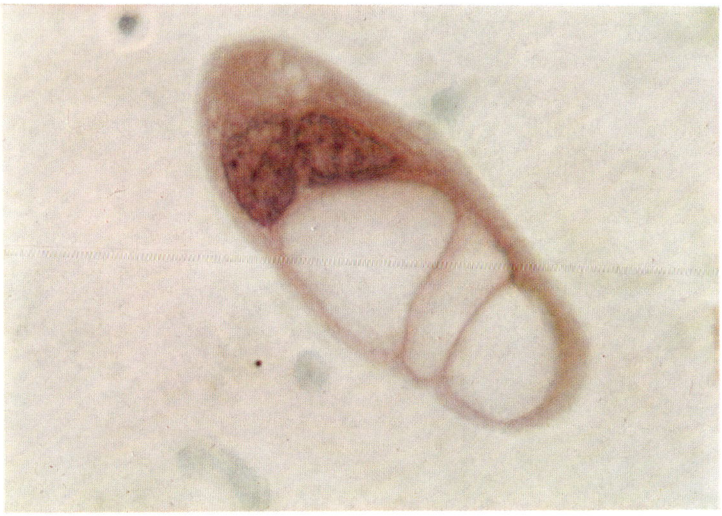

b. High magnification

1. Nuclei: Round, elliptical or triangular. Slightly hyperchromatic.
2. Cytoplasm: Lacy The borderlines of cluster cells are distinct.
3. Prominent vacuoles.

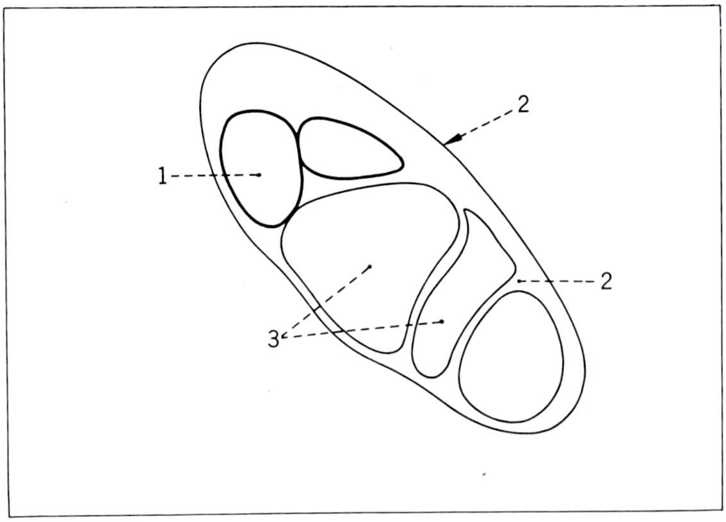

123. Squamous cells — Urethra

They are similar to squamous cells from other parts of the body, but no sign of keratinization is noticed. Basophilically stained.

Nuclei : Round. Medium sized. Slightly vesiculated. Uniform in size. Chromatin is evenly distributed.

Cytoplasm : Abundant and thin.

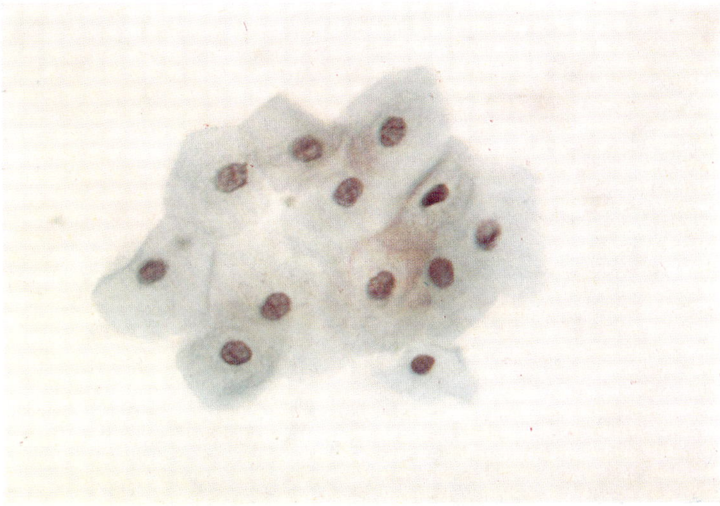

124. Prostatic cells

These cells were obtained by prostatic centesis. The cell is small and looks pyknotic.

Nucleus : Round or elliptical. The border is smooth and clear. Chromatin is comparatively abundant, hyperchromatic and dense.

Cytoplasm : Scanty, compared with the size of the nucleus. Thick and polygonal. The border may be distinct. Sometimes deeply stained (right picture).

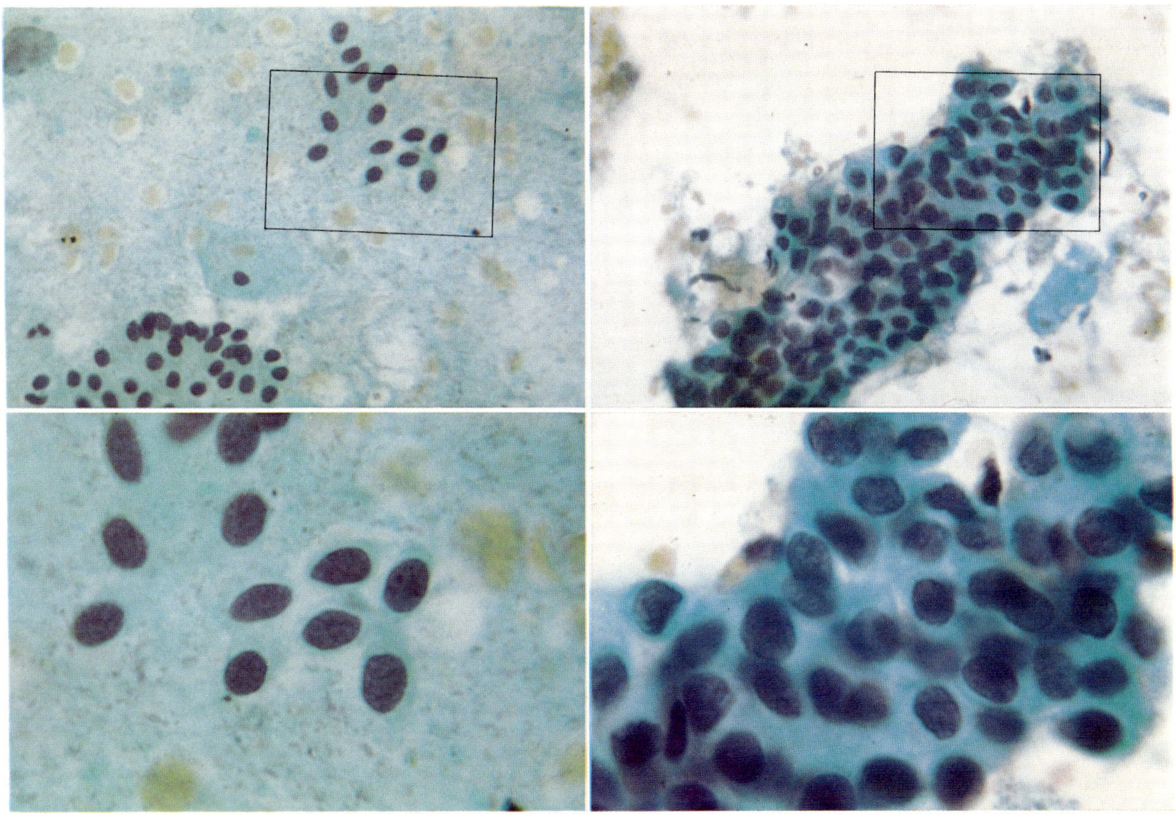

IV. Benign Atypical Cells in the Urine

The causes of benign atypia are 1) inflammation (cystitis, pyelonephritis and others), 2) presence of calculi lithangiuria, 3) papilloma, 4) hormonal changes (pregnancy, estrogen treatment and others), 5) radiation therapy and 6) cytomegalic inclusion disease.

Papilloma

Many cells exfoliate. The nuclei are mostly round or elliptical and uniform in size. The cells are elongated, columnar and vary in shape. Papilloma should be suspected when many of such cells are observed, but the cytodiagnosis is not always easy.

Hormonal effects

Navicular cells (*cf.* p. 54) can be observed in pregnancy.

The urinary smear of the male who has been given estrogen treatment for prostatic carcinoma is similar to the vaginal smear of the female who has been given the same hormone in menopause.

1. Numerous cells appear, and superficial cells especially increase.
2. The nucleus is pyknotic.
3. Eosinophilically stained.
4. Cytoplasmic granules are observed.

So-called "glycogenic cells" may be present in the prostatic secretion during estrogen treatment. They are round or elliptical and the cytoplasmic border is thickened. A yellowish-brown perinuclear accumulation is observed. This is glycogen (*cf.* pp. 54 and 56). The nucleus is pyknotic and often eccentric.

Cytomegalic inclusion disease

In this disease, the epithelial cells of the salivary gland and other organs are enlarged and numerous nuclear and cytoplasmic inclusions are observed. It is a viral disease that is seen especially in the newborn infant. The cells from the renal tubules, gastric mucosa and oral mucosa are cytodiagnostically important. A large cell of 20–30 microns with a large nucleus can be seen in the smear of urinary sediment. A homogeneous, basophilic inclusion with a smooth border is noticed at the center of the nucleus. Sometimes these inclusions are observed in the cytoplasm. In such cases there are more than two and they are small and strongly basophilic. They are located at one side of the cell and half are arranged in a concentric circle near the border.

125. Papilloma of the urinary bladder (Urine)

1. Cells with round or elliptical and uniformly sized nuclei.
2. Nuclei appear in a group. Chromatin is slightly increased in some nuclei.
3. Cytoplasm is faintly visible and sometimes the shape is almost columnar.

(Picture by courtesy of Dr. M. Takahashi)

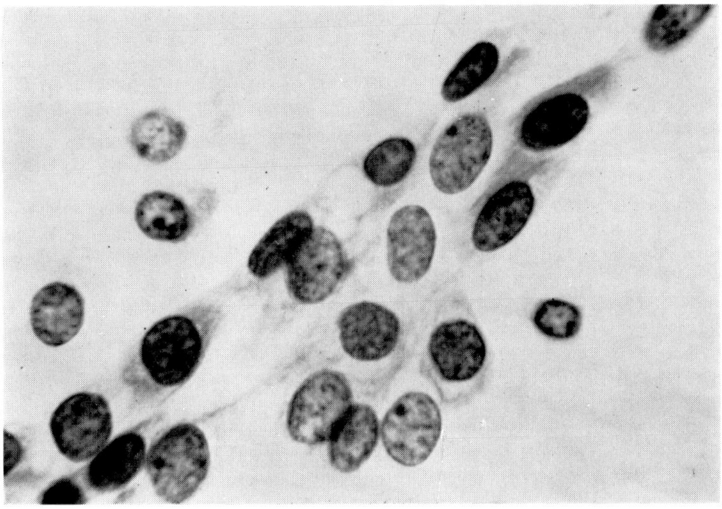

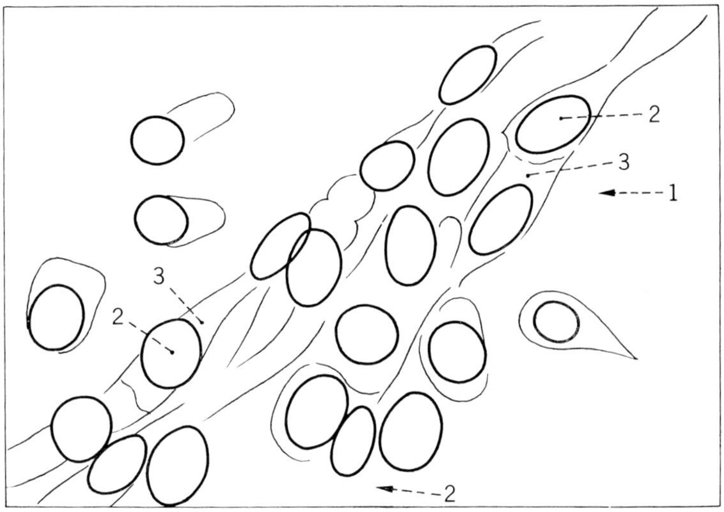

V. Characteristics of Malignant Cells of the Urinary Tract

Transitional cell carcinoma

The cells show almost the same characteristics as those of epidermoid carcinoma cells from other parts of the body : 1) increase in the nuclear-cytoplasmic ratio, 2) variety in nuclear size and shape, 3) hyperchromatism, 4) increase and uneven distribution of chromatin, 5) marked nucleoli, 6) overlapping cells, 7) vacuolation and 8) irregular cell shapes.

Malignant papilloma

Diagnosed by presence of columnar or elongated malignant cells.

Renal carcinoma

Cytodiagnosis of renal carcinoma is difficult, and renal carcinoma can be detected by smears only when it is so advanced that the cells exfoliate into the urine. Even then the cells look benign at first, and the usual yardsticks for malignancy may not be applicable.

Prostatic carcinoma

Mainly adenocarcinoma. The exfoliated cells are sometimes so small that they may be overlooked, but on other occasions they are large, irregularly shaped and show definite malignant features.

In smears obtained by prostatic massage, the cells are markedly collective and look as if they had been cast in a mold The cytoplasm is very scanty and only seen around the cell clusters. The nuclei are strongly hyperchromatic and the chromatin is unevenly distributed. Sometimes prominent nucleoli are observed.

126. Transitional cell carcinoma of the urinary bladder (Urine)

a. Low magnification

Two large cells are observed. The nuclei are markedly hyperchromatic.

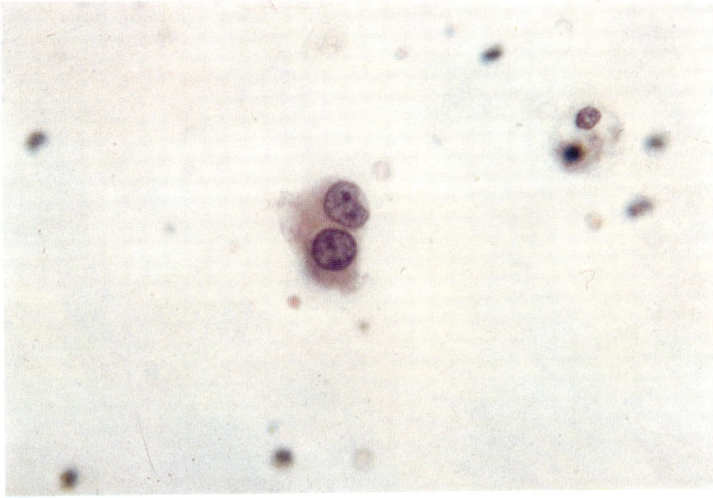

b. High magnification

1. A round nucleus. Markedly hyperchromatic. The nuclear membrane is very thick. Chromatin is unevenly distributed.
2. An irregularly shaped nucleus. Not very hyperchromatic, but chromatin has accumulated around the border.
3. A marked nucleolus.
4. Scanty, lacy and irregularly shaped cytoplasm.

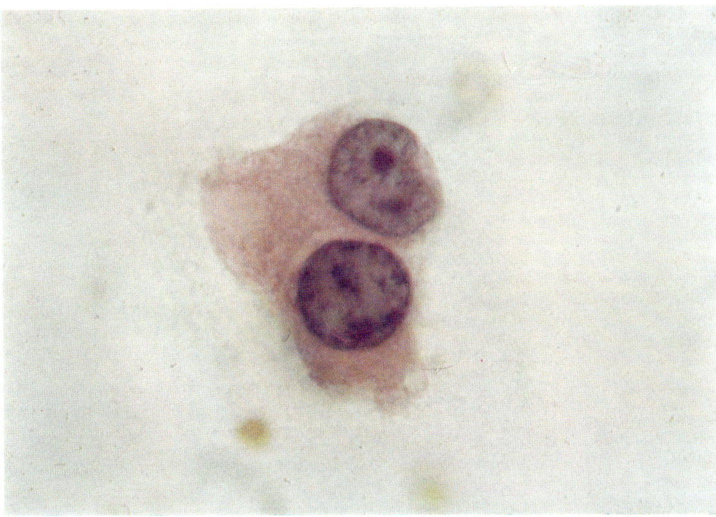

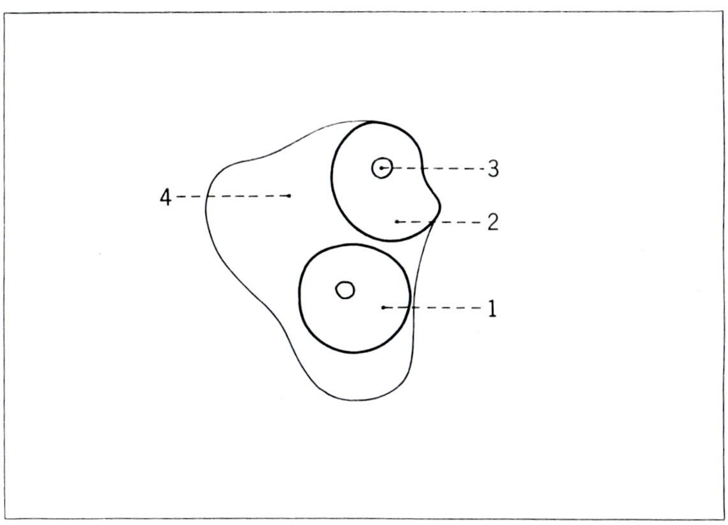

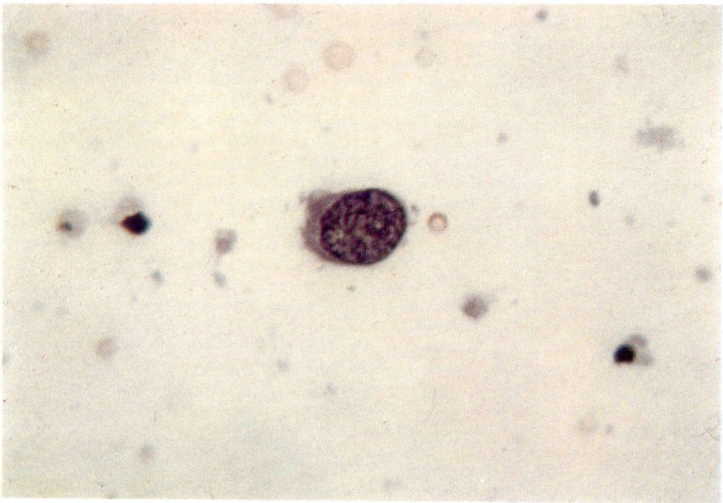

127. Transitional cell carcinoma of the urinary bladder (Urine)

a. Low magnification

An extremely large and hyperchromatic nucleus is present. The cytoplasm is barely visible. Other smaller but hyperchromatic malignant cells are also scattered in the smear.

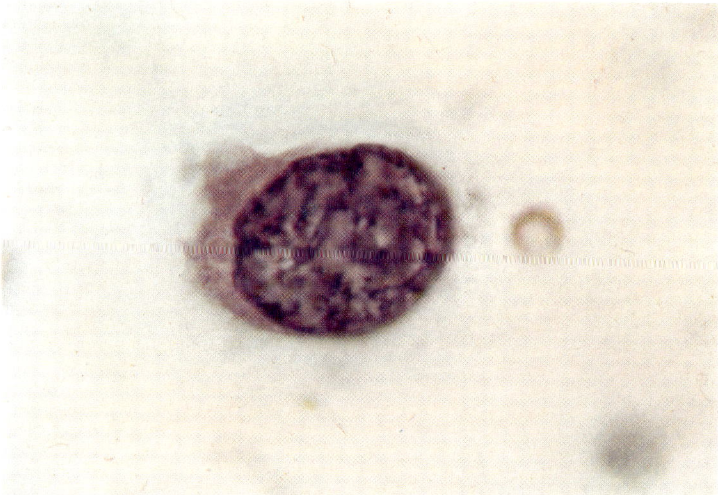

b. High magnification

1. The nucleus is markedly enlarged. The border is slightly irregular. Chromatin is irregularly clumped in masses.
2. Cytoplasm is barely visible.

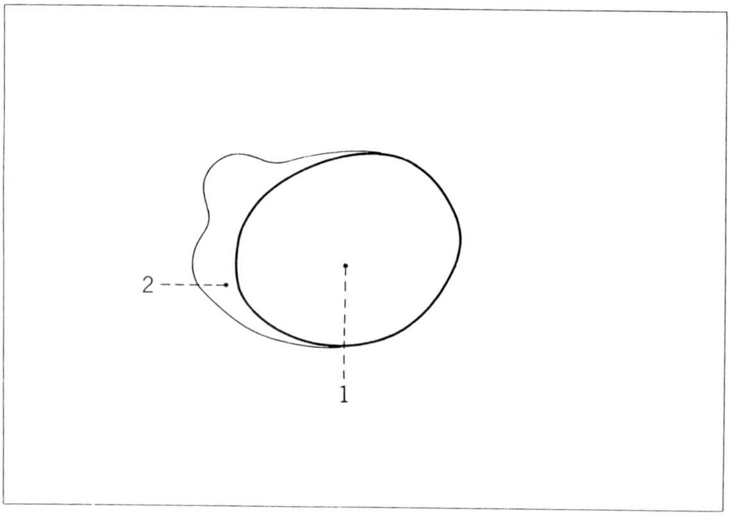

128. Transitional cell carcinoma of the urinary bladder (Urine)

a. Low magnification

A group of cells which have round and hyperchromatic nuclei is seen at the center.

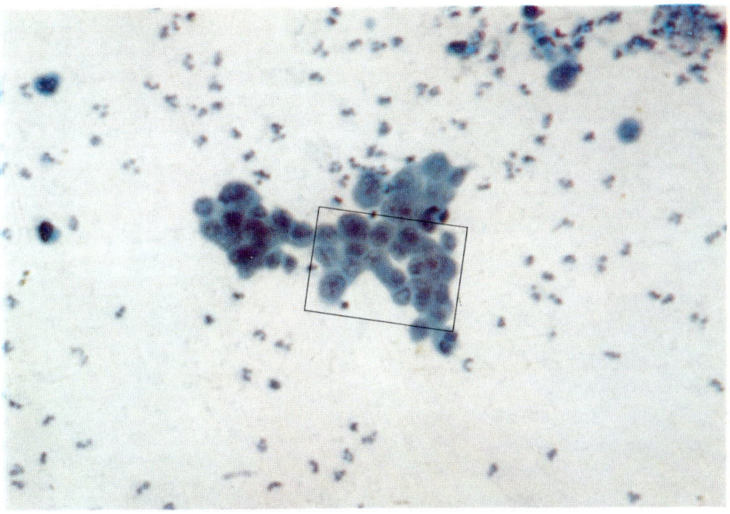

b. High magnification

1. Nuclei reveal coarse and irregularly granular chromatinic clumps. The nuclear border looks fairly distinct.
2. Vague nucleoli are recognizable.
3. Cytoplasm Lacy appearance and its border is fairly well preserved.

(Pictures by courtesy of Dr. Y. Sakai)

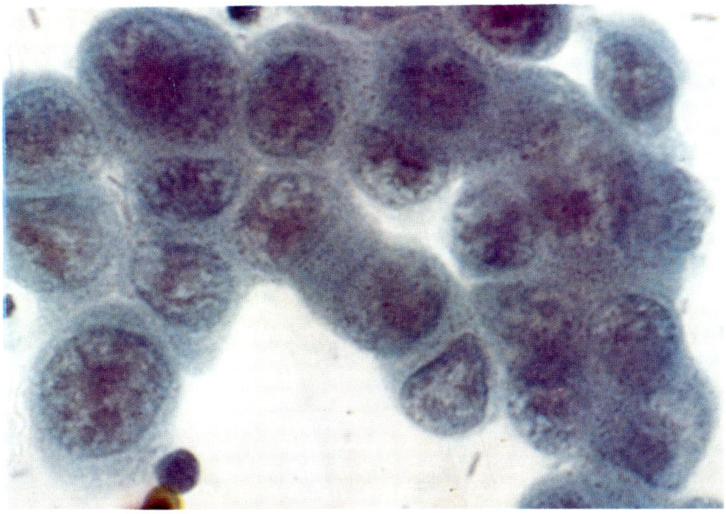

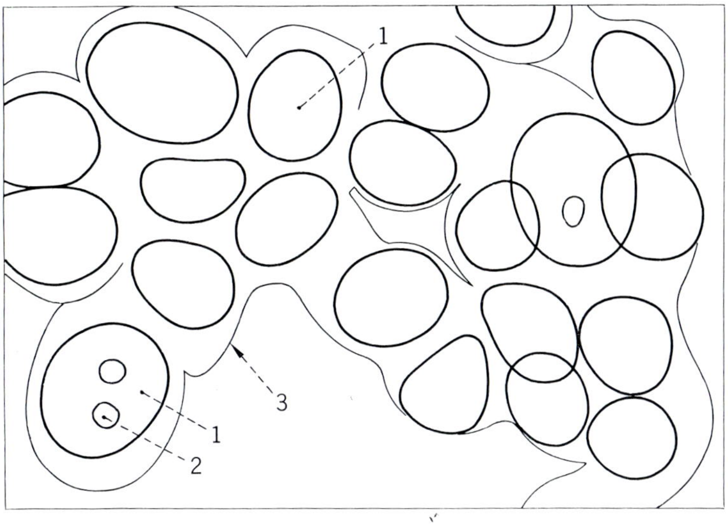

129. Transitional cell carcinoma of the urinary bladder — Radiation effects (Urine)

a. Low magnification

Large, vacuolated cells are seen among numerous polymorphonuclear leukocytes.

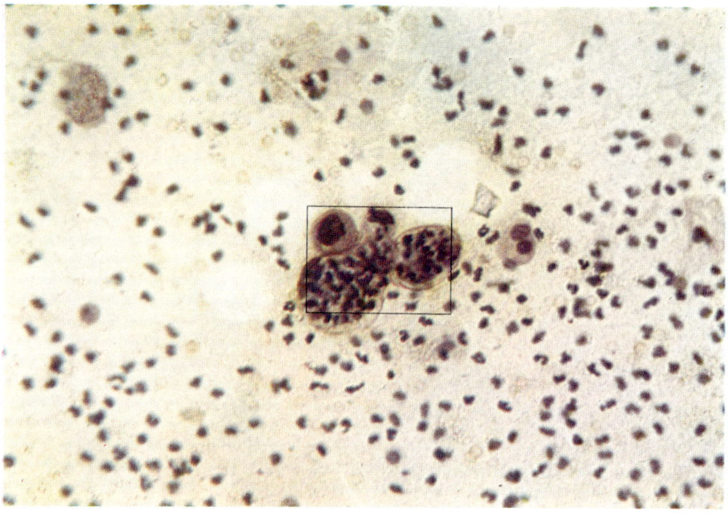

b. High magnification

1. A round, enlarged nucleus. Chromatin pattern is not clear and the nucleus looks structureless.
2. An irregularly crescent-shaped nucleus. Hyperchromatic and structureless. It is pushed to the cytoplasmic border by a large cytoplasmic vacuole.
3. A marked cytoplasmic vacuole. Numerous polymorphonuclear leukocytes are included.
4. This nucleus is also pushed to the cytoplasmic border and flattened by a large vacuole.
5. Fairly prominent nucleolus.
6. Well preserved cytoplasmic border.

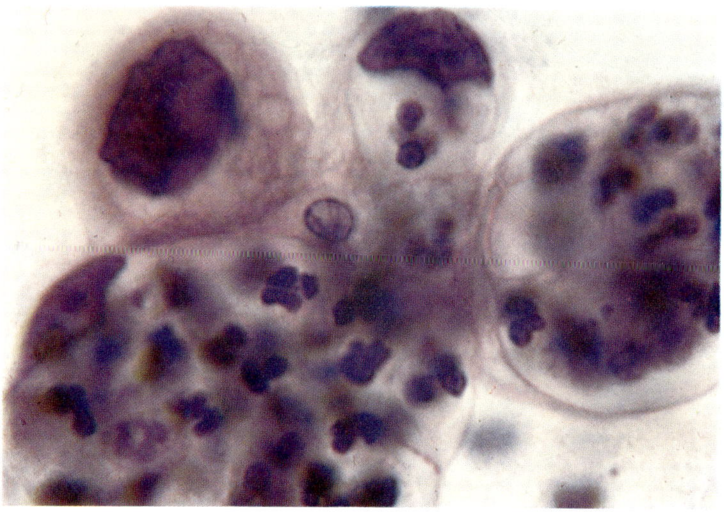

130. Metastasis from rectal carcinoma to the urinary bladder (Urine)

a. Low magnification

A columnar shaped cluster of malignant cells is present in the center.

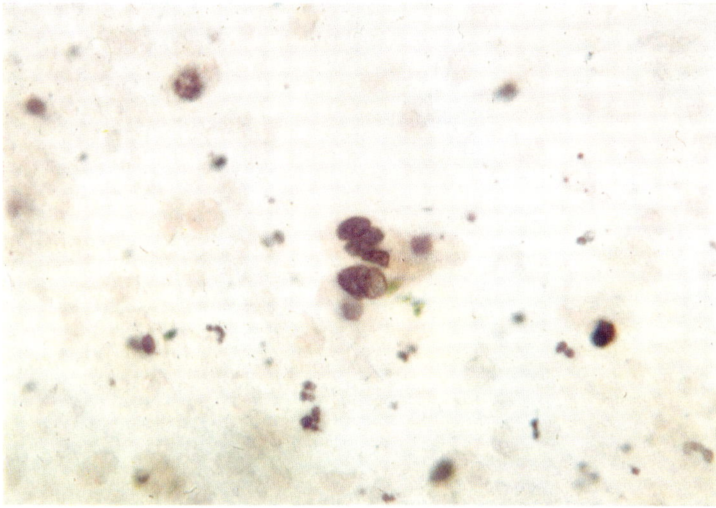

b. High magnification

1. Triangular or rectangular malignant nuclei showing hyperchromatism and irregular chromatinic distribution.
2. Characteristically columnar cytoplasm. Eosinophilically stained.
3. Flat surface of the cells toward the luminal space suggests that these cells may be derived from glandular tissue.

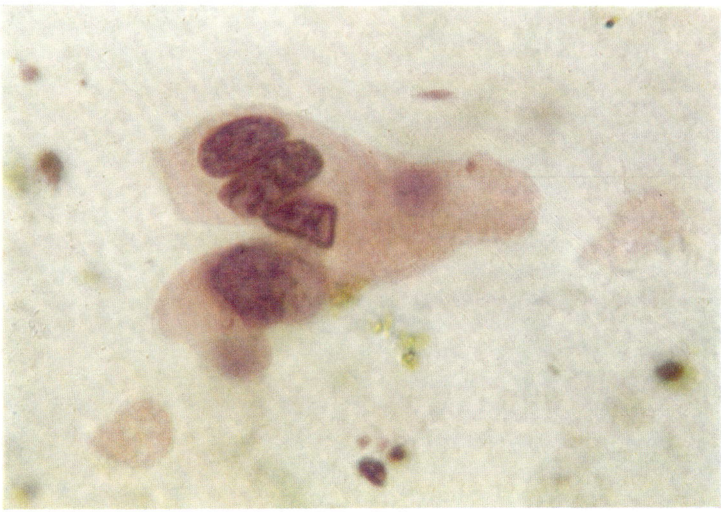

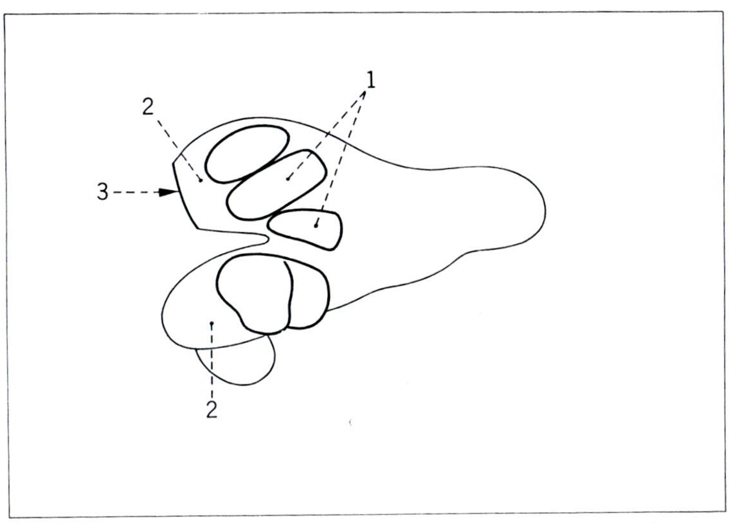

CYTOLOGIC EXAMINATIONS OF THE EFFUSIONS

I. Anatomy of the Body Cavities

Body cavities are potentially mesodermal spaces lined with pleura, peritoneum or pericardium, and these cavities are named the pleural cavity, the peritoneal cavity and the pericardial sac, respectively. Each membrane is divided into the parietal layer covering the outer wall and the visceral layer covering the inside organs.

Each is a single flat epithelial cell layer called the mesothelium. Even in the normal state, the body cavities contain a small amount of fluid, which, like a lubricant, helps the parietal and visceral layers to move easily. In the pathological state, this fluid increases, its properties change or it is replaced with air or blood.

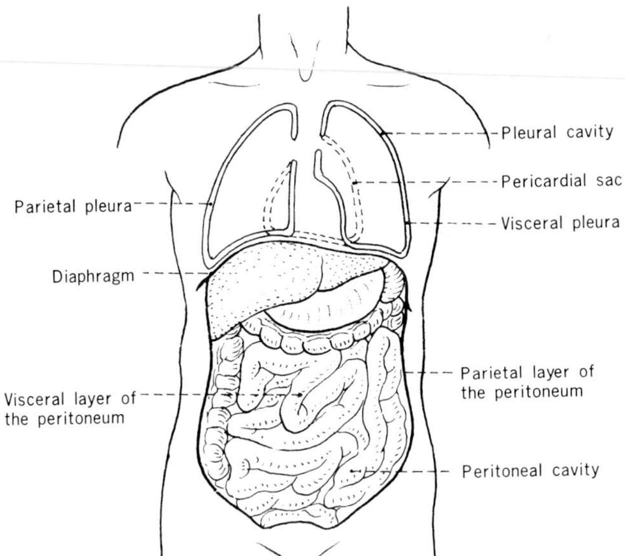

Fig. 55 Body cavities.

The objects of the smear test are mainly ascites in the peritoneal cavity and pleural effusions in the pleural cavity.

Causes of the effusions

1) Inflammation

Causes exudation. The exudate contains much protein and the specific gravity is over 1020. It contains various kinds of leukocytes and mesothelial cells.

2) Circulatory disturbances

Cause transudation. The transudate contains less protein and the specific gravity is about 1012. It may contain a few cells. The most common causes are an increase in pressure due to stagnation of the venous circulation and an increase in capillary permeability.

3) Primary or metastatic malignant tumor

Causes exudation. The definite mechanism is unknown.

II. Benign Cells in Fluid in the Body Cavities

Mesothelial cells (Fig. 56)

1) In normal states

The mesothelium consists of a single layer of flat epithelial cells. When an effusion stagnates, separation of mesothelial layer takes place and these cells become stratified into several layers. The cells become cuboidal and readily exfoliate.

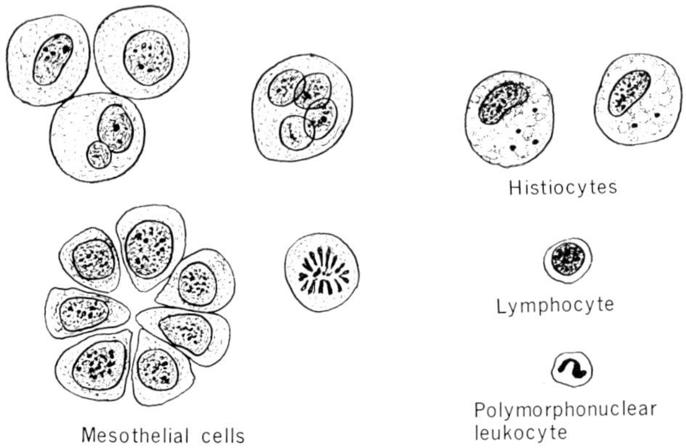

Fig. 56 Benign cells.

2) Exfoliation

Cells sometimes form a rosette-like or papillary structure, or are separate on other occasions. The cells forming those structures are thought to have exfoliated singly first, continued to grow for a while in the effusion, and then formed the structures.

3) Nucleus

1. Round, kidney-shaped or elliptical.
2. Located almost in the center.
3. Uniform in size.
4. The border is thick and distinct.
5. Fairly hyperchromatic. Finely reticulated. Chromocenters may be seen.
6. A nucleolus is sometimes present.
7. Mitosis can be noticed. This indicates that the exfoliated cell proliferates for a time in the effusion, but this has little diagnostic meaning.
8. Occasionally multinucleated.

4) Cytoplasm

1. Round or elliptical and homogeneous.
2. The border is comparatively distinct.
3. Stained both eosinophilically and basophilically.
4. Sometimes vacuolated. This is a sign of degeneration.
5. The nuclear-cytoplasmic ratio is higher than in normal epithelial cells.

131. Mesothelial cells

a. Low magnification

Three mesothelial cells are observed among scattered polymorphonuclear leukocytes.

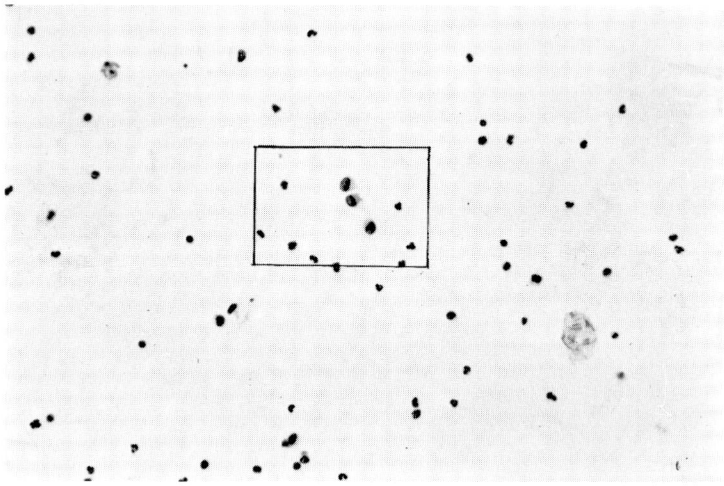

b. High magnification

1. Nuclei: Round or elliptical. Uniform in size. The borders are distinct. Chromatin is more abundant around the borders.
2. Cytoplasm: Comparatively distinct and homogeneous.
3. Polymorphonuclear leukocytes.

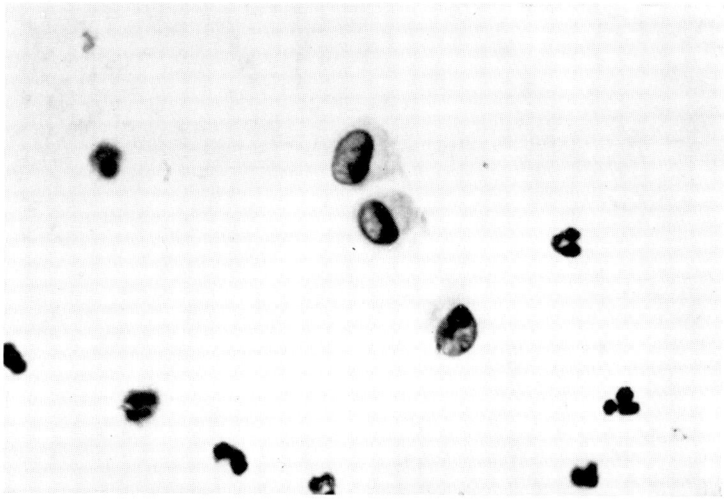

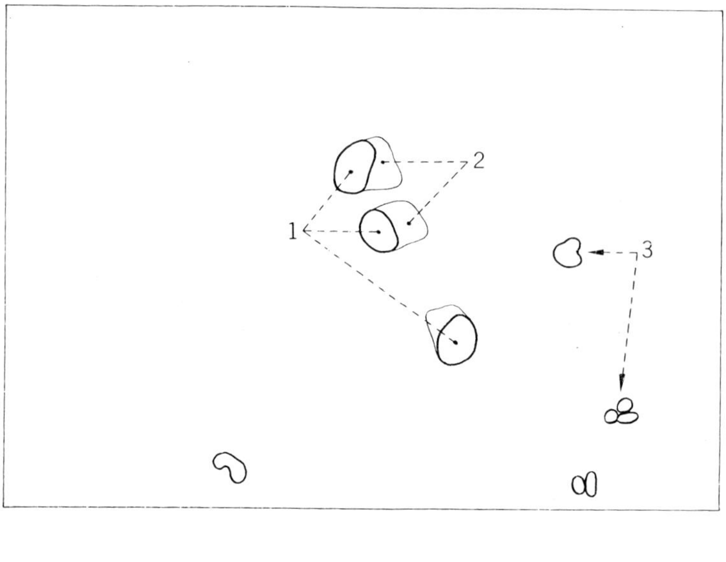

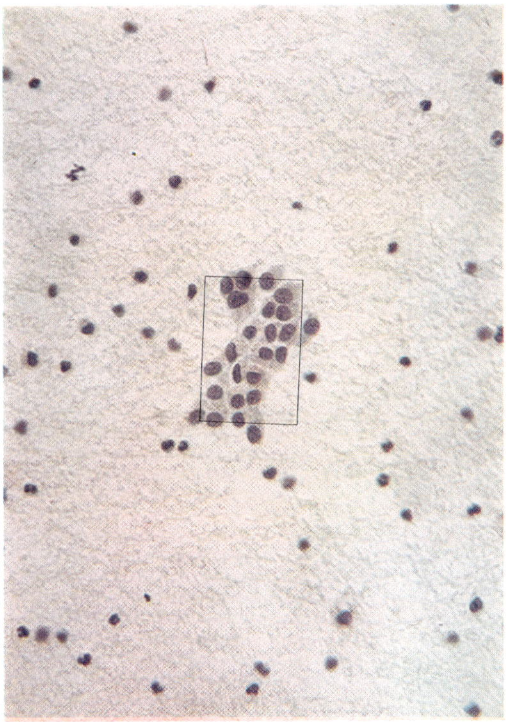
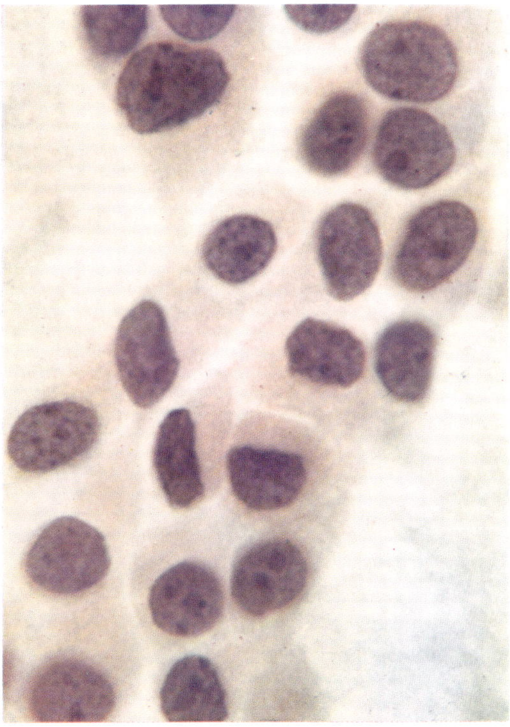

132. Mesothelial cells
A group of mesothelial cells.
Nuclei : Mostly round or elliptical. Chromatin is finely granular and evenly distributed.
Cytoplasm : Various shapes. Generally distinct and homogeneous. Cellular borders are mostly distinct.

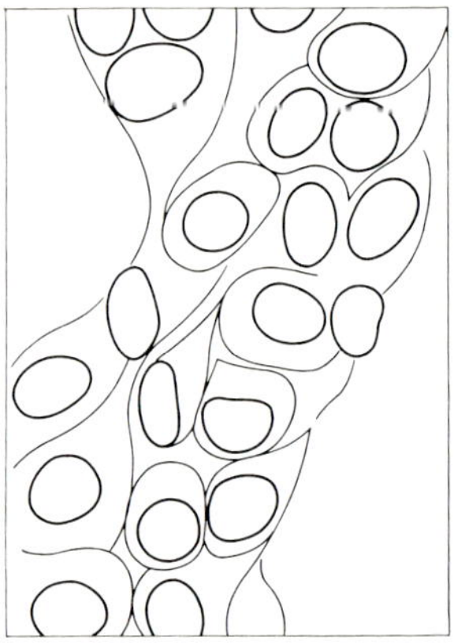

Histiocytes (Fig. 56)

May have essentially almost the same characteristics as those of histiocytes in other parts of the body

1) Nucleus

Large, round, elliptical or kidney-shaped. Eccentrically located. Chromatin is finely granular Sometimes multinucleated. Mitosis can be observed.

2) Cytoplasm

Round or elliptical in shape. Finely vacuolated or foamy. Phagocytotic.

Other cells (Fig. 56)

Polymorphonuclear leukocytes, lymphocytes and plasma cells may be also present.

III. Atypism of Benign Cells

Benign atypia of mesothelial cells and histiocytes are caused by chronic inflammation, congestive heart failure, cirrhosis of the liver, tuberculosis and pulmonary infarction. They may also be caused by poor fixation of the smear or prolonged effusion.

Fig. 57 Atypia of benign cells.

Enlargement of the cells

This may be due to poor fixation, prolonged effusion or irradiation. The nucleus looks cloudy without any chromatin pattern.

Abnormal clustering of the cells

Cirrhosis of the liver, chronic inflammation or prolonged effusion should be suspected when arranged in a rosette-like, pseudoglandular, alveolar or papillary formations of cells. A pseudoglandular structure of mesothelial cells may be artificially formed in the specimens prepared by cell block method.

Signet-ring cell like change

This is observed as a stage of degeneration of a cell or in the later stage of cirrhosis of the liver. It is probably due to a change in a histiocyte.

All these benign cells should be carefully distinguished from malignant cells. A diagnosis should be made prudently after the consideration of the present history and the clinical symptoms.

Differential diagnosis by supravital staining: As a rule, mesothelial and tumor cells in effusions are not stained supravitally by neutral red or Janus green, while histiocytes and leukocytes are generally well stained, though the required time for staining is not always the same.

133. Mesothelial cells — Benign atypia

a. Low magnification

A cluster of cells arranged in a semicircle is observed.

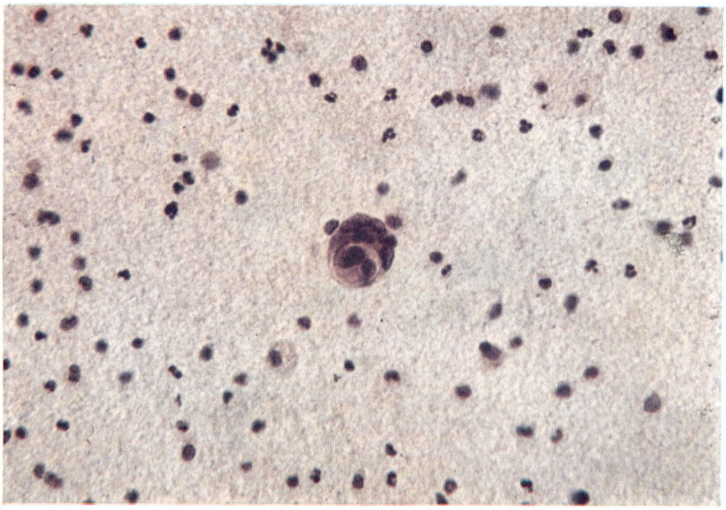

b. High magnification

1. Several elliptical nuclei are arranged in a horseshoe shape.
2. Two or three round nuclei are overlapping at the center.
3. Every nucleus is hyperchromatically stained and markedly enlarged. Chromatin is coarsely distributed.
4. Cytoplasm: A large vacuole is observed. The borders of individual cells are indistinct, and the border of the cell cluster only vaguely distinguished.

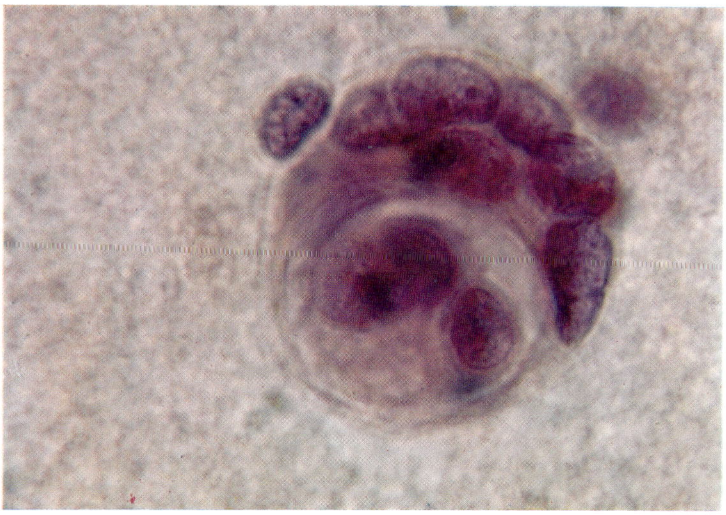

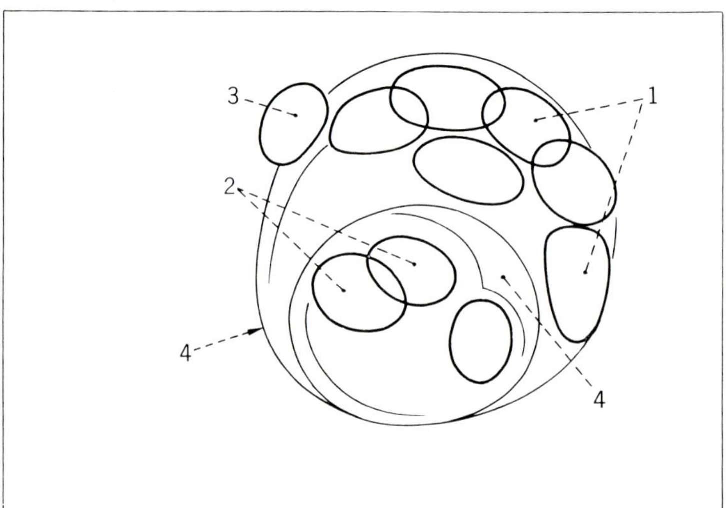

134. Mesothelial cells — Benign atypia

a. Low magnification

Cells arranged in a rosette-like formation are seen in the center.

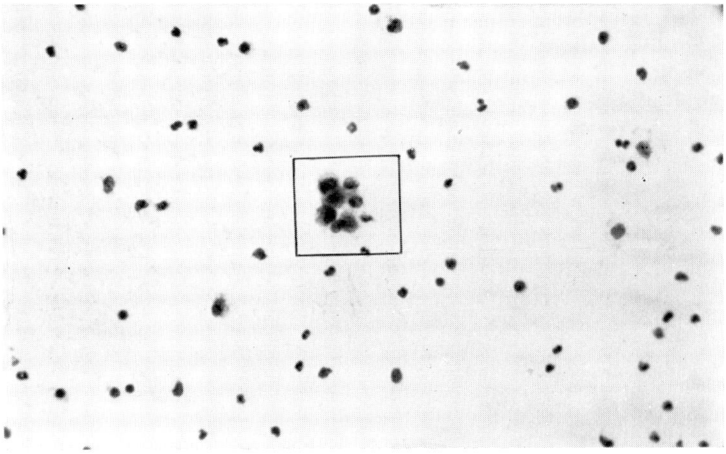

b. High magnification

1. Several cells form a round cluster.
2. Nuclei: Round, elliptical or kidney-shaped. Rather hypochromatic. Chromatin is accumulated around the border, so the nuclear membrane appears thickened.
3. Marked mitotic figure.
4. The borders of individual cells are only faintly distinguished. The cytoplasm looks round or elliptical.

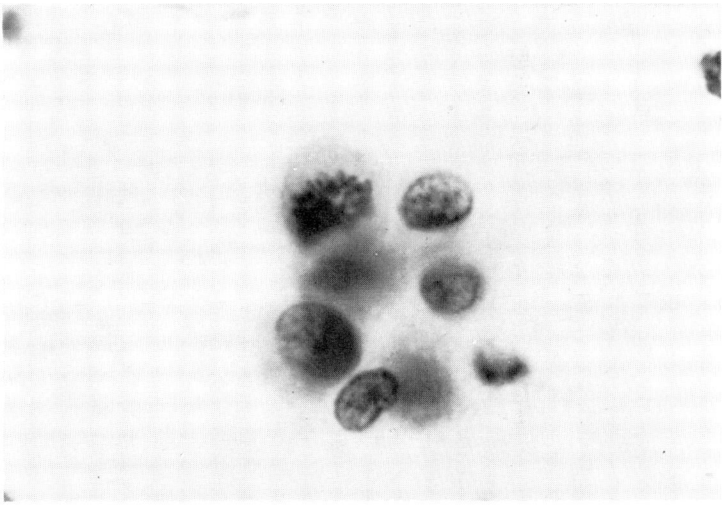

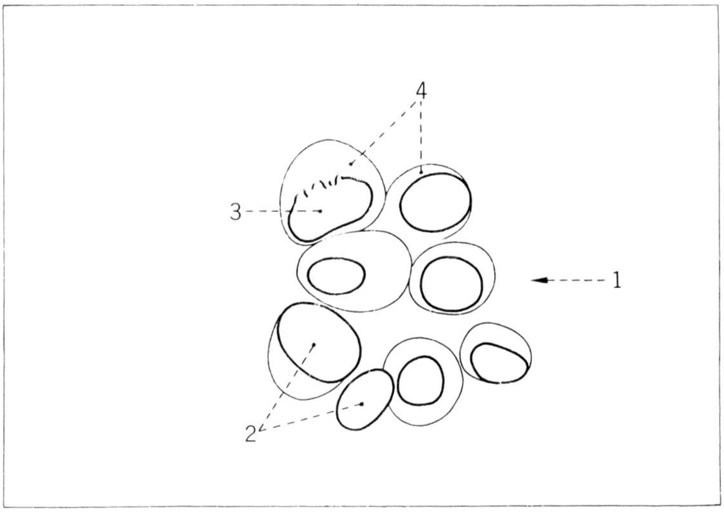

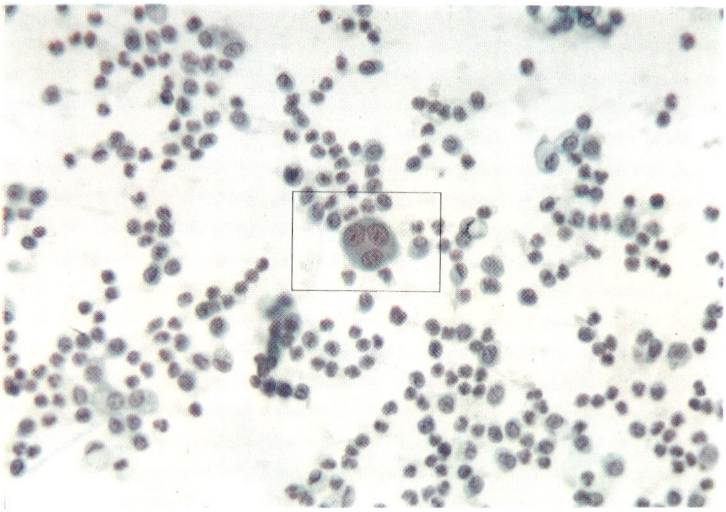

135. Theca cell tumor of the ovary (Ascites)

a. Low magnification

Among abundant large atypical cells a trinucleated giant cell is found.

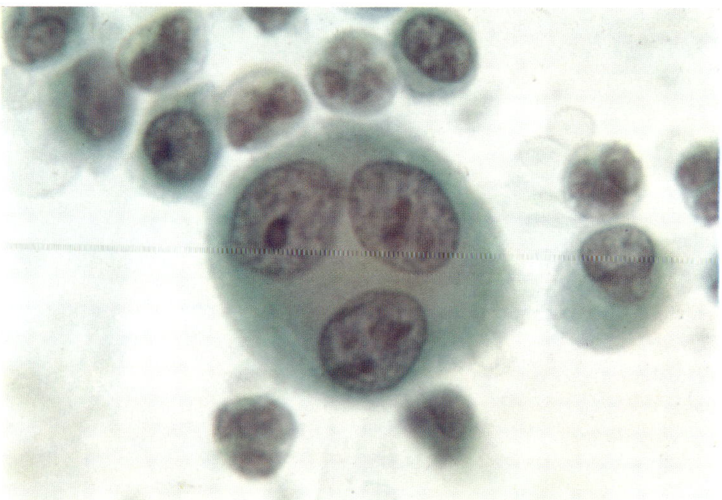

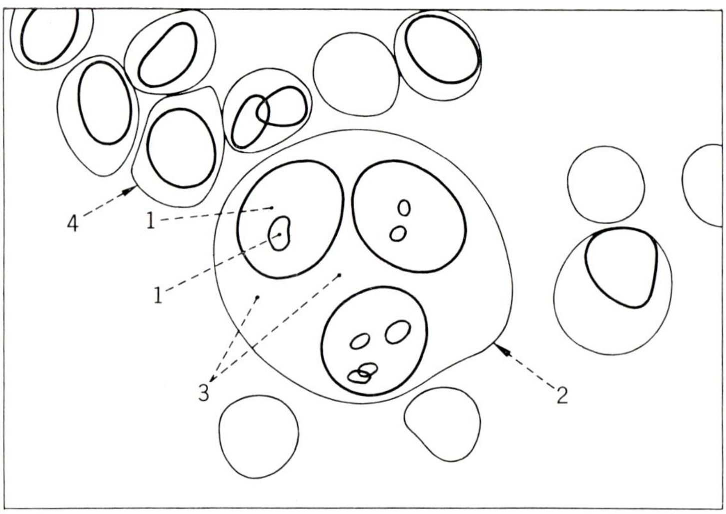

b. High magnification

1. Large, round nucleus showing coarse granulation of chromatin and prominent nucleolus. The nuclear border is clear.
2. Cytoplasm looks thickened and the border is comparatively clear.
3. Staining reaction in the center of the cytoplasm appears different from the peripheral area.
4. The same type of atypical cells. The cytoplasm is more vague.

136. Pseudomucinous cystadenoma of the ovary (Ascites)

a. Low magnification

A glandular cellular cluster was accidentally found in the ascitic fluid in a case of pseudomucinous cystadenoma.

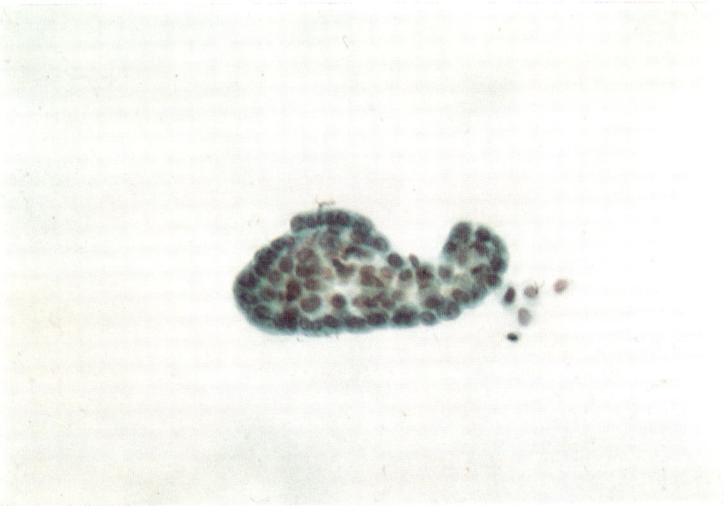

b. High magnification

1. Columnarly arranged glandular cells the nuclei of which look merely moderately hyperchromatic. Those nuclei arrange regularly and are regular in size and shape. No individual cellular border is recognizable.
2. Irregularly arranged cells are seen inside of the cluster. The nuclei are round, elliptical, triangular or polygonal in shape.

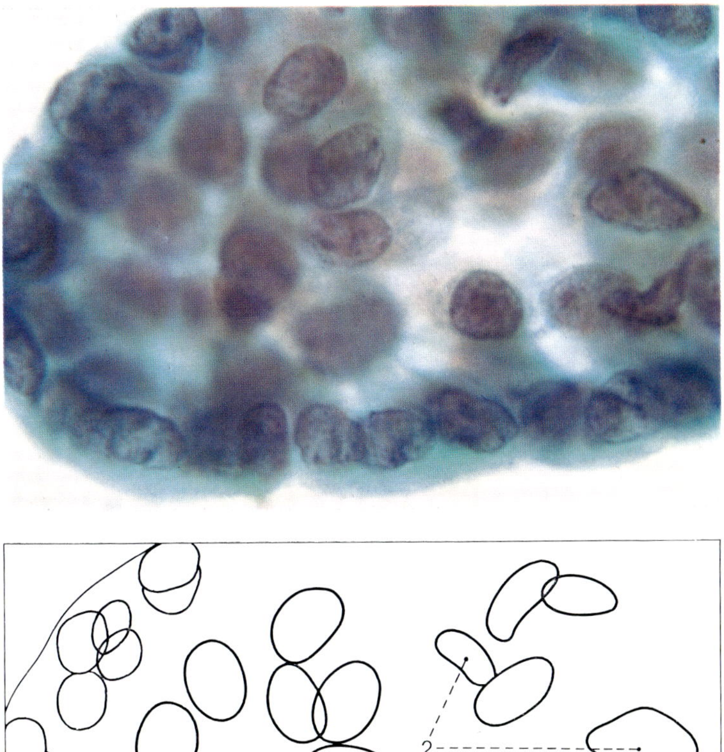

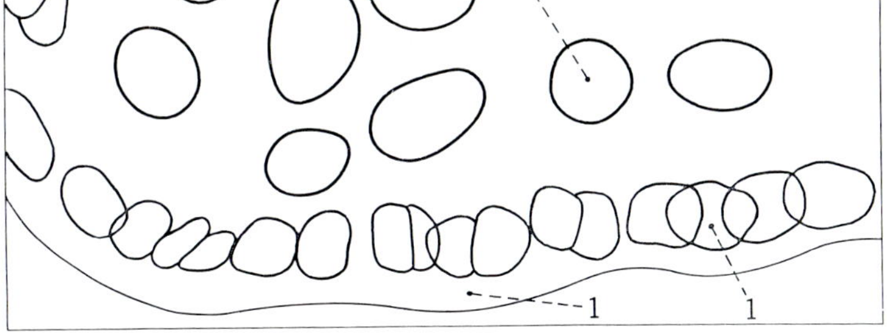

IV. Diagnosis of Malignant Cells

When malignant cells appear in effusions

Both primary and metastatic carcinoma should be suspected. Primary carcinoma, however, is limited only to mesothelioma, and this is a rare tumor. Thus, metastatic carcinoma may be suspected in most cases, and the primary focus must be in a fairly advanced stage. Smear tests of effusions, therefore, are not used for the early detection of carcinoma, which is the original purpose of cytodiagnosis. Instead, cytology serves as a guide for prognosis of the patients and determining treatment.

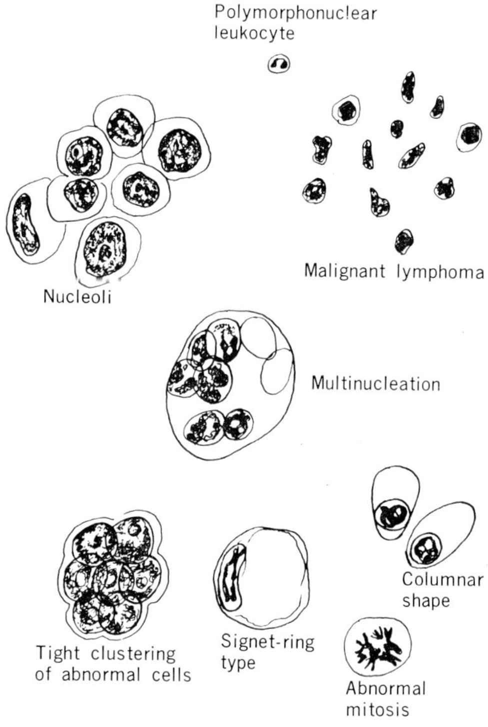

Fig. 58 Various types of malignant cells.

Characteristics of metastatic malignant cells in effusions

Since effusions themselves may have the nature of a tissue culture medium, the cells exfoliated in them rarely become necrotic or cornified and can live and proliferate for a time even after exfoliation. Therefore, the criteria for malignant cells coming directly from a malignant tissue are often not applicable, and distinction of the types of malignancy may be difficult for the same reason.

1) Appearance of large malignant cells

These are irregularly or bizarrely shaped and often form dense clusters. In malignant lymphoma, however, they appear separately and do not form clusters.

2) Changes in the nucleus

1. Considerably varied in size and shape.
2. Hyperchromatic.
3. Chromatin accumulates around the border.
4. Chromatin is very unevenly distributed. Large clumping of chromatin granules may be observed. Fission of the nuclear membrane is occasionally noticed.
5. Abnormal nucleoli : Attention should be paid to the appearance of one or more large nucleoli. The nucleolar-nuclear ratio may be approximately 1 : 3, while it may be 1 : 6 in a benign cell. Try to differentiate such nucleoli from the one or two small nucleoli which are sometimes observed in mesothelial cells.
6. Abnormal mitosis: Abnormal mitotic figures are one of important diagnostic criteria of malignant tumor. One should be able to distinguish abnormal mitosis from that of a benign cell which is sometimes observed in effusions (*cf.* Fig. 58 and p. 209).
7. Multinucleation : Do not confuse a multinucleated histiocyte with a multinucleated malignant cell.

3) Changes in the cytoplasm

1. Size and shape: Not uniform. Generally scanty, compared with the size of nuclei; therefore, the nuclear-cytoplasmic ratio is high.
2. Cytoplasmic vacuolation: This is a characteristic of mucus-producing adenocarcinoma. However, a benign cell can degenerate and be vacuolated in the exudate, and also cells exfoliated from epidermoid carcinoma are occasionally vacuolated. Therefore, vacuolation is not always very significant for determining the type of carcinoma.
3. Necrosis or cytolysis is rarely observed; thus the cytoplasm is often fairly abundant and well preserved. It is seldom stained eosinophilically because it is barely keratinized.

Diagnosis of mesothelioma

This is a rare tumor that affects the pleura and peritoneum and rarely diffuses beyond the primary cavity. The cells exfoliated from mesothelioma are essentially similar to benign mesothelial cells. The tentative diagnostic criteria are as follows.

1. Chromatin is increased and unevenly distributed.
2. Nucleoli are more easily observed.
3. Multinucleated.
4. Cell size varies.
5. Peculiar mulberry-shaped clustering of cells. Such criteria, besides clinical observations, are sometimes helpful in making a diagnosis of mesothelioma.

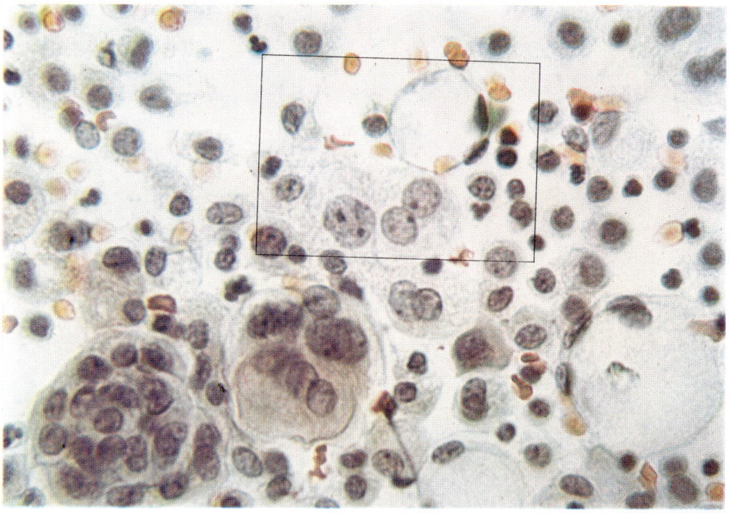

137. Mucus-producing adenocarcinoma of the uterine cervix (Ascites) (*cf.* pp. 101 and 103)

a. Low magnification

Numerous abnormal cells are observed. They are scattered or form rosette-like structure. In these structure, large and small nuclei overlap one another, and the cytoplasm is seen only in the periphery. The borders of individual cells are hardly visible.

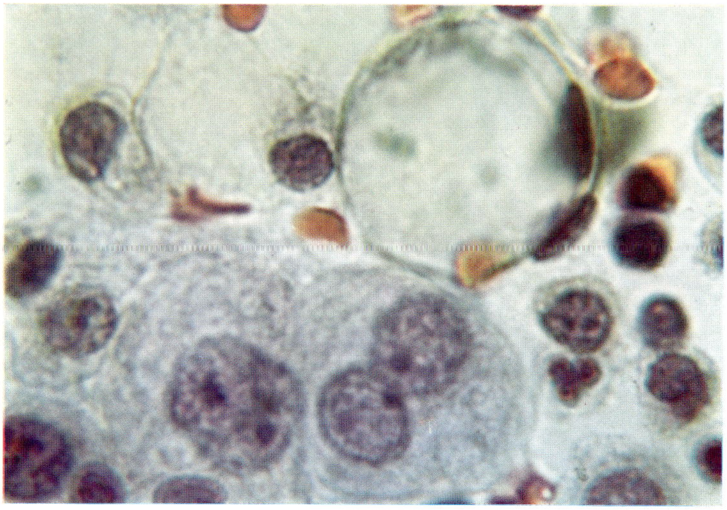

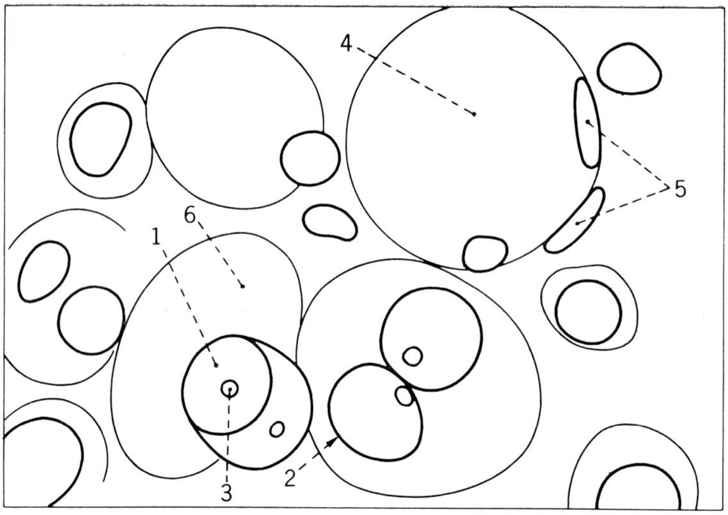

b. High magnification

1. Large cells. The nuclei are roundish but not uniform in size.
2. The nuclear membrane appears thick.
3. Chromatin is unevenly distributed. Large nucleoli are observed.
4. A large cytoplasmic vacuole.
5. The nuclei are pushed to one side by the vacuole and flattened. Very hyperchromatic.
6. The cytoplasm looks thin and lacy

138. Endometrial adenocarcinoma (Ascites)

a. Low magnification

A large cell cluster is observed among numerously scattered round cells.

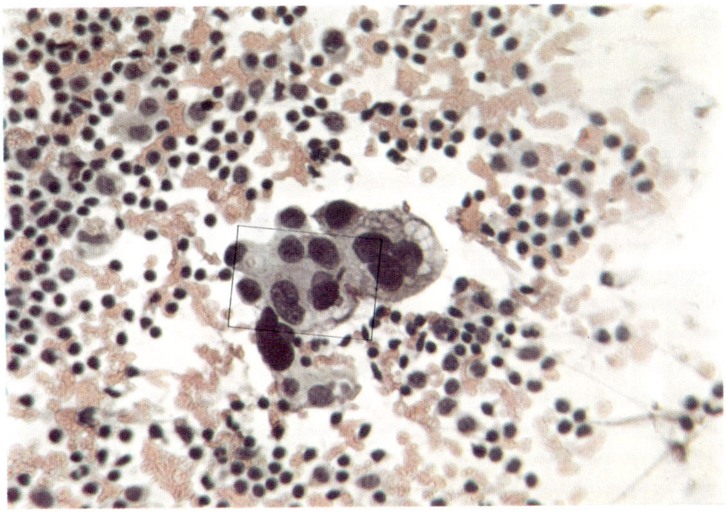

b. High magnification

1. Overlapping large nuclei showing marked hyperchromatism and prominent nucleoli.
2. Characteristically lacy cytoplasm.
3. Cytoplasmic border is visible, though no individual cellular border is seen.
4. Faint vacuoles.

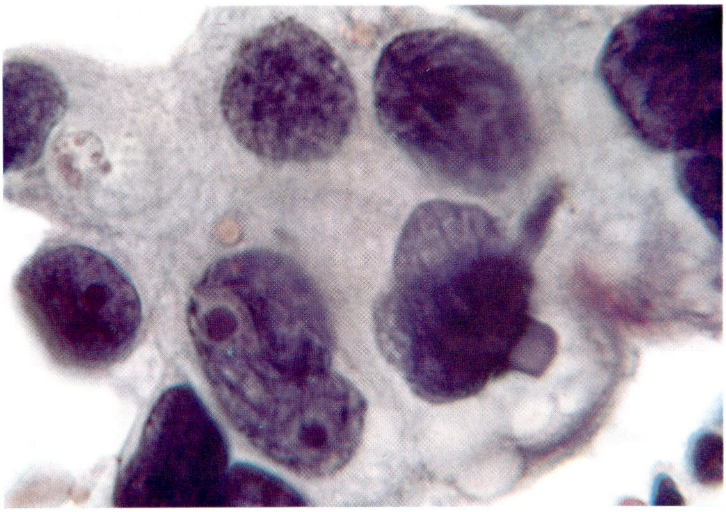

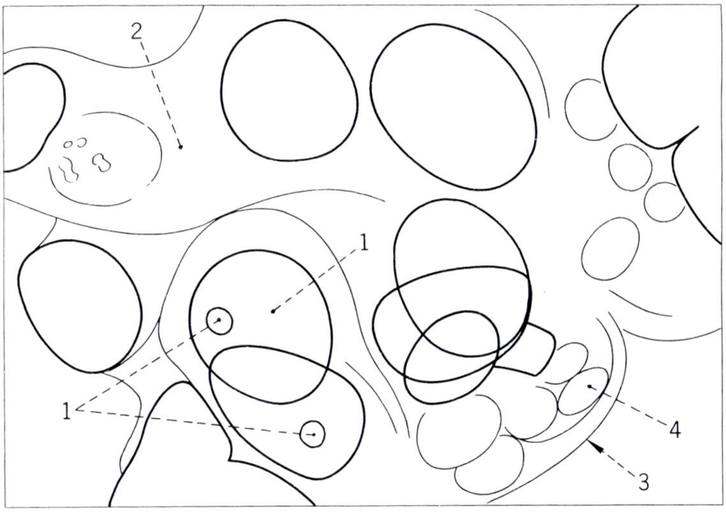

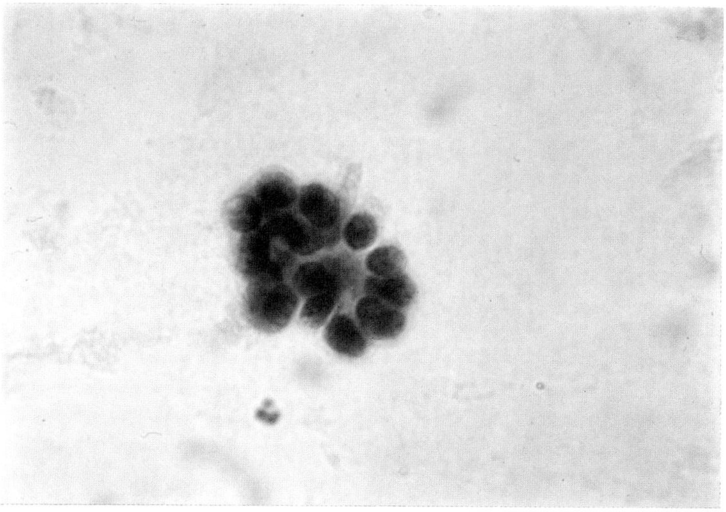

139. Papillary adenocarcinoma of the ovary (Ascites)

a. Low magnification

A peculiar round papillary clustering of cells is observed.

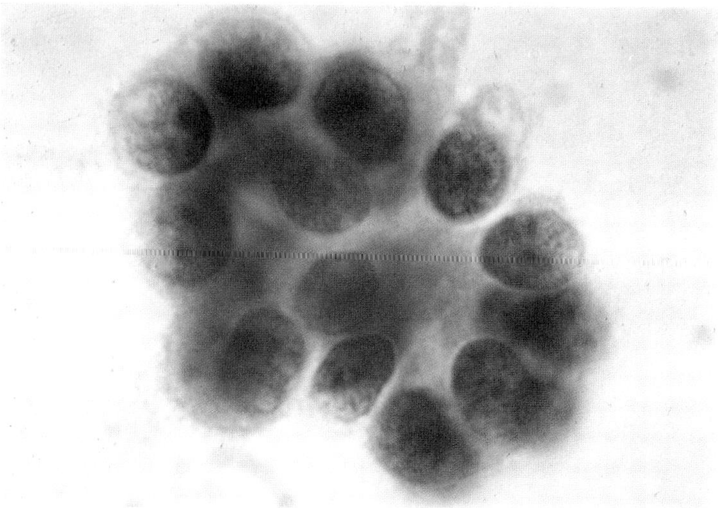

b. High magnification

1. Nuclei: They are dense and overlap one another. It is difficult to distinguish separate nuclei. However, variety of size, irregular shapes and hyperchromatism are seen.
2. Cytoplasm: Seen in small amounts peripherally.
3. Distribution of nuclear chromatin is uneven and to one side.
4. Cytoplasmic vacuoles are observed.

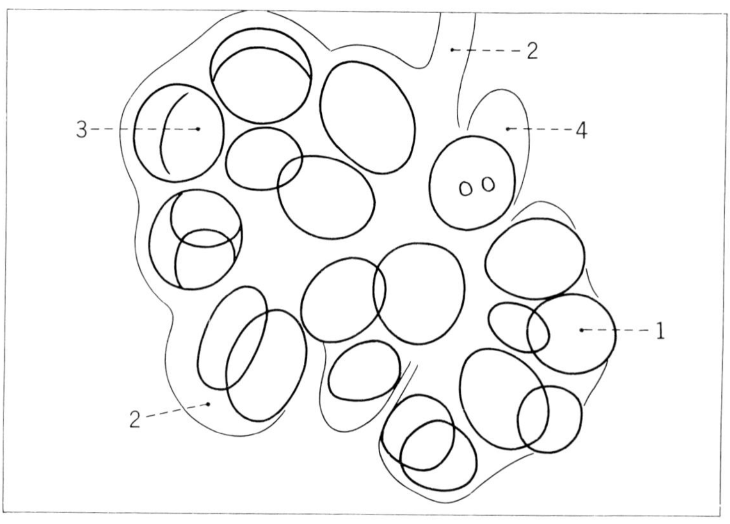

140. Papillary adenocarcinoma of the ovary (Pleural effusion)

a. Low magnification

A few cell groups are scattered. The component cells comparatively overlap each other.

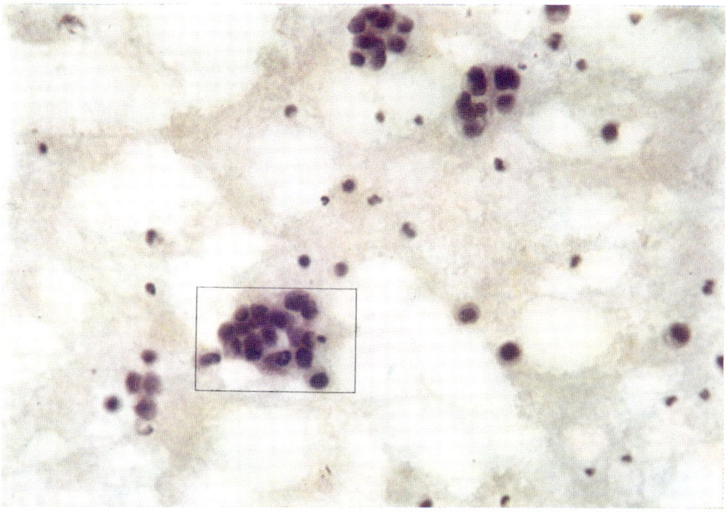

b. High magnification

1. The nuclei are irregularly shaped and hyperchromatic They overlap each other.
2. Chromatin is unevenly distributed.
3. Abnormal mitosis is noticed.
4. The cytoplasm is vague. The border is faintly visible.

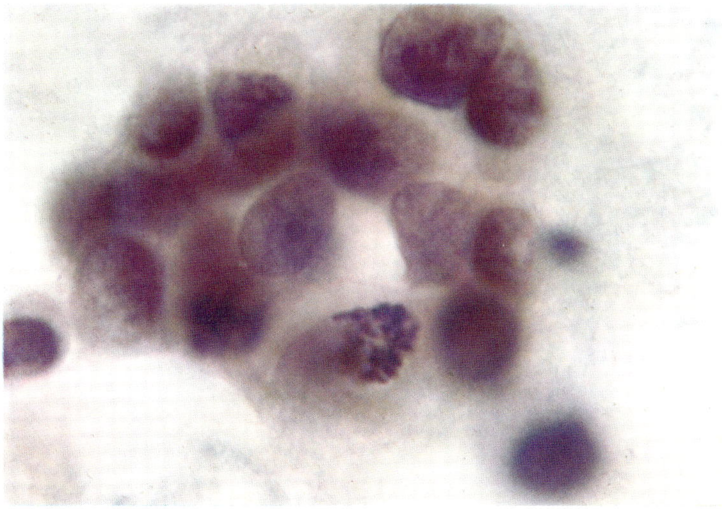

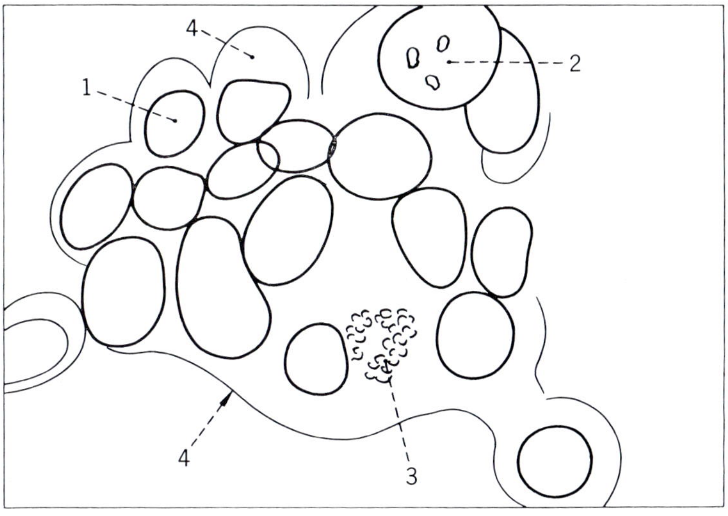

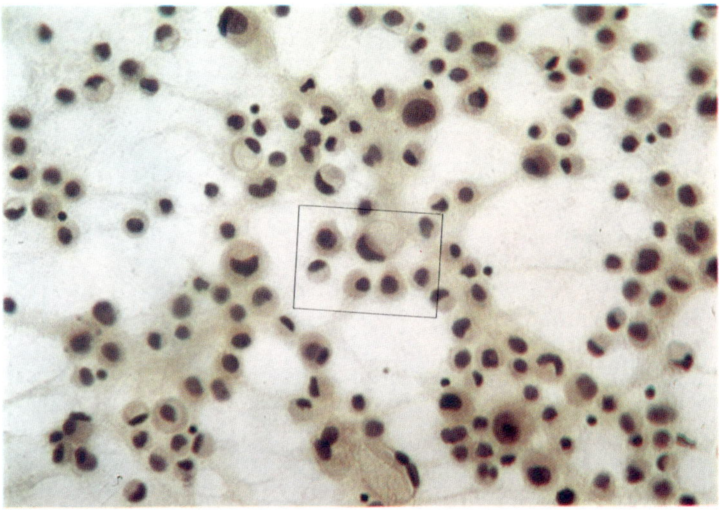

141. Duct cell carcinoma of the breast (Pleural effusion)

a. Low magnification

Large cells appear separately disseminated in the pleural effusion. The nuclei are very hyperchromatic, and not uniform in size and shape.

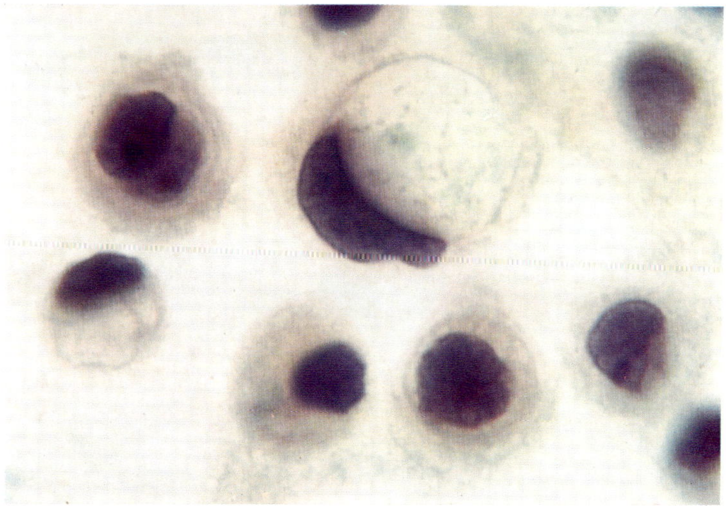

b. High magnification

1. The nuclei vary in size. Irregularly shaped, round, crescent-shaped or polygonal. Generally hyperchromatic.
2. Chromatin is greatly increased and unevenly distributed. No detailed chromatin pattern can be recognized in the hyperchromatic nuclei.
3. The nucleus is pushed aside by a cytoplasmic vacuole and assumes a crescent shape.
4. The cytoplasm is unexpectedly abundant, vesicular and thin.

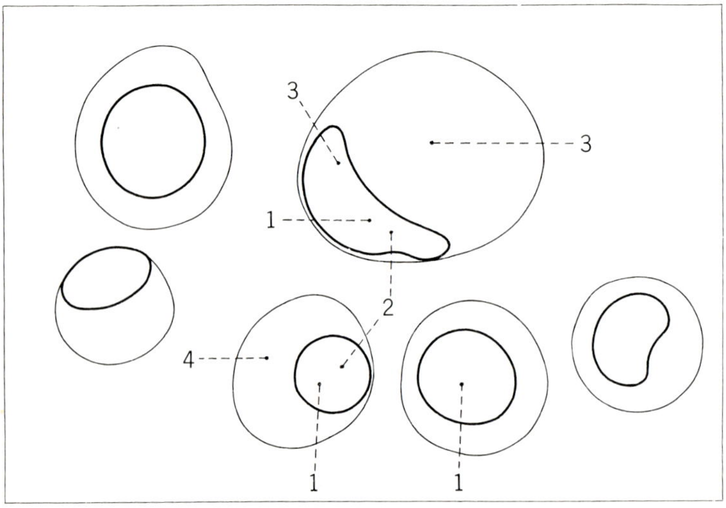

142. Terminal bronchiolar carcinoma (Pleural effusion)

a. Low magnification

There are groups consisting of several cells loosely organized. Some cells are vacuolated.

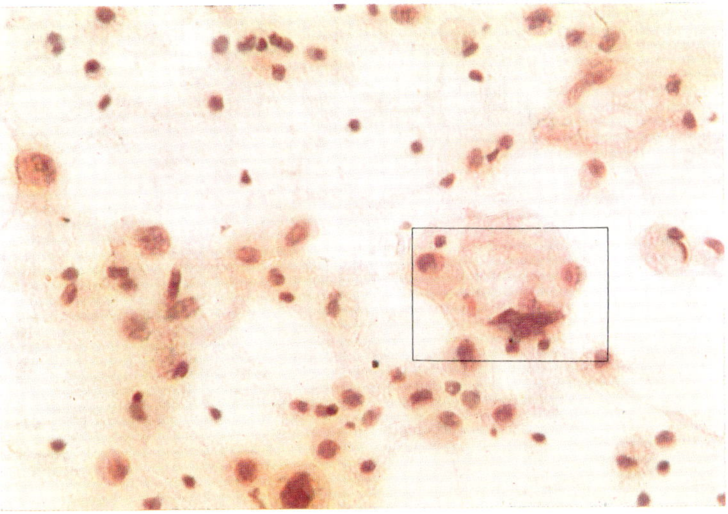

b. High magnification

1. Nucleus : Large, markedly irregularly shaped and hyperchromatic.
2. A nucleus is vaguely observed.
3. Chromatin is unevenly distributed and the nuclear border is distinct.
4. The cell is vacuolated.

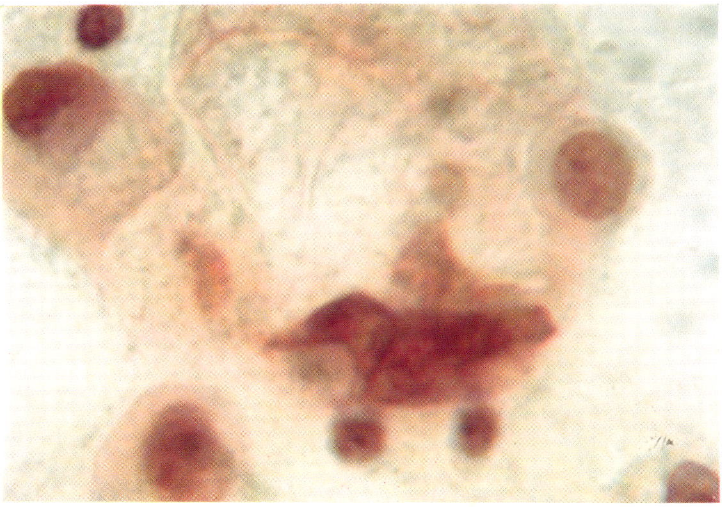

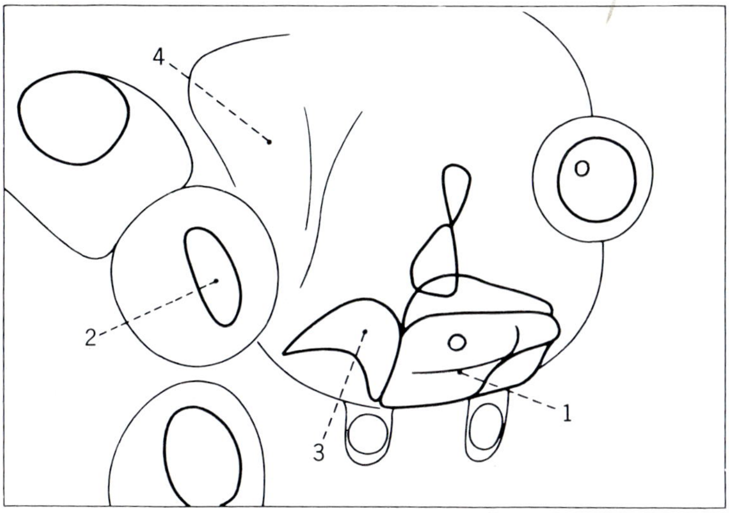

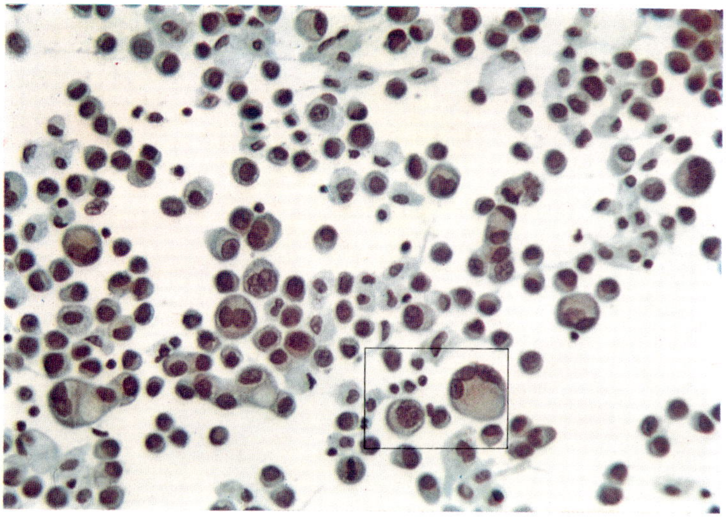

143. Gastric adenocarinoma (Ascites)

a. Low magnification

There are scattered large hyperchromatic cells.

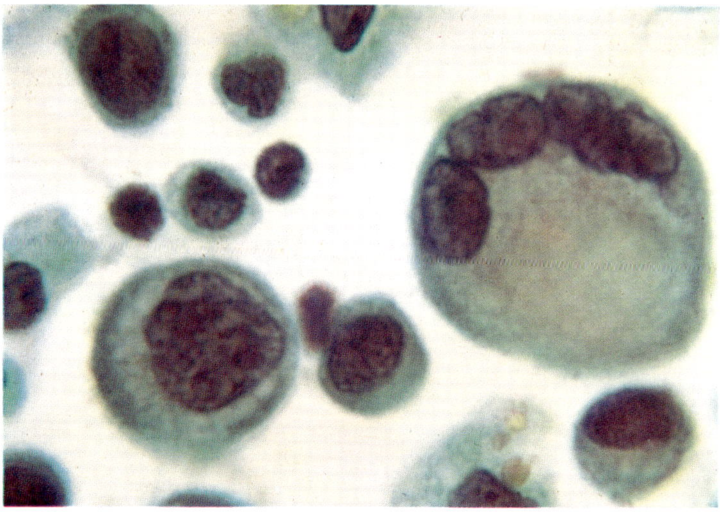

b. High magnification

1. Among several malignant cells, a signet-ring like cell is observed.
2. Large, elliptical, hyperchromatic nuclei. Chromatin is abundant and the pattern is indistinct.
3. Cytoplasm: Distinct and the borders are comparatively clear. Vesiculated and reticulated.
4. Thickened cytoplasmic border.
5. About five nuclei arrange peripherally and overlap each other.
6. Clear nuclear rim.

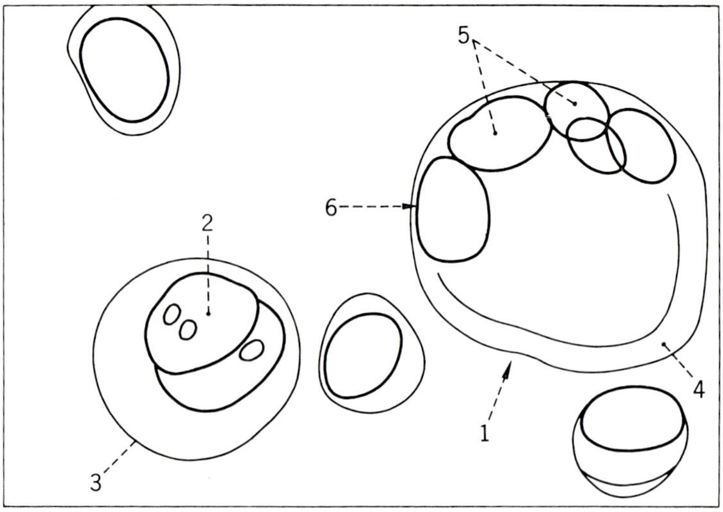

144. Colonic carcinoma (Ascites)

a. Low magnification

Hyperchromatic cells are scattered throughout the field. A yeast-like group is observed nearby.

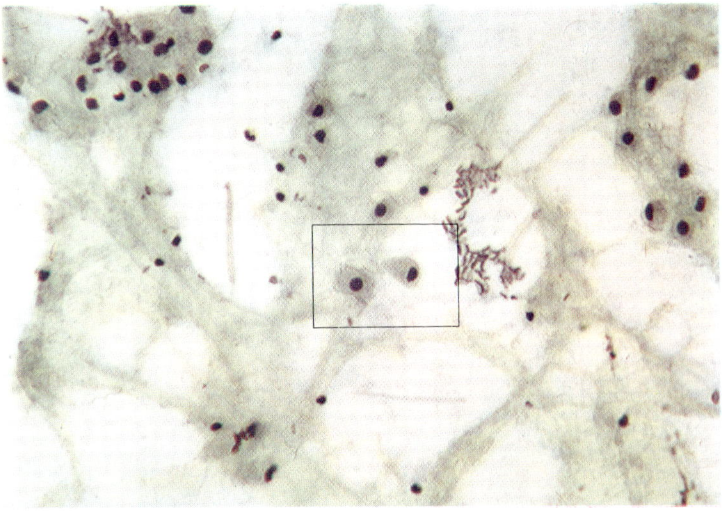

b. High magnification

1. The yeast-like group is a fungus with double refractile borders. This is often observed in protracted effusions or when the patient's general condition worsens.
2. Each cell has a large, round or elliptical and hyperchromatic nucleus. The nucleus is often eccentrically located.
3. The cytoplasm is abundant and reticulated, and the nuclear-cytoplasmic ratio is not very high. More rectangular in shape.

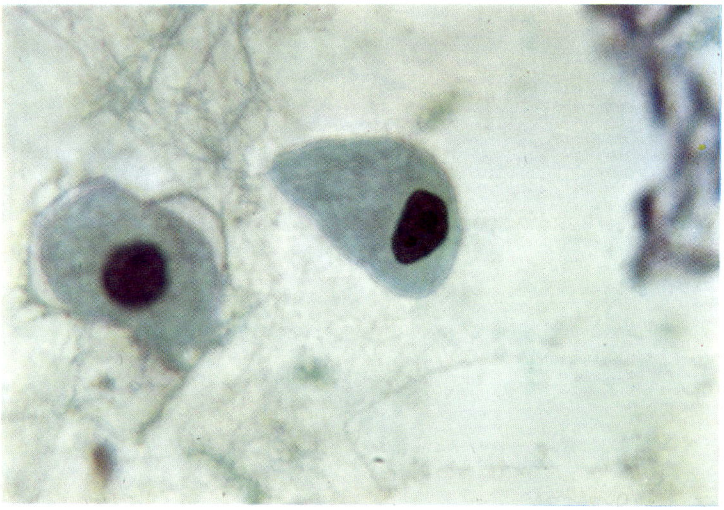

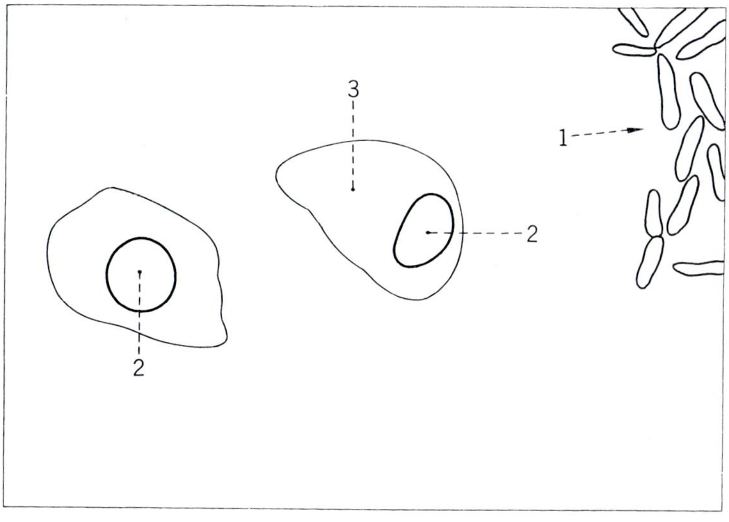

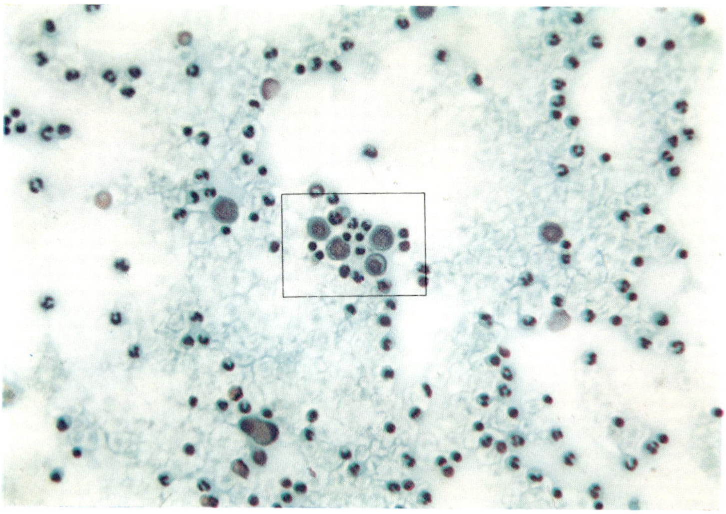

145. Prostatic carcinoma (Ascites)

a. Low magnification

Isolated, comparatively small abnormal cells are present in the center.

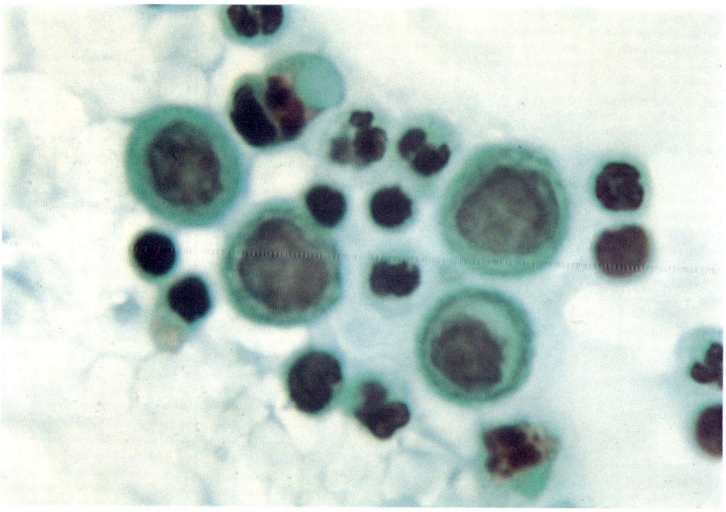

b. High magnification

1. Nuclei: Not too hyperchromatic compared with leukocytes. They are polygonal in shape and vary in size. No definite nucleolus is found.
2. Distinct and thickened cytoplasmic border.
3. Leukocytes.

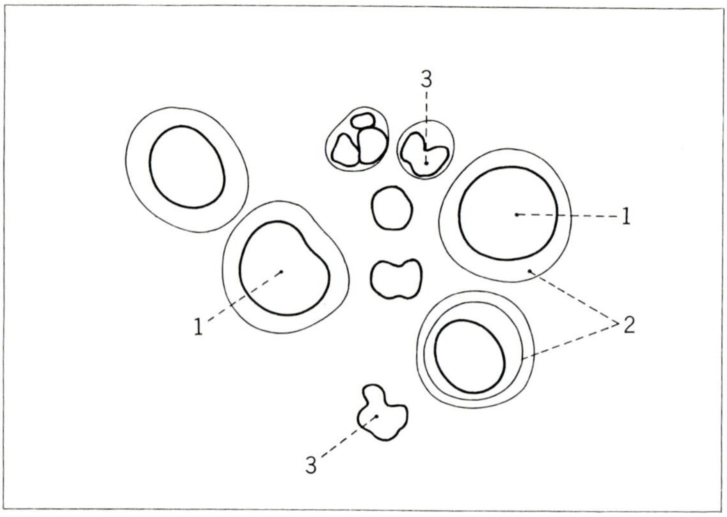

146. Bronchogenic carcinoma — Prominent mitotic figure (Pleural effusion)

a. Low magnification

Beautiful picture of mitosis.

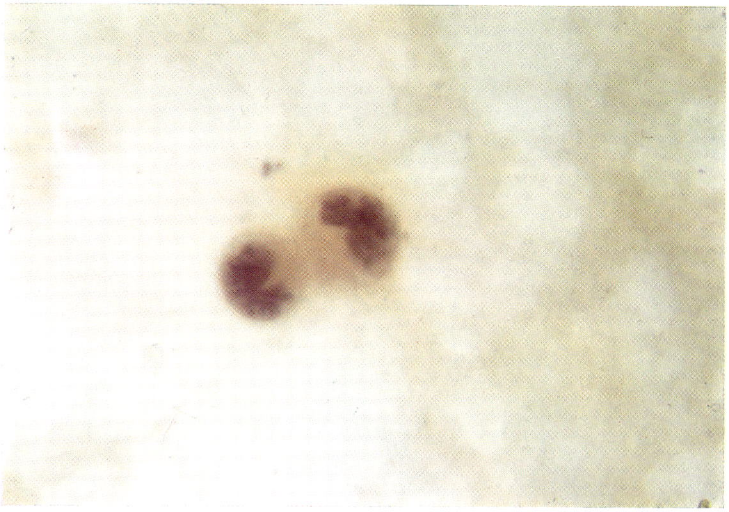

b. High magnification

Cell division in stage of telophase showing that the cytoplasm now constricts in the middle, forming two cells of almost equal size. In both cells, the nuclear membrane may reappear to surround the chromosomes and nucleoli return to the nucleus.

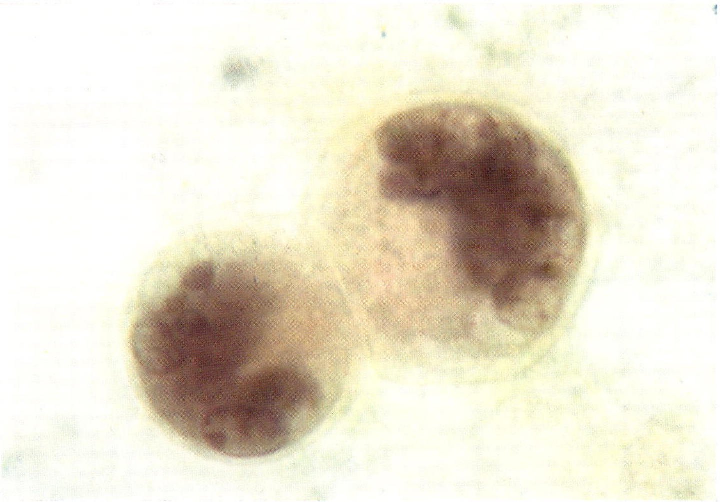

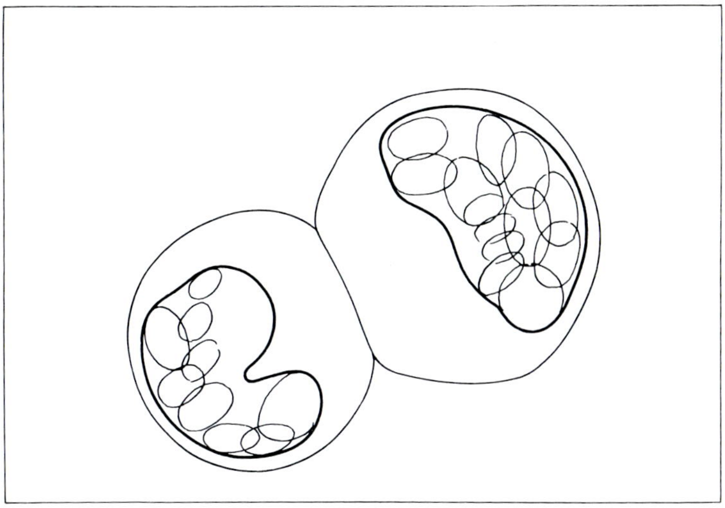

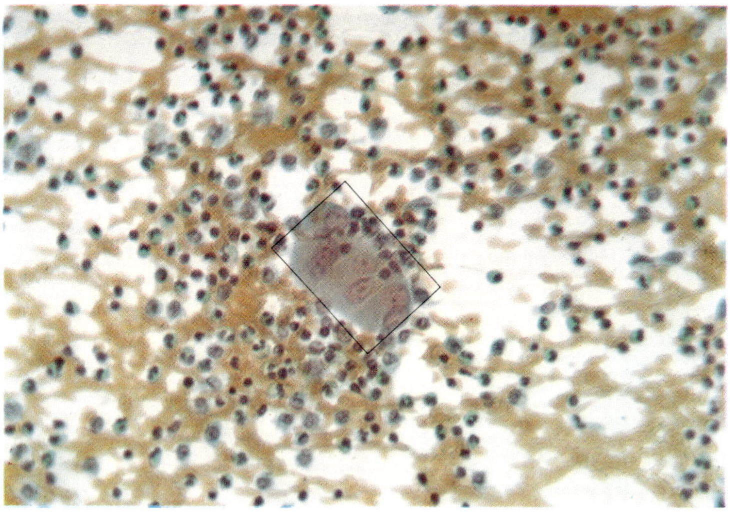

147. Osteogenic sarcoma of the right femur (Pleural effusion)

a. Low magnification

A large multinucleated cell is found among numerous leukocytes and erythrocytes in the background.

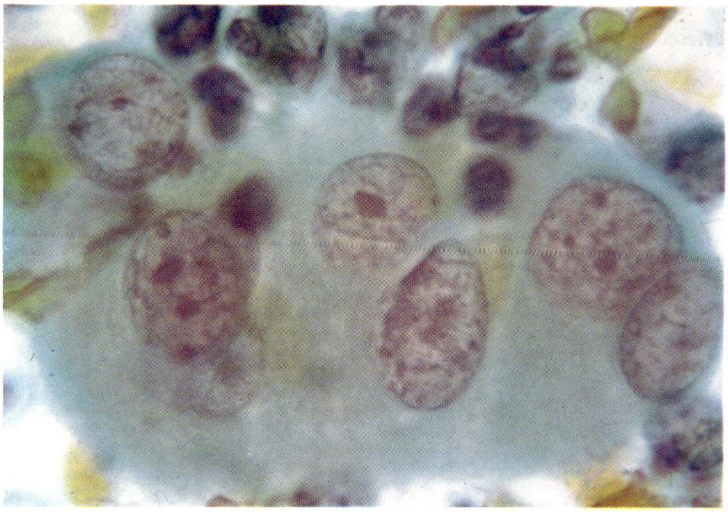

b. High magnification

1. Large, round or elliptical nuclei reveal coarse chromatinic granules and nucleoli. Chromatin content is rather little in amount.
2. Cytoplasm appears thick and the border looks distinct.

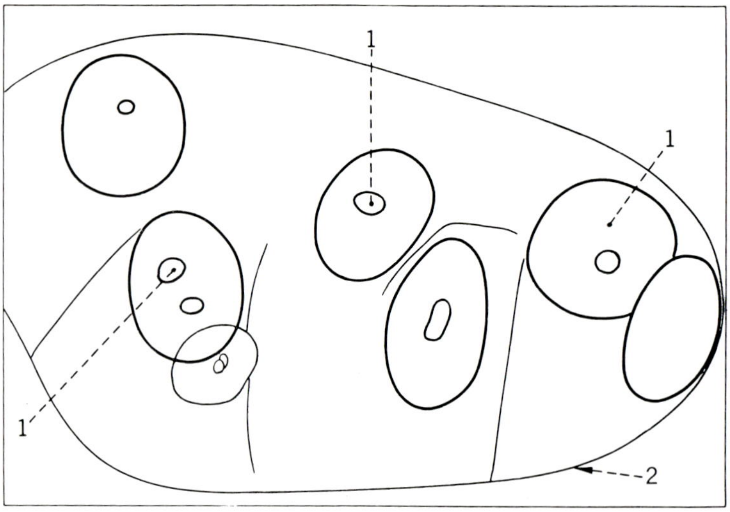

148. Malignant melanoma (Ascites)

a. Low magnification

Large dark-brown cells are scattered throughout the field. The cells are round, and the nuclei are also round or elliptical.

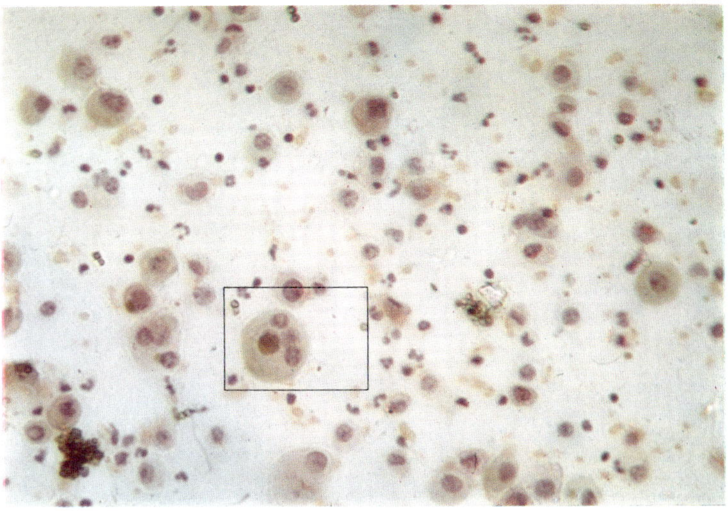

b. High magnification

1. Characteristic cytoplasm: Dark-brown melanin pigment is contained. The whole cytoplasm looks dark. The border is distinct.
2. Nucleus: Large and round. Hyperchromatic.
3. Chromatin is coarsely granular and unevenly distributed.
4. A conglomerated melanin pigment in the cytoplasm.

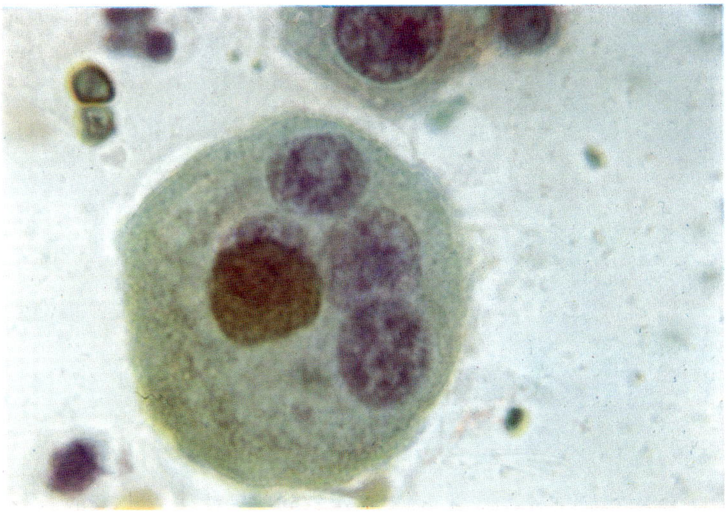

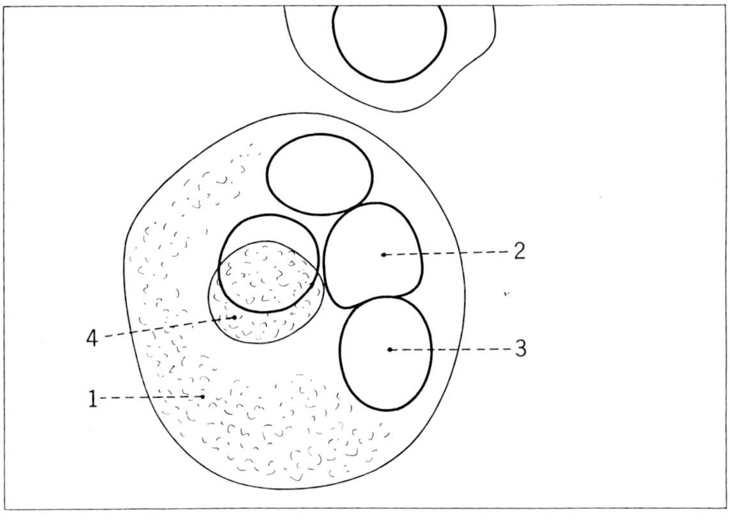

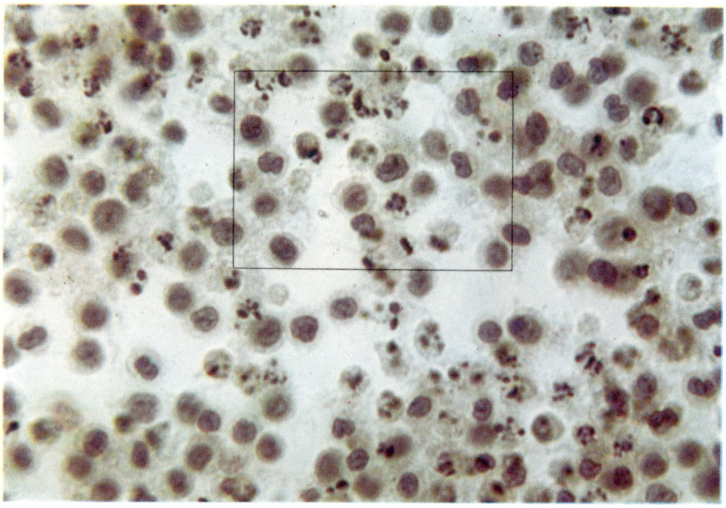

149. Reticulum cell sarcoma (Pleural effusion)

a. Low magnification

The most striking feature is that the cells appear scattered like sesame seeds. Close observation reveals that they are hyperchromatic nuclei of various sizes and irregular shape. There is little cytoplasm.

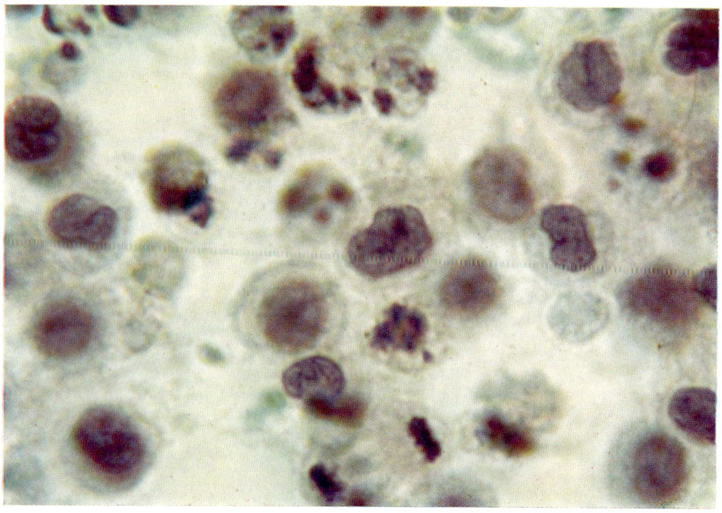

b. High magnification

1. Nuclei: Not uniform in size. Irregular and various in shape. Markedly hyperchromatic. Nuclear structures are almost absent.
2. Cytoplasm: Little or no cytoplasm. The border is vague and indistinct. The nuclear-cytoplasmic ratio is very high.
3. Cells overlap one another. Therefore, when one cell is in-focus the other ones are out-of-focus.
4. Scattered cells showing karyorrhexis due probably to irradiation.

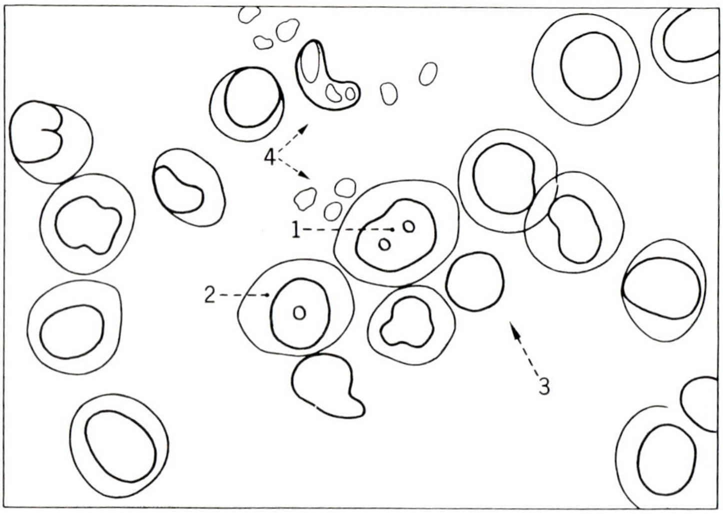

CYTOLOGIC TECHNIQUES

I. Clinical Method of Obtaining Cytologic Specimens and Their Preparation

Collection of materials from the female genital tract and their preparation

Vaginal and cervical smears

1) How to obtain specimens

Various instruments may be used, such as cotton swab, wooden scraper, pipette, platinum aese and small spoons, but a fine cotton swab method is most advisable (Fig. 59).

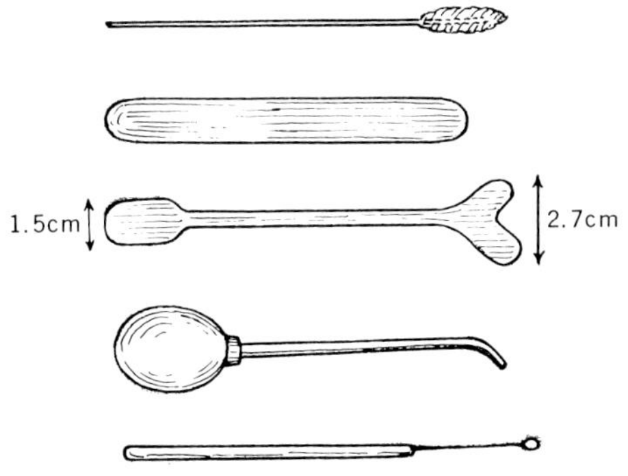

Fig. 59

The advantages are :

1. It does not injure the mucous membrane.
2. It is not painful to the patient.
3. Many cotton swabs can be easily prepared, sterilized and stored for use. They are inexpensive and are disposable.
4. They can be used for obtaining vaginal specimens from both the posterior fornix and the cervical canal.
5. They provide well preserved specimens in their natural state. For hormonal evaluation, the specimens should contain freshly shed squamous cells. The site for obtaining smears is, therefore, limited to the lateral vaginal wall.

Precautions :

1. Advise the patient not to insert tampons or any medication into the vagina within the 24 hours before taking the specimen.
2. Do not make pelvic examinations before obtaining specimens.

3. Instruments for obtaining specimens : For example, a dry vaginal speculum. If the vaginal discharge is scanty, however, and it is difficult to introduce the speculum, it may be moistened with a small amount of normal saline solution. The cotton swab may be lightly dipped in normal saline solution.
4. Collect an adequate amount of specimen by inserting the cotton swab into the posterior vaginal fornix. Sometimes, however, secretion in this area is insufficient. In such case, move the swab all around the portio vaginalis to obtain sufficient material (Fig. 60).
5. After obtaining the vaginal specimen, wipe the area around the external os of the cervix with a cotton ball to remove excess secretions. Then introduce the swab carefully into the cervical canal and rotate it and take cervical specimens (Fig. 61).
6. If the patient is a virgin and insertion of a speculum is not desirable, a fine pipette can be used instead of the swab (Fig. 62).

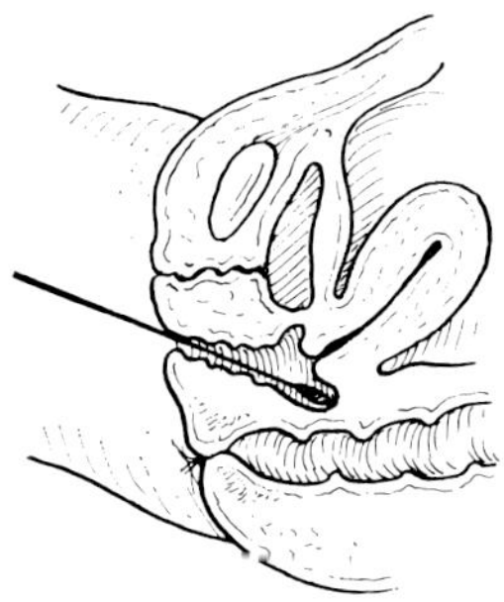

Fig. 60

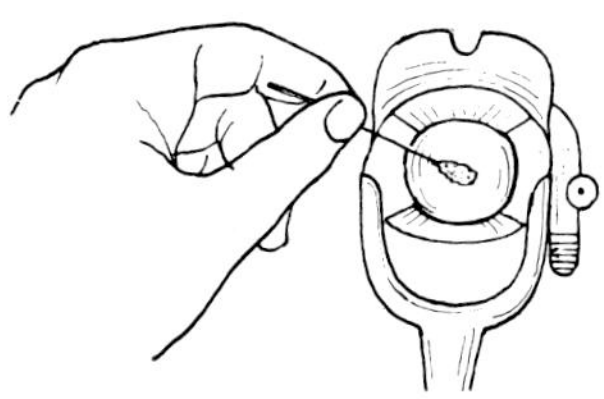

Fig. 61

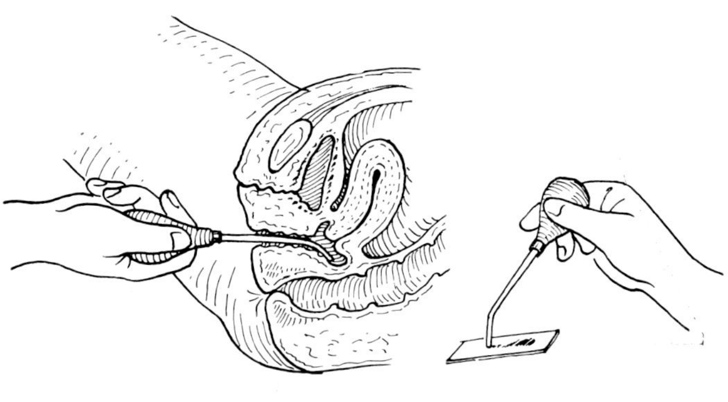

Fig. 62

2) How to prepare specimens

1. Vaginal and cervical smears may be placed on separate glass slides, but it is more convenient to put the vaginal smear in the center of the glass slide and the cervical smear to one side. To distinguish between vaginal and cervical smears it is easy for an experienced examiner, but it is sometimes difficult for a beginner. Therefore, it is convenient for the beginner to know

the location of each smear on the slide. This also has the advantage that vaginal and cervical smears can be easily compared on the same slide (Fig. 63).

2. When applying specimens to a glass slide, it is important to rotate the swab lightly on the slide so that the material is evenly smeared. Leave a little space between the two types of smears to avoid mixing.
3. Write the patient's name or number with a diamond marking pencil at the vacant end of the slide on the same side on which the specimen has been smeared. A label will be pasted here after the slide has been completed (Fig. 63).

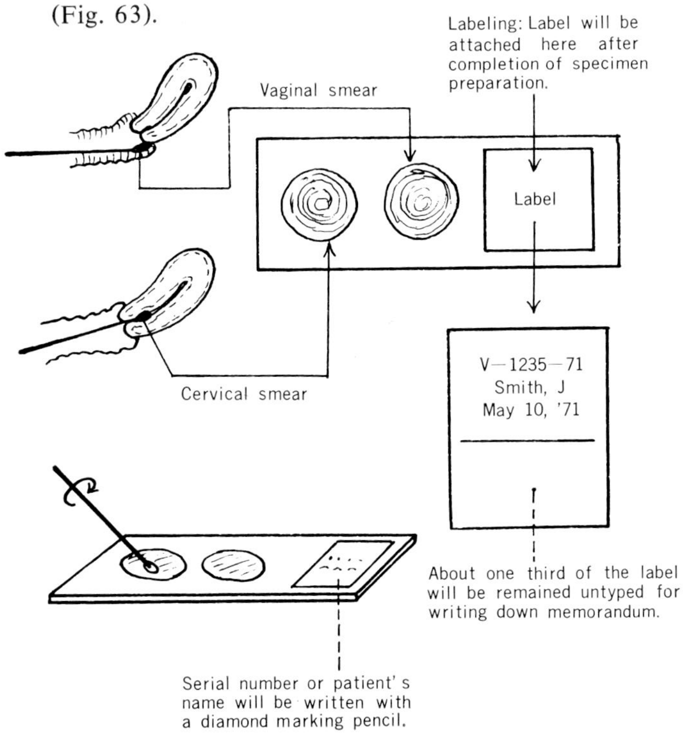

Fig. 63

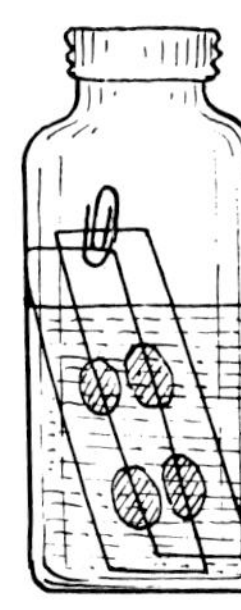

Fig. 64

Incising the slide properly identifies the smear and easily tells which is the smeared side, even after the slide has been soaked in xylol or other solutions for staining. Some glass slides are frosted at one end. An ordinary pencil can be used to label such slides.

4. A small paper clip is sometimes attached to the labeled end of a glass slide to prevent the smeared surface of the slide coming into contact with another slide when several slides are placed in a fixation bottle. Such clips are, of course, unnecessary when the bottle is divided into sections (Fig. 64).

Endometrial smears

1) How to obtain specimens

Obtaining endometrial specimen is not as easy as that of vaginal and cervical material. The most desirable method is one which can be performed sanitarily in the out-patient department without disturbing the patient.

Various methods have been devised, such as: 1) insertion of a brush into the endometrial cavity and rotating it gently, 2) insertion of a specially designed fine cannula, washing the endometrium with a sterile isotonic solution and recovering the solution containing cells, 3) scratching the endometrium gently with a fine curette and smearing the material on a glass slide. The simplest and most accurate method is to introduce a fine cannula with small holes at the tip into the uterine cavity and aspirate the endometrium with the syringe attached to the other end of the cannula.

1. As shown in the figure, the cannula is about 4 mm across and 21 cm long, and the top end is slightly curved like a metal urinary catheter. Four or five small holes are placed at the end. A stylet which fits the cannula snuggly and closes the top holes is inserted into the cannula (Fig. 65).

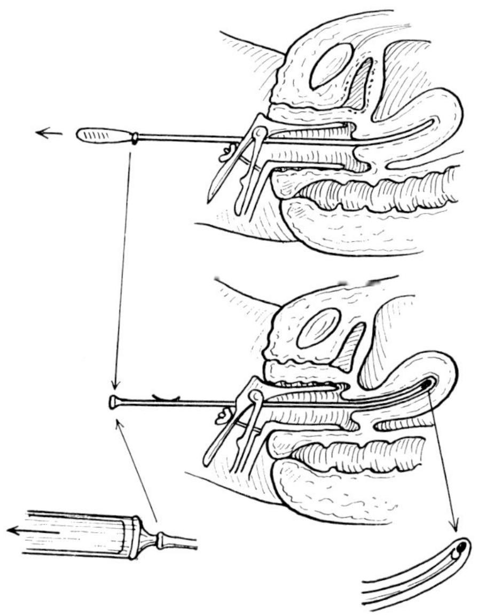

Fig. 65

2. Using a speculum, expose the portio vaginalis of the patient in the lithotomy position on the examining table. After determining the version and the flexion of the uterus, then introduce the sterile cannula gently through the external os of the cervix. If necessary, pull the anterior lip of the portio vaginalis lightly with a tenaculum forceps while inserting the cannula. This usually does not hurt the patient
3. Slight resistance should be felt when the inserted cannula reaches the uterine fundus. It should not be forced any further. Then pull the internal tube out gently and attach a 20 cc syringe to the end. Move the top end of the cannula to and fro in the uterine cavity while aspirating the endometrium lightly. If the top end makes close contact with the endometrium, fairly

strong resistance should be felt when the injector is pulled. Pull the cannula slightly to separate the top end from the endometrium, and a small amount of endometrial tissue will be aspirated into the cannula. Repeat this procedure, moving the top of the cannula little by little to aspirate various parts of the endometrium. The cannula should occasionally be pulled out and the aspirated material should be pushed out onto a clean glass slide so that the top holes will not be clogged.

2) How to prepare specimens

The aspirated material is often blood or mucus-containing small tissue fragments. Therefore, place another glass slide over the smeared slide and crush the specimen lightly by rubbing the two slides together, so that the material is evenly spread on the slides (Fig. 66). If excess blood or mucus is expressed, it should be wiped off with gauze or tissue paper. Write the patient's name with a diamond marking pencil on one end of smeared side of each slide and put both the slides into a fixative.

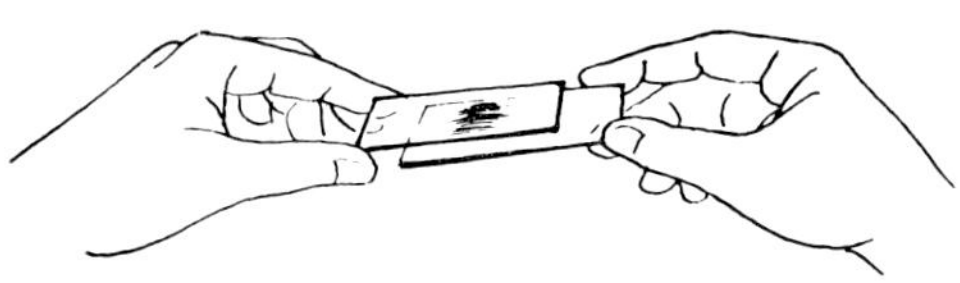

Fig. 66

Aspiration smears are inclined to thicken; therefore, try to make smears as thin as possible. Excess material on a slide will cause difficulty in microscopic observation. If there is too much material for two slides, more slides may be smeared.

Collection of materials from the respiratory tract and their preparation

Sputum smears

1) How to obtain specimens

1. Early morning sputum is the best material.
2. Normal expectoration is the best method for obtaining specimens.
3. Let the patient expectorate into a 100 cc wide-necked bottle filled with 50–70% ethyl alcohol or 2% formalin-Ringer's solution (Fig. 67).
4. When natural expectoration is difficult: Dissolve 5–7 mg of α-chymotrypsin in 3–5 cc of 15% hypertonic saline solution (or a 10% saline solution in inflammation of the lower respiratory tract) containing 20% propylene glycol. Put the mixture in a hot vaporizer and have the patient inhale the vapor. Use the expectorated sputum within 24 hours of the inhalation (Fig. 68).

2) How to prepare specimens

Place the sputum in a Petri dish and examine the contents carefully against a black background. Select blood, pus or a tissue fragment, if there is any. Place a small amount between two glass slides, and grind with a circular motion. Spread the material as evenly as possible on the two slides so that the smears are about as

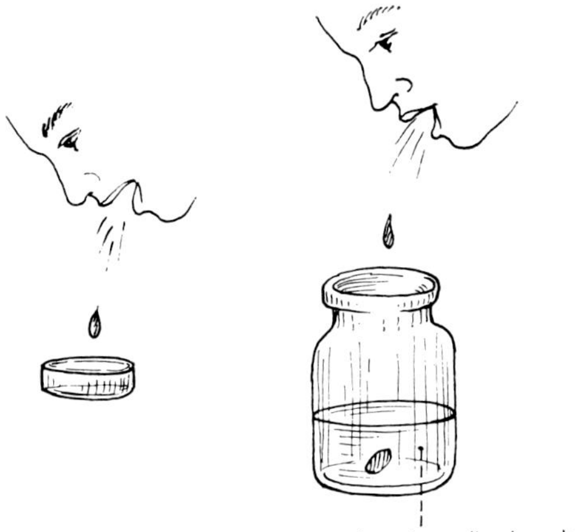

Fig. 67

Fig. 68

thin as blood smears (*cf.* p. 217 : how to prepare endometrial aspiration smears).

Place the slides in a fixative immediately. At least four smears should be made for examination. The patient's name or number should be clearly marked with a diamond marking pencil on one end of each slide.

Bronchial washings or brushing smears

1) How to obtain specimens

One can aspirate secretions using an aspirating tube and a bronchoscope or inject 1–2 cc of normal saline solution and aspirate it after light washing. Another method is to brush the focus directly or scrape it with a small metal scraper under roentgenoscopy. With this method it is essential to accurately determine the location of the focus beforehand by bronchography.

2) How to prepare specimens

Place the washings in a 50 cc centrifuge tube and centrifuge at 1500 rpm for 30 minutes.

Remove as much as possible of the supernatant and make smears from the sedi-

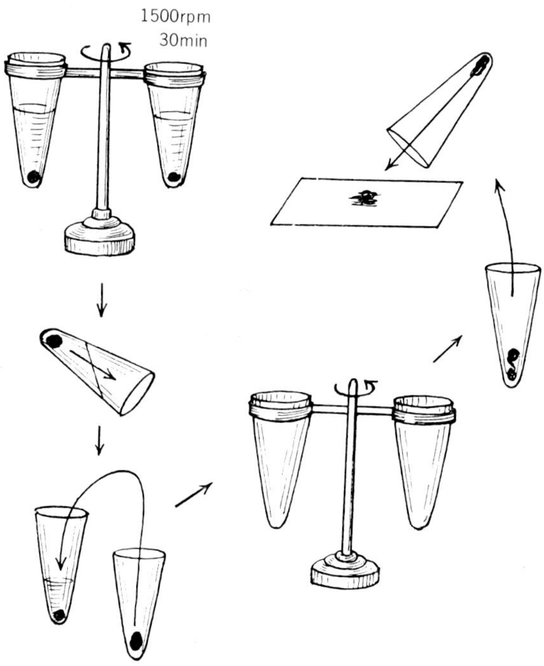

Fig. 69

ment in ordinary way (Fig. 69). A small amount of albumin to the glass slide may help to keep the smear attached during fixation and staining, but it is not absolutely necessary. Scraped specimen should be treated in the same way as sputum, *i.e.*, the brush can be applied directly to the slide.

Collection of materials from the digestive system and their preparation

1) Oral cavity

Wipe with a cotton swab in the same way as in obtaining the vaginal specimen.

2) Esophagus

Insert a Levine tube, inject normal saline solution and aspirate the solution. Or scrape the focus gently and carefully under esophagoscopy. Before brushing, insert a gastric sound first, then a small brush through the sound to brush the focus. Wash the brush in normal saline solution, centrifuge the solution and smear the sediment.

3) Stomach

1. Gastric irrigation is often used. The patient should fast on the examination day. If possible, irrigate the stomach the previous evening and keep it empty and clean.

   The patient should be in a sitting or lateral decubitus position. Insert a sound. If the patient gags, anesthetize the throat lightly with a xylocaine spray. When the sound reaches the stomach, inject normal saline solution and wash the stomach until the washings become clean.

Then inject about 300 cc of α-chymotrypsin solution (7 mg/ 1/10M acetic buffer solution) and repeat the aspiration and injection alternately. This is quite tedious; therefore, a mechanical method has been devised by the Cancer Institute of Japan. Collect as much of the washings as possible and centrifuge them at 2000 rpm for 10 minutes. Smear the sediment on a glass slide to which albumin has been applied (Fig. 70).

2. Abrasive ballon method. The patient should be fasting from the previous evening. Instruct him to swallow a balloon covered with a fine net. When it reaches the pylorus, remove the gastric juice first and enlarge the balloon by inflating it with air. Then pull the balloon up to the cardia and remove the air. Push it down again to the pylorus, inflate it and pull it up to the cardia. After repeating this procedure several times, remove the balloon. Then wash the stomach with normal saline solution, collect the washings and make smears from the sediment.

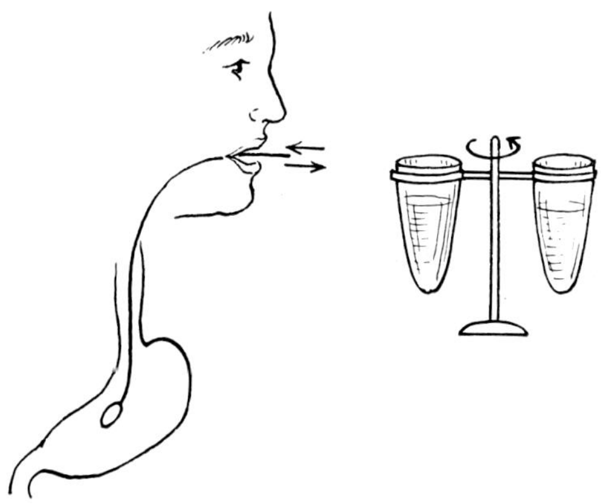

Fig. 70

3. Rotary brush method. This is similar to the abrasive balloon method. Insert a firm, two-winged brush into a polyethylene tube with a metal point. After inserting the tube into the stomach, push the brush out of the tube. Brush the gastric wall by rotating the brush. Remove the brush from the stomach, dampen it with normal saline solution and apply it directly to a slide.
4. Collection of cells under gastroscopy. Push sound wrapped with cotton, a small curette or small forceps from the tip of a gastroscope and scrape the wall of the stomach. Inject washing solution directly onto the focus to accelerate the exfoliation of cells.

Collection of materials from the urinary tract and their preparation

1) How to obtain specimens

Voided or catheterized urine may be used.

Use urine voided early in the morning. During catheterization, move the catheter back and forth in the bladder and scrape the wall lightly to obtain more cells.

Ureteral catheterization has the advantage that the source of exfoliated cells can be located to some extent, because urine can be obtained separately from each kidney.

2) How to prepare specimens

As soon as it is voided, place the urine in a centrifuge tube and centrifuge it at 1500 rpm for 30 minutes. If there is not enough sediment, divide a large volume of urine into many centrifuge tubes and centrifuge the contents of each tube. Discard the supernatant and suspend the sediment again in a small amount of urine. Pool all material in one tube and centrifuge it to obtain the sediment (This is the same method as in centrifugation of bronchial washings; *cf.* pp. 218-219). Add two or three drops of an albumin solution to the sediment and smear it on a glass slide to which albumin has been applied. Place the slide gently in a fixative when the smear begins to get dry.

N. B.: If the specimen cannot be immediately smeared, add an equal amount of 95% ethyl alcohol or [40 cc of alcohol + 58 cc of 1/2 diluted Ringer's solution + 2 cc of formalin] to it for preservation. The other procedures are the same.

Collection of materials from the body cavities and their preparation

Material is collected by peritoneo- or pleuro-centesis, or directly from the peritoneal or pleural cavity during surgery. If possible, obtain the sediment immediately from the collected effusion by centrifugation. A quantity of 3.7% sodium citrate equivalent to one fifteenth of the amount of specimen may be added to prevent fibrin reduction. When immediate centrifugation is not possible, add an equal amount of 50% ethyl alcohol or [30 cc of alcohol + 68 cc of Ringer's solution + 2 cc of formalin] to the material. In such cases, however, protein in the exudate sometimes coagulates and makes the solution turbid and collection of cells difficult. Thus it may sometimes be impossible to obtain good smears.

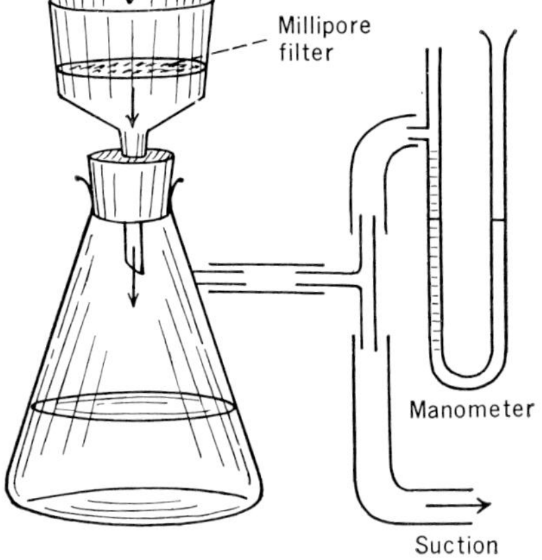

Fig. 71

The other smear-making procedures are the same as those for urinary smears. The Millipore filter method can be adopted when there is plenty material. Filter the material through a Millipore filter with a pore size of 0.5–5 μ. The cellular elements remain on the filter. Paste the filter onto a glass slide and examine it microscopically

Filter the material under normal pressure or negative pressure of about 25 mm Hg, and collect the cellular constituents on a Millipore filter (Fig. 71). The specimen should be fixed in 10% formalin. The filter can also be dissolved. In this case, filter about 10 cc of the material as described above. Place two albumin smeared slides close together and attach the filter face down on them Cut the filter between the two slides so that half of the filter is attached to each slide Place the slides in a fixative of acetone and alcohol (3 : 1) or 95% methyl alcohol and 95% ethyl alcohol (1 : 20) and fix for 2–3 hours. Then place in 95% ethyl alcohol for 10 minutes and stain.

II. Fixation of Specimen

Whether a smear is good enough for cytodiagnosis depends upon fixation. The first step to a good smear is rapid and proper fixation.

Proper fixation prevents cells from being deformed and degenerating and assists staining.

The most common method

The most common fixation is a solution of equal parts of ether and 95% ethyl alcohol. It has been used the most often since Papanicolaou and is most suitable for gynecological, prostatic and mammary specimens which can be smeared on slides immediately after collection.

The specimen should be fixed in the fixative as soon as possible before they get dry. When specimens are smeared on a glass slide, the excess water dries first, but it takes about 2–3 minutes for the specimen to dry completely. It is important to put the smear in the fixative immediately after the smear was made and before the material begins to dry (Fig. 72).

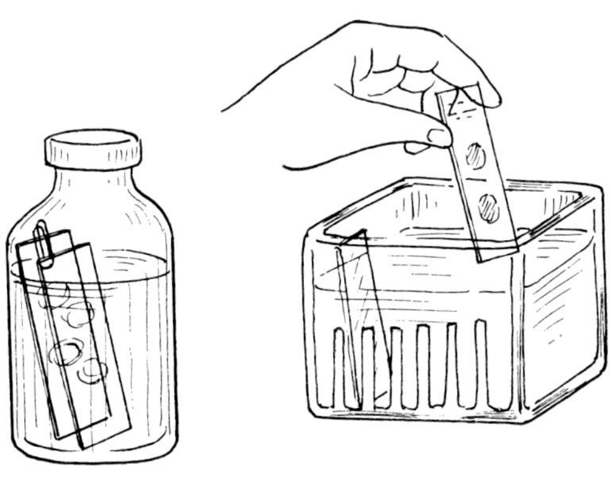

Fig. 72

Fixation time: At least 15 minutes are necessary, but it is best to fix the smear for more than one hour.

Sometimes smears can be left in the fixative for more than several weeks. This may cause some changes in staining reaction but this produces no significant diagnostic changes.

Fixative that is used repeatedly should be filtered as often as possible to prevent smears from being contaminated with floating debris which may cause misdiagnoses. The fixative container should be covered to prevent the ether and alcohol from evaporating.

Fixation by 95% ethyl alcohol without ether

Recently, fixation by ethyl alcohol alone without the highly inflammable and evaporable ether has been adopted. Smears are adequately fixed by this method, also. The procedure is the same as the ether-alcohol method. The use of methyl alcohol instead of ethyl alcohol has an economic advantage although the results are somewhat inferior.

Fixation by placing the smear into a solution of polyethylene glycol (M. W. 4000) and 95% alcohol (1 : 9) or dropping this solution on the smear is also recommended.

" Spray-cyte " method

Recently, Spray-cyte manufactured by Clay Adams has become popular. This is an aerosol spray fixative. It can be sprayed on the material as soon as smeared on a slide. This method is recommended because it is simple and the results are as good as those with the ether-alcohol method.

It should be sprayed evenly for several seconds from a distance of about 15–20 cm from the object. It dries within several minutes. Therefore, the smear can be stained right away (Fig. 73). Furthermore, since the spray is watersoluble, the smear can be placed in a hematoxylin solution immediately for staining ; thus the staining time is shortened.

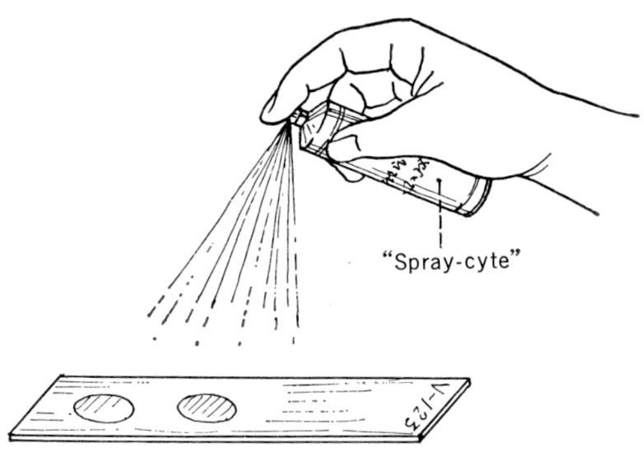

Fig. 73

Fixation of sputum, bronchial washings, effusions, gastric washings and urine

1) Sputum

As stated above (*cf.* p. 217), if possible, the specimen should be smeared while it is fresh and placed in the ether-alcohol fixative. If this is not possible, the sputum should be expectorated into a wide-necked bottle filled with 50% ethyl alcohol or 2% formalin-Ringer's solution (*cf.* p. 217). The smears made from such specimen are satisfactory.

2) Liquid specimen such as washings, effusions and urine

It should also be smeared while fresh. Centrifuge the specimen to collect the sediment, smear it on a glass slide and fix it in ether-alcohol. If possible, divide all of the specimen into several 50 cc centrifugation tubes. Centrifuge each tube, discard the supernatant and suspend the sediment in a small quantity of the specimen. Pool the material from all the tubes and centrifuge again. Remove as much as possible of the supernatant and smear the sediment on a glass slide.

The smeared slide is usually fixed in ether-alcohol.

When the material cannot be made into smears immediately after collection, add an equal quantity of 50 or 95% ethyl alcohol or an alcohol-Ringer-formalin solution (*cf.* p. 221) to the material and store in a refrigerator until use.

III. Treatment of Unfixed and Air-dried Smears

The cells of unfixed and air-dried smears are often deformed and are hard to diagnose accurately. In such cases, rehydrate the smear first and then stain it by Papanicolaou method (Bonime, R.G.: Obstet. & Gynec. **27,** 783, 1966).

1) Rehydration of an air-dried smear

Soak in a solution of equal parts of glycerin and distilled water for 3 minutes. Smears which have been dry for more than 10 days should be left in this solution longer.

2) Removal of the rehydrating solution

① 95% ethyl alcohol 15 sec → ② 95% ethyl alcohol 15 sec → ③ Distilled water 15 sec

Or place the smear in gently running water for 90 seconds.

3) Fixation

95% ethyl alcohol — 15 min.

4) Staining procedure

① 90% ethyl alcohol 15 sec → ② 70% ethyl alcohol 15 sec → ③ 50% ethyl alcohol 15 sec

→ ④ Distilled water 15 sec → ⑤ Harris hematoxylin 6 min → ⑥ Distilled water 10 dips

→ ⑦ 0.5% hydrochloric acid solution 1-2 quick dips → ⑧ Distilled water 15 sec → ⑨ 1.5% ammonia solution (in 70% ethyl alcohol) Dip until the smear is stained blue.

→ ⑩ 50% ethyl alcohol 15 sec → ⑪ 70% ethyl alcohol 15 sec → ⑫ 95% ethyl alcohol 15 sec

→ ⑬ OG-6 2 min → ⑭ 95% ethyl alcohol 10 dips → ⑮ 95% ethyl alcohol 10 dips

→ ⑯ EA 50 (36) 3 min Use EA 65 for urine and sputum. → ⑰ 95% ethyl alcohol (three baths) 10 dips in each → ⑱ 100% ethyl alcohol 15 sec

→ ⑲ Xylol (three baths) The slides may be left for a while in the last bath. → ⑳ Coverslipping

IV. Staining Techniques

Usually Papanicolaou's method is used for staining smears. Some people may recommend Giemsa's stains or a simple hematoxylin-eosin method. In cytodiagnosis, however, Papanicolaou's stains should be used as the standard method, and the examiner should get used to cytodiagnosis by this method.

Papanicolaou's stains

Advantages

1. The nucleus is properly stained. This is essential for observation of the nuclear detail and differentiating between benign and malignant cells.
2. The cytoplasm is clearly stained. This is essential for observation of cytoplasmic structures and to note the difference between these and the nuclear structures.
3. The cytoplasm is transparent. This is essential for observing underlying cells when cells are overlapping.

Preparation of staining solutions

1) Harris hematoxylin

Used to stain the nucleus. The nuclear membrane and chromatin are stained dark-blue or reddish-purple, and the nucleolus red, pink or orange. Similar staining reactions can be expected from other hematoxylin solutions, too, if the staining time is adjusted properly. The advantage of Harris hematoxylin is that it is easily prepared and allows to age rapidly, so that it can be used after 24 hours.

a. How to prepare the stock solution :

Hematoxylin	5 gm
95% ethyl alcohol	50 cc
Ammonium alum	100 gm
Distilled water	1,000 cc
Mercuric oxide	2.5 gm

1. Dissolve hematoxylin in alcohol.
2. Place distilled water in a beaker and dissolve ammonium alum in it by heating. Heat the solution until it boils.
3. Add the hematoxylin solution to the alum solution and heat again to boiling.
4. Remove heat and add Mercuric oxide immediately. Stir gently until the solution becomes dark reddish-purple.
5. Plunge the beaker into a cold water bath to cool it rapidly.
6. The cooled solution should be filtered, placed in a dark bottle, and left for at least 24 hours to allow to age.

b. How to use the stock solution : Use the stock solution in full-strength or dilute it with an equal volume of distilled water. Filter the solution each time before using. Cork the bottle and store in a dark cool place.

2) Orange G (Papanicolaou OG-6)

Orange G and EA solutions are used to stain the cytoplasm. Both are alcoholic solutions and the alcohol makes the cytoplasm transparent.

OG-6 stains the cytoplasm containing keratin, yellow or orange. Keratin may be present in benign cells, but it is often abundant in a cell exfoliated from epidermoid carcinoma. The cytoplasm stains a beautiful orange with OG-6, in contrast to the hyperchromatic nucleus which stains a dark reddish-purple with hematoxylin. This helps to make a diagnosis.

How to prepare the solutions :

This is a 95% alcoholic solution, but Orange G powdered dye is not readily soluble in alcohol. Therefore, prepare the aqueous solution first and dilute it with alcohol.

1. Place 10 gm of Orange G in a beaker. Add distilled water until the total amount is 100 cc, and stir to dissolve. Bottle, stopper and leave in a dark cool place for several days.
2. How to prepare OG-6

10% Orange G aqueous solution	50 cc
95% ethyl alcohol	950 cc
Phosphotungstic acid, C.P.	0.15 gm

Place the above in a dark bottle, stopper and store in a dark cool place. Filter the solution before use.

The best color balance using this prescription is obtained when dyes from the National Aniline & Chemical Company are used. Therefore, strictly speaking, the color balance should be adjusted when the stain of any other company is used.

3) EA- (eosin, light-green and Bismarck brown) alcoholic solutions

The staining solutions originated by Papanicolaou are marked EA. EA-36 (The commercial solution of the same composition is named EA-50) is for vaginal smears and EA-65 is for other smears. EA-65 contains one half of the amount of light green contained in EA-36 ; therefore EA-65 is better for thicker smears if the green color is inclined to be too dark.

Eosinophilic cytoplasm → stains well with eosin → pink to yellow in various degrees

Basophilic cytoplasm → stains well with light green → pale blue to greenish-blue in various degrees

Generally, cornified superficial cells are eosinophilic and the other cells are basophilic. However, the staining is changed whenever the smear is dry, the pH of the secretion is different or the smear is too thick or thin ; therefore, the color of the cytoplasm is often not a very reliable factor in the smear test.

a. How to prepare EA-36 :

1. Prepare the following aqueous solutions with distilled water.
   2% light green S. F., yellowish
   10% Bismarck brown Y
   Store in a dark cool place for several days.
2. Prepare the following alcoholic solutions (stock solutions).
   0.1% light green stock solution (A)

2% light green aqueous solution	50 cc	Mix
95% ethyl alcohol	950 cc	

0.5% Bismarck brown stock solution (B)

10% Bismarck brown aqueous solution	5 cc	Mix
95% ethyl alcohol	95 cc	

0.5% eosin stock solution (C)

Eosin Y	5 gm	Dissolve
95% ethyl alcohol	1000 cc	

3. To prepare EA-36 solution

0.1% light green stock solution (A)	450 cc
0.5% Bismarck brown stock solution (B)	100 cc
0.5% eosin stock solution (C)	450 cc
Phosphotungstic acid	2 gm
Saturated lithium carbonate solution	10 drops

Mix well. Store in a stoppered, dark bottle. Filter before use without changing the concentration.

The saturated concentration of lithium carbonate is about 1.33% at 20°C. Therefore, put at least 0.2 gm in 10 cc of distilled water at normal temperature, stir well, decant the supernatant following the sedimentation of the lithium carbonate and filter before using.

b. The preparation of EA-65 and its differences from EA-36 :

1. Preparation of EA-65 solution

0.05% light green stock solution (D)	450 cc
0.5% Bismarck brown stock solution (B)	100 cc
0.5% eosin stock solution (C)	450 cc
Phosphotungstic acid	6 gm

Mix well. Seal the container and store it in a dark cool place. Be sure to filter the solution before use.

2. The difference is that EA-65 contains 0.05% light-green stock solution while EA-36 contains 0.1% solution :

2% light green aqueous solution	25 cc	Mix (D)
95% ethyl alcohol	975 cc	

Other solutions to be prepared

1. 0.5% collodion solution : Dissolve 5 gm of collodion in 1000 cc of equal parts of ether and 95% alcohol.
2. 0.25% hydrochloric acid solution : Pharmacopoeial undiluted hydrochloric acid is at about 30% concentration. Add distilled water to 0.88 cc of undiluted hydrochloric acid to make 100 cc.
3. 1.5% ammonia solution : Add an ammonia solution to 70% ethyl alcohol to make 1.5 % ammonia solution (A commercial undiluted ammonia solution is about 28%. Therefore, add 70% ethyl alcohol to 5.4 cc of an undiluted ammonia solution to make 100 cc)
4. Distilled water.
5. Ethyl alcohol — 50%, 70%, 80%, 95% and 100%.

*EA 50 (36) may be purchased from Ortho Pharmaceutical Corporation, Raritan, N. J.

6. Equal parts of 100% ethyl alcohol and xylol.
7. Xylol.

How to stain

1) Standard method (Table 4)

This is similar to Papanicolaou's original method and is used by most laboratories. Cells are stained strongly first, then bleached in a hydrochloric acid solution. The nuclei are properly stained while excess stain is removed from the cytoplasm. The acid is washed off by running water, and the nuclei are stained deep blue.

2) Alternate method (Table 5)

This method is suitable for staining urinary and gastric washing smears. It is a rapid method and has the advantage of staining smears in a shorter time. However, it is not good for thick smears because hematoxylin remains in the cytoplasm making the nucleus indistinct and the cytoplasm less transparent.

Precautions using Papanicolaou's staining procedure

1) Fixation

1. Fixative (ether-alcohol): Filter occasionally. When it gets old, floating cellular fragments may attach to other smears and cause misdiagnoses. Therefore, do not use old fixative. The container should be covered as tightly as possible. Ether evaporates readily.
2. A fixative containing ether is very inflammable.

2) Staining

1. The 80% ethyl alcohol of ③ of the staining procedure should be filtered every day. Cellular fragments detached from slides are often present and attach to other slides to cause misdiagnoses.
2. Harris hematoxylin should be filtered every day. Ordinary No. 2 filter paper may be used.
3. To test the condition of the hematoxylin solution, apply a drop of the solution to filter paper. If a light colored watery ring forms around a dark-blue spot, the solution is old and diluted. It should be discarded.
4. Baths for staining solutions should not be washed with neutral detergents. They should be washed only with distilled water and air dried. Staining baths should be cleaned every time the solution is changed.
5. The distilled water of ⑥ of the staining procedure should be changed every day.
6. When slides are transferred from one solution to another, no matter whether the slides are stained individually or collectively in a slide carrier, the previous solution should be drained as completely as possible. Especially in the process of distilled water ⑥ →hematoxylin ⑦ →distilled water ⑧, running water ⑩ (Standard method) → 50% ethyl alcohol ⑪, or 95% ethyl alcohol ⑭ → orange G ⑮ → 95% ethyl alcohol ⑯ → EA ⑰ → 95 % ethyl alcohol, slant the carrier to drain the solution in the process shown by arrows. Then place them on a paper towel or filter paper for 2–3 seconds to absorb the remaining solution before transferring them to the next bath.

Table 4 Standard method.

→ Gynecological smears begin here. (at step 3)

(1) Remove a slide from the ether-alcohol fixative, transfer it to the slide carrier, then soak in 95% alcohol.	→ (2) 0.5% collodion-ether-alcohol solution 2 min	→ (3) 80% ethyl alcohol 5 dips or 10-15 sec. The slide may be left for a while.
→ (4) 70% ethyl alcohol 5 dips or 10-15 sec	→ (5) 50% ethyl alcohol 5 dips or 10-15 sec	→ (6) Distilled water 5 dips or 10-15 sec
→ (7) Harris hematoxylin (Add an equal amount of distilled water to the stock solution) 6 min	→ (8) Distilled water (two baths) 5 dips in each	→ (9) 0.25% hydrochloric acid solution 2 dips
→ (10) Running water (gently) 6 min	→ (11) 50% ethyl alcohol 5 dips or 10-15 sec	→ (12) 70% ethyl alcohol 5 dips or 10-15 sec
→ (13) 80% ethyl alcohol 5 dips or 10-15 sec	→ (14) 95% ethyl alcohol 5 dips or 10-15 sec	→ (15) Orange G 90 sec
→ (16) 95% ethyl alcohol (two baths) Dip the slide in and out slowly. Don't soak for extended periods.	→ (17) EA-36 (50) 90 sec	→ (18) 95% ethyl alcohol (three baths) Dip the slide in and out slowly. Don't soak for extended periods.
→ (19) 100% ethyl alcohol (two baths) 5 dips or 10-15 sec. When many slides are stained together, use two baths.	→ (20) Equal parts of 100% ethyl alcohol and xylol 5 dips or 10-15 sec	→ (21) Xylol 5 dips or 10-15 sec
→ (22) Xylol 5 dips or 10-15 sec	→ (23) Xylol. The slide may be left here.	→ (24) Coverslipping

(Modified from the tables 11-8 to 11-11 of "A Manual of Cytotechnology" 2nd edition, 1963. National Committee for Cancers in Medical Technology)

Table 5 Alternate method (Rapid method).

① Remove a slide from the ether-alcohol fixative, transfer it to the slide carrier, then soak it in 95% ethyl alcohol.	② **0.5% collodion-ether-alcohol solution 2 min**	→ Gynecological smears begin here. ③ **80% ethyl alcohol 5 dips or 10-15 sec** The slide may be left for a while.
④ **70% ethyl alcohol 5 dips or 10-15 sec**	⑤ **50% ethyl alcohol 5 dips or 10-15 sec**	⑥ **Distilled water 5 dips or 10-15 sec**
⑦ **Harris hematoxylin** The hematoxylin stock solution + glacial acetic acid (4cc/100cc of the stock solution) **45 sec**	⑧ **Distilled water (three baths) 5 dips in each**	⑨ **50% ethyl alcohol 5 dips or 10-15 sec**
⑩ **1.5% ammonia solution (1.5% in 70% ethyl alcohol) 1 min**	⑪ **70% ethyl alcohol 5 dips or 10-15 sec**	⑫ **70% ethyl alcohol 5 dips or 10-15 sec**
⑬ **80% ethyl alcohol 5 dips or 10-15 sec**	⑭ **95% ethyl alcohol 5 dips or 10-15 sec**	⑮ **Orange G 75 sec**
⑯ **95% ethyl alcohol (two baths)** Dip the slide in and out slowly in each bath. Don't soak it for extended periods.	⑰ **EA-65 3 min**	⑱ **95% ethyl alcohol (three baths)** Dip the slide in and out slowly in all baths. Don't soak it for extended periods.
⑲ **100% ethyl alcohol 5 dips or 10-15 sec**	⑳ **Equal parts of 100% ethyl alcohol and xylol 5 dips or 10-15 sec**	㉑ **Xylol 5 dips or 10-15 sec**
㉒ **Xylol 5 dips or 10-15 sec**	㉓ **Xylol** The slide may be left here.	㉔ Coverslipping

(Modified from the tables 11–8 to 11–11 of "A Manual of Cytotechnology" 2nd edition, 1963. National Committee for Cancers in Medical Technology)

7. Orange G and EA solutions should be filtered or changed at least once a week. It is sometimes permissible to change one half the first week and all of the solution the second week.
8. Xylol ㉑ – ㉓ should be filtered when cellular fragments are seen floating in it.
9. If possible, change all solutions except the stains once a week or after every 250–300 slides. Wash the containers well with distilled water and then dry them.
10. Slide or slide carriers should be placed gently in each solution. A gynecological smear may be dipped gently about 5 times, because the cells are comparatively hard to dislodge from the slide. The other smears should be left in the solution for 10–15 seconds without being dipped frequently.
11. All the containers, especially those with 100% alcohol and xylol should be covered when not in use.

Other Precautions: Sputum and effusions should be handled with care, because they are occasionally infectious. All containers, instruments and slides used with them should be soaked in a germicidal solution, *i.e.*, a cationic soap. Use adequate paper towels and newspapers and dispose of them frequently to avoid contamination of the working area. Don't forget to disinfect the hands and fingers.

Rapid staining methods (especially for gynecological smears)

Rapid method using Giemsa's solution

Various characteristic staining procedures other than Papanicolaou's are used for smear tests. It is not the purpose of this book to describe all of them, but one or two quick methods recommended to clinicians will be described.

Giemsa's method is quick and easy and gives comparatively good results. It is a simple method that can be performed anywhere, and only requires Giemsa's solution and alcohol. Therefore it is convenient in laboratories with limited facilities.

1) Solution

1/10 Giemsa's alcohol solution

Mix 1 cc of Giemsa's stock solution in 9 cc of 90% ethyl alcohol. The alcohol may be ethyl alcohol or solution of 8.0 cc dehydrated alcohol, 0.5 cc phosphate buffer (pH 6.4) and 0.5 cc distilled water.

2) Staining

1. Soak a moist smear in ordinary ether-alcohol fixative for more than 15 minutes. Or rehydrate a dried smear (*cf.* p. 226) and fix in 95% ethyl alcohol.
2 Dip in 95% ethyl alcohol about 5 times.
3. Place the smear on a flat surface. Apply about 10 drops of the Giemsa's alcohol solution with a glass pipette fitted with a rubber bulb and watch the solution gradually evaporating and the stain changing from blue to reddishpurple. Most of the Giemsa's solution covering the surface of the smear will evaporate in 7–10 minutes, and the surface will remain slightly damp (Fig. 74). Next, dip the smear in acetone.

4. Acetone — 5 dips.
5. 95% ethyl alcohol — The smear is covered with concentrated Giemsa's solution as a result of evaporation. Place the smear in alcohol and agitate it to wash the Giemsa solution off the surface.
6. 100% ethyl alcohol — several dips.
7. Equal amounts of 100% ethyl alcohol and xylol — 5 dips.
8. Xylol — 5 dips. It may be left here for a while until mounting.
9. Coverslipping.

(Cytology Service, Cancer Institute Hospital, Tokyo)

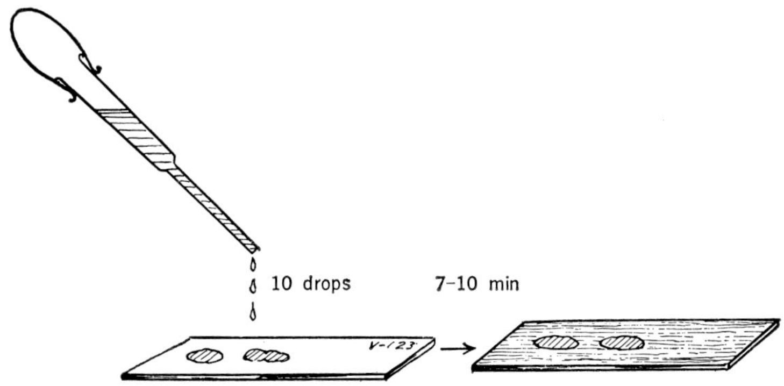

Fig. 74

Ink stain

1) Staining solution

Prepare a stock solution (A) by mixing 10 drops of blue Sheaffer Ink, U.S.A., one drop of red ink and 10 cc of watei and store in a refrigerator. Sometimes a stock solution (B) made of 10 drops of blue ink and one drop of red ink without water is used.

2) Procedure

1. Drop stock solution (A) immediately on an unfixed vaginal smear and leave about ten minutes. Or drop stock solution (B) on a vaginal smear and leave for two or three minutes. Or apply 1/20 stock solution (B) diluted with distilled water and leave for about ten minutes.
2. Remove the staining solution and mount the smear in a drop of glycerin.
3. Examine under the microscope immediately.
4. Wash the smear gently with water and fix it by the regular procedure. It may be stained by Papanicolaou's method.

3) Precautions

1. Prepare a new staining solution each time.
2. Stain the smear at a low temperature (*e.g.* in a refrigerator).
3. Keep the slide in a horizontal position and stain it by dropping the staining solution onto it.
4. Mount in glycerin.

V. Restaining Old Faded Smears

Precious and interesting smears often become faded because they are exposed to strong light during photography or are left on the microscope stage for long periods of time. First, such smears are treated by the following method.

① Xylol Until the mounting medium is softened and the coverslip can be removed. 12 hours or more	➡ ② Xylol 3 dips	➡ ③ Xylol 3 dips
➡ ④ Equal parts of xylol and 100% ethyl alcohol 3 dips	➡ ⑤ 100% ethyl alcohol 3 dips	➡ ⑥ 95% ethyl alcohol Until the eosin is removed. about 1 min
➡ ⑦ 80% ethyl alcohol 3 dips	➡ ⑧ 70% ethyl alcohol 3 dips	➡ ⑨ 50% ethyl alcohol 3 dips
➡ ⑩ Distilled water 3 dips	➡ ⑪ 0.5% hydrochloric acid solution Until the hematoxylin is removed. about 1 min	➡ ⑫ Running water 5 min 30 sec for a urinary smear
➡ ⑬ 50% ethyl alcohol 3 dips	➡ ⑭ 70% ethyl alcohol 3 dips	➡ ⑮ 80% ethyl alcohol

Then restain by Papanicolaou's method. The staining results, however, may be inferior to the original stain.

VI. Mounting with a Coverslip

1) Remove excess xylol beforehand so that bubbles will not form after the xylol evaporates.

2) A coverslip may be easily placed by warming the slide to about 37°C beforehand, as the medium spreads well at this temperature.

3) How to place a coverslip

1. Apply a drop of mounting medium of proper viscosity to the slide. Just enough should be used ; not too much or too little. The proper amount can only be learned by experience.
   HSR and Permount are recommended.
2. Bring one of the longer sides of the cover glass into contact with the drop of mounting medium. After confirming that the medium is spread over the whole length of the side, press the cover glass gently on the slide. Press the cover glass lightly with the tip of a pair of forceps to spread the medium to the edges of the coverslip. If the medium reaches the edges of the coverslip only when it is pressed hard, and retracts when pressure on the coverslip is released, too little mounting medium has been used. On the other hand, if there is too much mounting medium, it will be forced out from under the coverslip. When only some medium is forced out, apply filter paper to the edges to absorb the excess. The medium forced out on the slide should be carefully wiped off with gauze dampend in xylol (Fig. 75). Mounting medium should be diluted properly with xylol before use. However, if the mounting medium is too diluted, air will enter after it dries and will damage the slide.

   An easier method of coverslipping is to place a drop of mounting medium

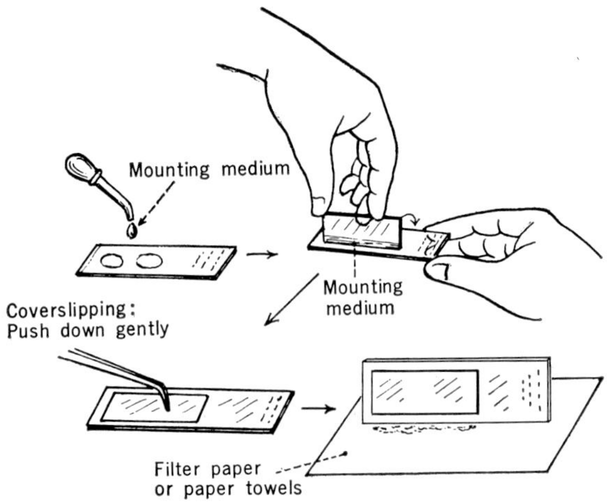

Fig. 75

on the coverslip and lower the inverted slide onto the coverslip. This has the advantage of giving the operator the relatively large slide to manipulate instead of the much smaller coverslip.

3. The covered slide should be left flat for half a day. If it has to be examined immediately, try to keep it in a horizontal position as much as possible during the microscopic examination.

VII. Labeling

Any form of labeling can be used as long as it is used consistently.

As shown in the figure, a label may be attached to the right-hand side of the slide where the patient's name has been previously written with a diamond marking pencil when the smear was obtained. It is best to indicate the kind and number of the smear and the date when it was prepared. If possible, write the patient's full name on the next line. The kinds of smears may be indicated as follows: *V* for vaginal-cervical smears, *EC* or *EM* for endocervical or endometrial smears, respectively, *Sp* for sputum, *B* for bronchial washings, *G* for gastric smears, *U* for urinary smears, *A* for ascites, *P* for pleural effusions and *Br* for breast secretions, *e.g.* V-1235-71, P-2563-68 and Sp-263-65. The middle number is the serial number of the smear, and the last number is the year when it was made.

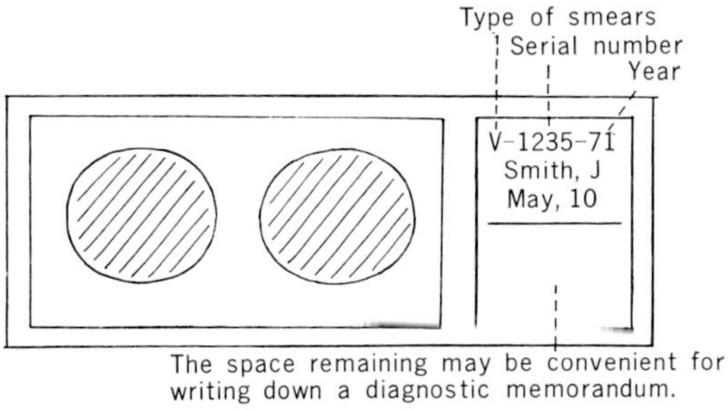

Fig. 76

VIII. How to Examine Smears

However good the smear may be, it is of no use unless properly examined.

1. Unlike observations of tissue slides, observations of cell smears are concerned with individual cells. Therefore, as shown in the figure, the slide should be examined systematically from one end to the other. There are two ways to move the slide during microscopic examination : vertically and horizontally. According to the author's experience, the vertical movement causes less eye fatigue. This method of microscopic observation is called "screening" (Fig. 77). A ×10 objective and a wide-view ocular (×10, ×12.5 or ×15) are recommended for the screening.

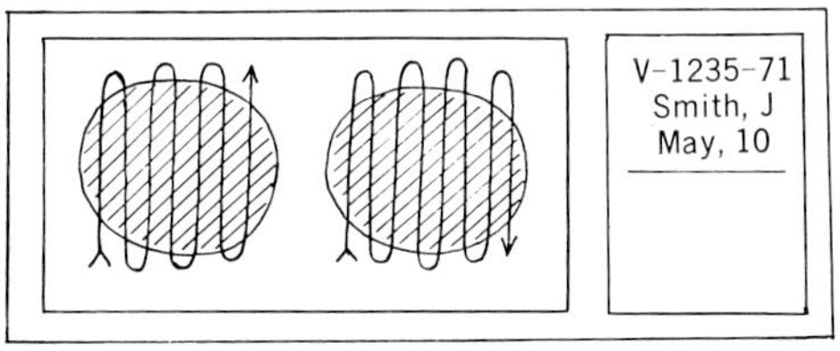

Fig. 77

2. An ink mark may be placed on the slide whenever a malignant, interesting or noteworthy cell is observed. Without this mark, it is often time-consuming to identify the same cell again. It is almost impossible to do this on smears of effusions with numerous cells.

   The mark should always be made in a uniform way. If placed beneath the cell, it will be observed above the cell through the microscope; therefore, the cell in question can easily be found beneath the mark. First, place the cell in the center of the microscopic field. Make a mark with an ordinary blue ink pen slightly below the center of the spot lighted by the condenser (Fig. 78). Recently, an instrument to make such marks has been used*.

   Sometimes a group of noteworthy cells may be observed. These cells may be enclosed in circle. Anyone can easily draw this circle by observing the cells under low magnification (×4 objective).

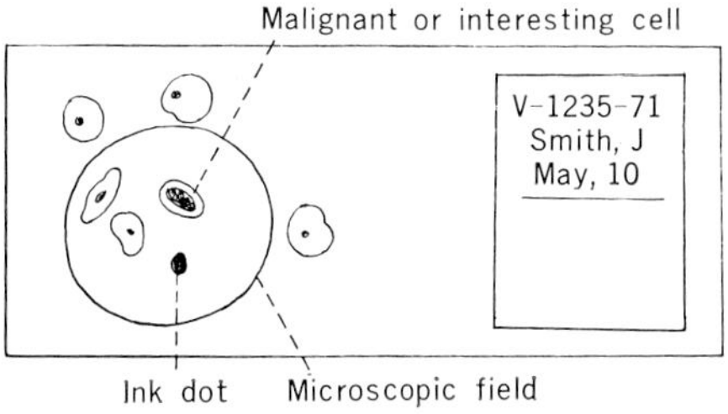

Fig. 78

* Zieler Instrument Co., Inc., 31st., James Ave., Boston, Mass.

IX. Self-irrigation Vaginal Smears

Recently a self-irrigation method to collect vaginal secretions has been devised for mass population screening of uterine carcinoma and has become popular. The patient inserts a special pipette into her vagina and aspirates the vaginal secretion into the pipette bulb. The pipette bulb is sent to a laboratory, and smears are made from this material and examined microscopically. This method may have a weak point, however, in that the sufficient endocervical material necessary for the early detection of carcinoma in situ is difficult to obtain.

Self-irrigation method

1. This should be performed about a week after the termination of menstruation. Sexual intercourse or vaginal douche should be avoided for 24 hours before irrigation. After menopause a woman can perform it any time.
2. A special plastic pipette, shown in Fig. 79, is inserted into the vagina. The individual sits on the front part of a toilet stool, removes the cap from the top of the pipette and inserts the pipette into the vagina up to the bulb.

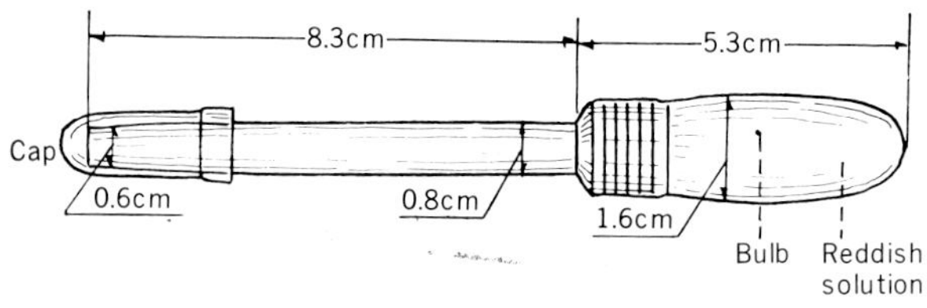

Fig. 79

3. She then presses the bulb with her fingers to inject an irrigating solution (normal saline solution containing 20% ethyl alcohol, and adjusted to a slightly alkalier pH against phenol red) into the vagina. The top of the pipette is moved gently in and out, while pressing the bulb, to mix the red solution with vaginal secretions.
4. Next, the bulb is released and the pipette is slowly extracted from the vagina while the bulb is being restored to its original shape. The solution which has been injected into the vagina and mixed with vaginal secretions is aspirated into the bulb by negative pressure. The red color should change to orange when the solution is well mixed with vaginal secretion.
5. The cap is replaced on the pipette. The individual fills in her age, name, address, history of pregnancy and marital state on a printed form, and inserts it with the pipette into an attached tube. She then sends or brings it to a laboratory.

How to make smears from the solution mixed with vaginal secretions in the pipette bulb

1) 50% Carbowax 1540 stock solution

Mix Carbowax 1540 (Union Carbide Chemical Co.) with an equal amount of distilled water.

2) Preparation of Carbofix

50% Carbowax stock solution	50 cc
Glacial acetic acid	20 cc
95% ethyl alcohol	430 cc

3) How to prepare vaginal smears

1. Separate the bulb part from the pipette and use it as a centrifuge tube. Or place the contents in an ordinary centrifuge tube. Centrifuge at 1200 rpm for two minutes.
2. Decant or aspirate the supernatant. Add Carbofix to the sediment at the top of the bulb and resuspend the cellular components. Add 5 cc of Carbofix if an ordinary centrifuge tube is used. Centrifuge at 1200 rpm for two minutes.
3. Remove as much of the supernatant as possible. Add an equal amount of Carbofix to the sediment and mix well.
4. Apply one or two drops of this mixture to a glass slide, spread thin and dry in air in a horizontal position for about 30 minutes.
5. The dried smear should be soaked in 95% ethyl alcohol for at least 10 minutes, and then stained by Papanicolaou's method.

4) A satisfactory irrigation smear should have 50–100 squamous cells in a microscopic field at ×125. The smear has no diagnostic value if less than 25–30 cells are observed (H. J. Davis).

X. The Best Way to Report Cytodiagnostic Findings and Inherent Precautions

A cytodiagnostic report should be made with consideration for clinicians some of whom have little knowledge of cytodiagnosis. Therefore, make all diagnoses carefully and avoid expressions which may cause misunderstanding.

1. It is usually best to make a diagnosis according to Papanicolaou's classification and add some explanatory remarks to it.

Class I	Negative. No atypia observed.
Class II	Negative. Some atypia observed.
Class III	Suspicious. But no definite sign of malignancy.
Class IV	Positive. Signs of malignancy.
Class V	Positive. Definitely malignant.

   Don't write only "Class II". Write, for instance, "Class II. No sign of malignancy is observed, but some atypia is noticed. It is considered to be an inflammatory change."
2. If possible, evaluate the hormonal state of a vaginal smear. Write, for instance, "Class I. Negative. Most of the squamous cells present are superficial It probably corresponds to the late follicular phase". This would be quite helpful to the clinician. Especially, when the clinical diagnosis is primary sterility or functional amenorrhea, etc. and the patient is young, a "negative" cytodiagnosis alone has little clinical value.
3. If the specimen is insufficient for a diagnosis or if it is obscured by strong bleeding or other contaminants, do not write "Class I. Negative", but write down clearly that the material is insufficient and request more material.
4. In the case of a malignant tumor, indicate the type of tumor, if possible, besides "Class IV" or "Clase V". However, it is risky to report mere assumptions without grounds for your belief.
5. The diagnosis of Class III should be made with great care and thought. It should be made only when the evidence nearly eighty percent positive. Try to prevent misunderstanding by the clinician by clearly describing the grounds of the diagnosis. A diagnosis above Class III often requires histological support, and this requirement should be written in the cytodiagnostic report.

INDEX